AF676423

Intelligent and Computer Technologies Application in Construction

Intelligent and Computer Technologies Application in Construction

Editors

Hongling Guo
Jia-Rui Lin
Yantao Yu

MDPI • Basel • Beijing • Wuhan • Barcelona • Belgrade • Manchester • Tokyo • Cluj • Tianjin

Editors

Hongling Guo
Tsinghua University
Beijing, China

Jia-Rui Lin
Tsinghua University
Beijing, China

Yantao Yu
The Hong Kong University of Science and Technology
Hong Kong, China

Editorial Office
MDPI
St. Alban-Anlage 66
4052 Basel, Switzerland

This is a reprint of articles from the Special Issue published online in the open access journal *Buildings* (ISSN 2075-5309) (available at: https://www.mdpi.com/journal/buildings/special_issues/Intelligent_Construction).

For citation purposes, cite each article independently as indicated on the article page online and as indicated below:

LastName, A.A.; LastName, B.B.; LastName, C.C. Article Title. *Journal Name* **Year**, *Volume Number*, Page Range.

ISBN 978-3-0365-8150-7 (Hbk)
ISBN 978-3-0365-8151-4 (PDF)

© 2023 by the authors. Articles in this book are Open Access and distributed under the Creative Commons Attribution (CC BY) license, which allows users to download, copy and build upon published articles, as long as the author and publisher are properly credited, which ensures maximum dissemination and a wider impact of our publications.
The book as a whole is distributed by MDPI under the terms and conditions of the Creative Commons license CC BY-NC-ND.

Contents

About the Editors

Hongling Guo

Hongling Guo is an Associate Professor at the Department of Construction Management at Tsinghua University. He obtained his bachelor's and master's degrees and his first PhD in Management Science and Engineering from Harbin Institute of Technology as well as his second PhD in Construction Management from The Hong Kong Polytechnic University, and then worked as a researcher and visiting lecturer at The Hong Kong Polytechnic University for three years. He focuses on employing advanced information technologies to improve the efficiency of construction engineering and management. His research interests include intelligent construction, virtual construction, BIM, and digital construction safety management.

Jia-rui Lin

Jia-rui Lin is an Assistant Professor at the School of Civil Engineering at Tsinghua University. He received his bachelor's degree and PhD from Tsinghua University. His research interests include intelligent design, construction process modelling, building information modeling (BIM), machine learning, and digital twin.

Yantao Yu

Yantao Yu is an Assistant Professor at the Department of Civil and Environmental Engineering at The Hong Kong University of Science and Technology. She received both her bachelor's and master's degrees from Tsinghua University and her PhD from The Hong Kong Polytechnic University, funded by the Hong Kong PhD Fellowship. Her research interests include smart construction, construction robotics, and occupational safety and health.

buildings

MDPI

Editorial

Intelligent and Computer Technologies' Application in Construction

Hongling Guo [1,*], Jia-Rui Lin [2] and Yantao Yu [3]

1 Department of Construction Management, Tsinghua University, Beijing 100084, China
2 Department of Civil Engineering, Tsinghua University, Beijing 100084, China
3 Department of Civil and Environmental Engineering, The Hong Kong University of Science and Technology, Hong Kong 999077, China
* Correspondence: hlguo@tsinghua.edu.cn

Citation: Guo, H.; Lin, J.-R.; Yu, Y. Intelligent and Computer Technologies' Application in Construction. *Buildings* **2023**, *13*, 641. https://doi.org/10.3390/buildings13030641

Received: 20 February 2023
Accepted: 27 February 2023
Published: 28 February 2023

Copyright: © 2023 by the authors. Licensee MDPI, Basel, Switzerland. This article is an open access article distributed under the terms and conditions of the Creative Commons Attribution (CC BY) license (https://creativecommons.org/licenses/by/4.0/).

The construction industry is faced with many challenges, such as lagging productivity [1], labor sustainability [2], and environmental sustainability [3]. Intelligent construction provides a solution to these challenges. In the past two decades, significant efforts have been devoted to enhancing the construction project delivery process using intelligent and computer technologies. Examples include, but are not limited to, smart site supervision [4], construction robotics [5], automatic safety [6], and health management with the IoT [7]. This Special Issue aims to provide a platform to explore state-of-the-art knowledge, practical implementation, and cutting-edge innovations in the area of intelligent and computer technologies' application in construction. A total of fourteen original research studies have been published, with contributions from international research groups. All these contributions address the main topics of this Special Issue with an effective and targeted effort.

Al-Sarafi et al. [8] explored factors that affect the adoption of BIM in the Yemeni construction industry. The authors investigated five factors, i.e., technology, process, policy, people, and the environment, using partial least squares structural equation modeling (PLS-SEM). The multivariate results indicate that all factors influencing BIM adoption in Yemen are highly correlated in the measurement model. The insight of this study illustrates how factors influence the adoption of BIM and help develop BIM implementation strategies in other countries.

Lin et al. [9] developed maturity scoring tables for assessing intelligent construction management (ICM). A case study on two construction enterprises was conducted to validate the feasibility of the developed assessment system. The results show that the system can assess the maturity of these enterprises and derive appropriate improvement plans accordingly. This method paves the way for an effective and accurate improvement in ICM maturity.

Xu et al. [10] proposed an automatic approach that expands domain knowledge elements (DKEs) from unstructured text to achieve better safety and risk management in metro construction. The authors first obtained the connected knowledge elements with a co-word co-occurrence network (CCN) and pruned the weakly related subnetworks using association rule mining (ARM). Finally, a structure of DKEs could be obtained. The presented method automatically expands DKEs from a small body of prior knowledge while reducing expert bias, contributing to a refined knowledge structure that can guide safety training and aid knowledge-based safety risk management.

Li, Zhang, and Xu [11] aimed to determine the factors influencing the adoption of blockchain technology. The authors developed a technology–organization–environment framework and collected data from 244 practitioners using questionnaires. The hypothesis was validated using partial least squares structural equation modeling (PLS-SEM) and fuzzy-set qualitative comparative analysis (fsQCA).

Fei et al. [12] proposed an integrated schematic design method for reinforced concrete (RC) shear wall structures using generative adversarial networks (GANs). A cloud design platform was developed to provide a workable GAN application so as to address challenges in computer-aided design (CAD) drawing preprocessing and the high hardware and software requirements of users' computers. The experimental results show that the proposed method has a 97.3% accuracy in heterogeneous data conversion and can generate shear wall layout designs similar to those of qualified engineers.

Yan et al. [13] aimed to identify and analyze the key factors driving intelligent construction (IC) development and to produce general laws to guide IC development. The authors designed a five-stage method to obtain key driving factors and outlined general laws based on an empirical study in China.

Xu, Kang, and Li [14] designed a novel feature-based deep learning method for construction component classification. The presented method leverages local and global features and performs feature fusion through deep convolution to achieve robust classification. An experiment conducted on the construction dataset proved the efficiency of the proposed method. The method helps increase efficiency in construction digitization.

Wang et al. [15] recognized key construction scenes on highway bridges through a visual relationship detection-based method. The authors first identified five key construction scenes based on the underlying construction characteristics. Then, they formulated identification rules for these scenes. Finally, a novel construction scene identification model (CSIN) was built on these rules and vision-based techniques. The model's effectiveness was verified experimentally with an accuracy of 94%. This method helps to ensure safe construction through remote monitoring.

Guo et al. [16] developed a virtual simulation method that achieves automatic selection and localization of mobile cranes to improve the safety and efficiency of lifting operations. The authors first extracted the required information from building information modeling (BIM). Then, candidate locations and types of mobile cranes could be determined based on the crane capacity and simulation results. More specifically, three constraint checks and two efficiency optimizations were conducted. This study contributes to crane operation planning and automatic construction simulation.

Shen et al. [17] aimed to identify and prevent safety risks during construction. The authors developed a method that integrates the safety rule library, BIM, and natural language processing technology to identify risks and intelligently present results in a visual way. The findings and insights provide new information for construction safety management.

Aguilar et al. [18] proposed a framework for analyzing the acoustic behavior of rooms based on reverberation time (RT). The presented framework enables decision-making in the early design phase using BIM technology and Dynamo. The framework allows automatic evaluation of the RT each time after the modification of the BIM model, showing optimal solutions according to cost and optimum absorbent surface area.

Li et al. [19] investigated and analyzed factors that influence the development of intelligent construction (IC) in China. They developed a structural equation modeling (SEM) approach to identify the factors, examine their implications, and showcase the key means for successful IC development.

Zhao, Cao, and Liu [20] addressed problems in prefabricated component (PC) hoisting control. They proposed a novel framework that uses BIM and the Internet of Things (IoT) to structure a digital twin (DT) and adopts Dijkstra's algorithm to conduct hoisting route planning. In addition, long-range radio (LoRa) technology is also utilized for real-time information transmission to monitor the PCs' state. The proposed framework improves the intelligent management of prefabricated building construction.

Cao, Kamaruzzaman, and Aziz [21] proposed a review paper on BIM utilization in green building construction. They highlighted the advantages of BIM, discussed the potential application of BIM in different phases of green building construction, and revealed the barriers, challenges, and future research directions of BIM utilization in green building construction.

The guest editors would like to acknowledge all the authors for their scientific support and kind sharing of their knowledge. The editors would like to express their gratitude to the peer reviewers for their rigorous analysis of manuscripts as well as the managing editors of *Buildings* involved in this Special Issue for their continuous support.

Author Contributions: All authors contributed to every part of this editorial. All authors have read and agree to the published version of the manuscript.

Conflicts of Interest: The authors declare no conflict of interest.

References

1. Wu, M.; Lin, J.-R.; Zhang, X.-H. How Human-Robot Collaboration Impacts Construction Productivity: An Agent-Based Multi-Fidelity Modeling Approach. *Adv. Eng. Inform.* **2022**, *52*, 101589. [CrossRef]
2. Sungjin, K.; Soowon, C.; Daniel, C.-L. Dynamic Modeling for Analyzing Impacts of Skilled Labor Shortage on Construction Project Management. *J. Manag. Eng.* **2020**, *36*, 4019035. [CrossRef]
3. Opoku, D.-G.J.; Ayarkwa, J.; Agyekum, K. Barriers to Environmental Sustainability of Construction Projects. *Smart Sustain. Built Environ.* **2019**, *8*, 292–306. [CrossRef]
4. Edirisinghe, R. Digital Skin of the Construction Site. *Eng. Constr. Archit. Manag.* **2019**, *26*, 184–223. [CrossRef]
5. Carra, G.; Argiolas, A.; Bellissima, A.; Niccolini, M.; Ragaglia, M. Robotics in the Construction Industry: State of the Art and Future Opportunities. In *ISARC, Proceedings of the International Symposium on Automation and Robotics in Construction, Berlin, Germany, 20–25 July 2018*; IAARC Publications: Berlin, Germany, 2018; Volume 35, pp. 1–8.
6. Ding, L.; Fang, W.; Luo, H.; Love, P.E.D.; Zhong, B.; Ouyang, X. A Deep Hybrid Learning Model to Detect Unsafe Behavior: Integrating Convolution Neural Networks and Long Short-Term Memory. *Autom. Constr.* **2018**, *86*, 118–124. [CrossRef]
7. Yu, Y.; Li, H.; Yang, X.; Kong, L.; Luo, X.; Wong, A.Y.L. An Automatic and Non-Invasive Physical Fatigue Assessment Method for Construction Workers. *Autom. Constr.* **2019**, *103*, 1–12. [CrossRef]
8. Al-sarafi, A.H.; Alias, A.H.; Shafri, H.Z.M.; Jakarni, F.M. Factors Affecting BIM Adoption in the Yemeni Construction Industry: A Structural Equation Modelling Approach. *Buildings* **2022**, *12*, 2066. [CrossRef]
9. Lin, C.; Hu, Z.-Z.; Yang, C.; Deng, Y.-C.; Zheng, W.; Lin, J.-R. Maturity Assessment of Intelligent Construction Management. *Buildings* **2022**, *12*, 1742. [CrossRef]
10. Xu, N.; Zhang, B.; Gu, T.; Li, J.; Wang, L. Expanding Domain Knowledge Elements for Metro Construction Safety Risk Management Using a Co-Occurrence-Based Pathfinding Approach. *Buildings* **2022**, *12*, 1510. [CrossRef]
11. Li, C.; Zhang, Y.; Xu, Y. Factors Influencing the Adoption of Blockchain in the Construction Industry: A Hybrid Approach Using PLS-SEM and FsQCA. *Buildings* **2022**, *12*, 1349. [CrossRef]
12. Fei, Y.; Liao, W.; Zhang, S.; Yin, P.; Han, B.; Zhao, P.; Chen, X.; Lu, X. Integrated Schematic Design Method for Shear Wall Structures: A Practical Application of Generative Adversarial Networks. *Buildings* **2022**, *12*, 1295. [CrossRef]
13. Yan, X.; Zhou, Y.; Li, T.; Zhu, F. What Drives the Intelligent Construction Development in China? *Buildings* **2022**, *12*, 1250. [CrossRef]
14. Xu, Z.; Kang, R.; Li, H. Feature-Based Deep Learning Classification for Pipeline Component Extraction from 3D Point Clouds. *Buildings* **2022**, *12*, 968. [CrossRef]
15. Wang, C.; Lv, J.; Geng, Y.; Liu, Y. Visual Relationship-Based Identification of Key Construction Scenes on Highway Bridges. *Buildings* **2022**, *12*, 827. [CrossRef]
16. Guo, H.; Zhou, Y.; Pan, Z.; Zhang, Z.; Yu, Y.; Li, Y. Automated Selection and Localization of Mobile Cranes in Construction Planning. *Buildings* **2022**, *12*, 580. [CrossRef]
17. Shen, Q.; Wu, S.; Deng, Y.; Deng, H.; Cheng, J.C.P. BIM-Based Dynamic Construction Safety Rule Checking Using Ontology and Natural Language Processing. *Buildings* **2022**, *12*, 564. [CrossRef]
18. Aguilar, A.J.; de la Hoz-Torres, M.L.; Martínez-Aires, M.D.; Ruiz, D.P. Development of a BIM-Based Framework Using Reverberation Time (BFRT) as a Tool for Assessing and Improving Building Acoustic Environment. *Buildings* **2022**, *12*, 542. [CrossRef]
19. Li, T.; Yan, X.; Guo, W.; Zhu, F. Research on Factors Influencing Intelligent Construction Development: An Empirical Study in China. *Buildings* **2022**, *12*, 478. [CrossRef]
20. Zhao, Y.; Cao, C.; Liu, Z. A Framework for Prefabricated Component Hoisting Management Systems Based on Digital Twin Technology. *Buildings* **2022**, *12*, 276. [CrossRef]
21. Cao, Y.; Kamaruzzaman, S.N.; Aziz, N.M. Green Building Construction: A Systematic Review of BIM Utilization. *Buildings* **2022**, *12*, 1205. [CrossRef]

Disclaimer/Publisher's Note: The statements, opinions and data contained in all publications are solely those of the individual author(s) and contributor(s) and not of MDPI and/or the editor(s). MDPI and/or the editor(s) disclaim responsibility for any injury to people or property resulting from any ideas, methods, instructions or products referred to in the content.

buildings

MDPI

Article

A Framework for Prefabricated Component Hoisting Management Systems Based on Digital Twin Technology

Yuhong Zhao [1,2], Cunfa Cao [1,2] and Zhansheng Liu [1,2,*]

[1] Faculty of Urban Construction, Beijing University of Technology, Beijing 100124, China; zhaoyuhong@bjut.edu.cn (Y.Z.); caocunfa@emails.bjut.edu.cn (C.C.)
[2] The Key Laboratory of Urban Security and Disaster Engineering of the Ministry of Education, Beijing University of Technology, Beijing 100124, China
* Correspondence: liuzhansheng@bjut.edu.cn

Abstract: The hoisting of prefabricated components (PCs) is a key step during the construction of prefabricated buildings. Aiming at the problems existing in the control of PC hoisting, an innovative hoisting management system framework based on the digital twin (DT) is established in this paper. The system framework comprehensively utilizes the building information model (BIM) and Internet of Things (IoT) to establish a digital twin model (DTm) for PC hoisting control and uses Dijkstra's algorithm to conduct hoisting route planning according to the BIM data in the model. Meanwhile, long-range radio (LoRa) technology was used for data acquisition and transmission to monitor the movement state of the PCs in the hoisting process. By testing it in a prefabricated building project, the DT control method was conducted to realize the functions of real-time information collection, hoisting path planning and PC positioning, which proved the feasibility and effectiveness of the method. As a key technology to realize intelligent manufacturing, DT has been widely studied in academia. The DTm of the hoisting process of PCs is established in this study; it improves the level of intelligent management of prefabricated building construction and provides a new idea for intelligent building construction.

Keywords: digital twin; Internet of Things; prefabricated components; hoisting; building information model; long-range radio; Dijkstra's algorithm

Citation: Zhao, Y.; Cao, C.; Liu, Z. A Framework for Prefabricated Component Hoisting Management Systems Based on Digital Twin Technology. *Buildings* **2022**, *12*, 276. https://doi.org/10.3390/buildings12030276

Academic Editors: Hongling Guo, Jia-Rui Lin and Yantao Yu

Received: 2 January 2022
Accepted: 18 February 2022
Published: 1 March 2022

Publisher's Note: MDPI stays neutral with regard to jurisdictional claims in published maps and institutional affiliations.

Copyright: © 2022 by the authors. Licensee MDPI, Basel, Switzerland. This article is an open access article distributed under the terms and conditions of the Creative Commons Attribution (CC BY) license (https://creativecommons.org/licenses/by/4.0/).

1. Introduction

With the rapid development of society, the construction industry's labor-intensive construction noise and pollution, as well as other problems, are increasingly prominent. Prefabricated building is developing rapidly because of its advantages of factory prefabricated component production, fast construction site assembly and green environment protection [1]. At present, the scale of the global prefabricated building market is growing steadily year by year and the new construction area and market share of prefabricated building in China are increasing year by year. Although prefabricated buildings have many advantages, there are still many challenges in the construction site of prefabricated buildings, such as insufficient management of construction site information being unable to be timely transmitted, insufficient management of PC and low degree of visualization of the construction site [2]. Prefabricated building is the key link to realize the transformation of the construction mode from the traditional way to the industrial way. How to use intelligent management and advanced technology to solve the problems in the construction of traditional prefabricated buildings has attracted the attention of enterprises and scholars [3]. Zhong et al. [4] developed a multi-dimensional BIM platform by applying radio frequency identification (RFID) technology and BIM technology to enhance the real-time visibility and traceability of PCs. Zhao et al. [5] proposed a method based on the combination of cloud computing, BIM and IoT to solve the problem of delayed information transmission on the construction site. Through RFID, wireless sensor networks, LoRa, BIM and cloud computing, real-time dynamic assembly building project data are obtained from the construction site. Bortolini et al. [6] established a prefabricated building system model based on BIM 4D modeling for assembly logistics planning and control in the construction site, which is used for the delivery of sequence planning of PCs in the construction site and the collaborative work of all construction parties. Ko et al. [7] proposed a cost-effective material management and tracking system based on integrated RFID cloud computing services, which realizes the omni-directional automatic tracking of materials. Dave et al. [8] developed a communication framework based on the IoT to strengthen lean construction management and the use of tracking technology for RFID, GPS and other key components of the IoT and to track the whole process status of workers, materials and equipment. Valero et al. [9] introduced the application of RFID technology in tracking the status of PCs and Lee [10] designed a conceptual framework for prefabricated building construction management systems, integrating BIM, 3D laser scanning, IoT, cloud computing and other information technologies, to improve the efficiency and quality of prefabricated building construction management. Wang et al. [11] proposed a conceptual framework of an Intelligent Construction System for Prefabricated Buildings based on the IoT (ICSPB-IoT) and proved the feasibility of the realization of ICSPB-IoT through case studies. Professor Ma and his team, on the basis of overall building information management, also conducted research on prefabricated buildings and put more emphasis on intelligence on the basis of informatization [12]. He established an intelligent production planning and control system for residential parts [13] and built a material management system based on mobile terminals for the construction site [14].

Based on the above research studies and analyses, it was found that there are, mainly, the following deficiencies: (1) Research on prefabricated building construction focuses on the application of new technologies, such as IoT and BIM, for site management and information transmission, but does not apply the IoT and BIM in the hoisting process of PCs. (2) Although there are studies on the hoisting of complex objects, such as complex environment, there is a lack of research on the control method for the hoisting of PCs. (3) There is a lack of systematic methods for real-time monitoring and visual display of PCs in the hoisting process. This study analyzes the application of a variety of emerging technologies in the field of prefabricated buildings, studies the application ideas of DT technology in the construction management of prefabricated buildings and establishes a multi-technology combination of a PC hoisting management method based on DT. Meanwhile, the key technologies of hoisting planning and monitoring of the hoisting process are studied in combination with a hoisting example; the feasibility of the proposed method was verified.

2. Literature Review

2.1. Prefabricated Building Construction Management

In the field of prefabricated building construction management, scholars have conducted extensive research on construction management digitalization and intelligent upgrade. Liu et al. [15] and Liu et al. [16] used low-power wide-area network (LPWAN) in prefabricated building construction sites to solve the problems of small coverage, high energy consumption, and real-time information uploading and query in wireless network used in traditional prefabricated building construction. By comparing LoRa and the narrow-band IoT (NB-IoT), the most concerned LPWAN LoRa technology was found to have advantages such as low cost, flexible networking, being independent of operators, lower energy consumption and more mature industrial chain [17]. Combined with the characteristics of multiple terminal nodes and small data volume, it is believed that LoRa technology is more suitable for prefabricated building construction sites. Wang et al. [18] proposed an enhanced perception system based on IoT to solve the hoisting problem. Zhou et al. [19] proposed to use CPS-SMS to simulate and monitor the hoisting process based on the concept of information physical systems (CPSs), which overcame problems such as limited vision of underground space crane operators and complex construction environment. Lee et al. [20] aiming at the problem of the blind field of vision of operators in the process of tower crane hoisting, proposed to use laser element ranging to accurately locate the position of the object to be lifted, thus ensuring the safety of the tower crane hoisting process due to visual field and other factors. In order to deal with the problem of crane hoisting path planning in complex environments, Cai et al. [21] proposed a hybrid configuration space collision detection strategy based on an image collision detection algorithm and designed a master–slave parallel genetic algorithm. Through the verification of Compute Unified Device Architecture (CUDA) programming on the graphic processing unit, this method could efficiently generate a high-quality hoisting path in the complex environment.

2.2. Digital Twin

With the rapid development of the new generation of technologies represented by the IoT, big data, cloud computing, Blockchain and artificial intelligence [22,23], the application of DT theory to solve the problems existing in intelligent manufacturing has gradually become a research hotspot. Globally speaking, the digitization level of the engineering construction field is at a low stage, far behind the manufacturing industry, which indicates that the digitization development of the construction industry has broad prospects and space. *People's Daily* published an article on the broad prospect of DT and pointed out that, in the Yangtze River Delta integration Forum, relevant officials of the Ministry of Industry and Information Technology proposed to build a DT system for the full life cycle of products in key industries, which reflected China's confidence and support for the application prospect of DT.

The concept of "twin" originated from National Aeronautics and Space Administration (NASA), which predicted the flight status of the air vehicle through the accurate simulation test of the "twin" aircraft on the ground [24]. Professor Michael Grieves proposed the concept of "Conceptual Ideal for Product Lifecycle Management (PLM)" and pointed out that the digital model corresponding to the physical product could abstract the state of the physical product by simulating and testing the product's behavior [25]. It was not until 2011 that Professor Michael Grieves formally proposed the concept of DT, which includes physical entities, virtual digital models and the connection of data and information between them [26]. The DT conceptual model digitizes the concept of "twin", introduces virtual space for digital expression and establishes the connection between real space and virtual space, enabling them to interact in real time [27]. DT refers to the establishment of a virtual model of the real physical entity in a digital way, so as to map the physical model in the real world to the digital world, and to the use of a virtual model to simulate the behavior characteristics of the real physical entity by means of information exchange, fusion and iterative optimization [28,29]. Based on the physical entity, virtual digital model and three-dimensional architecture of interactive connection, Tao Fei et al. [29] introduced the twin database and service platform to establish the five-dimensional DTm. Wu et al. [30] proposed a conceptual modeling method of DT based on the five-dimension DT framework to represent the complex relationship between DT and their attributes and conducted class verification through the concept level modeling of DT for intelligent vehicles. Pei Wang et al. [31] proposed a big data virtual and real fusion framework for intelligent manufacturing based on DT and designed the framework comprehensively from three perspectives, such as the overall framework supported by the industrial Internet. Qiu et al. [32] enumerated a typical AR assembly system structure, analyzed key technologies and applications of AR in digital assembly and pointed out that DT technology is the development trend of intelligent assembly research in the future. The authors of [33] studied the availability of data resources in the manufacturing industry based on DT and took the available and unavailable data in the turning process as an example to conduct general form classification to ensure that every situation was covered. Many researchers have studied the application of DT in the production workshop, aerospace and other intelligent manufacturing fields and put forward the concepts of DT workshop and DT satellite, which reflect the promotion of the development of advanced manufacturing industry by DT [34–37].

Research on DT is also increasing in the field of architecture. China Information and Communication Institute pointed out that the "DT city" is the necessary way and means to realize the "smart city". The Chinese government is committed to making the Xiongan New Area a leading digital city in the world. Since the statement on DT cities was put forward in the 2018 outline of the plan for Xiongan New Area, DT cities have been mentioned in the fields of smart cities and community governance [38,39]. Yu et al. [40] proposed a method for health monitoring based on the DTm, which constructed a non-parametric Bayesian network to represent the dynamic degradation process of health state and the propagation of cognitive uncertainty. The authors of [41,42] studied the application of DT technology in steel structure building safety assessment and construction tensioning, respectively, verifying the feasibility of DT application through cases. In the field of building venue security, DT has been applied to the dynamic evacuation of personnel in the fire scene of the Winter Olympic Venues, which makes up for the shortage of traditional fire safety evacuations [43,44].

2.3. Research Emphasis and Novelty

Although information technology is widely used in the construction management of prefabricated buildings, the application of DT technology in the construction management of prefabricated buildings is seldom studied. Moreover, because of the complexity of construction and hoisting sites of PCs, it is impossible to collect and analyze the hoisting path and process information in real time. Additionally, because the management system of on-site construction hoisting cannot be combined with the existing virtual model layout, the management of construction hoisting still remains in the way of workers' communication and dialogue. Thus, the construction site process cannot be better displayed. To solve these problems, this study uses BIM, IoT and other technologies to establish a DT model for the hoisting management of PCs. Meanwhile, the model was used to conduct meaningful investigations and the results are as follows: (1) According to the DT model, the system framework of construction hoisting management was established to realize the real-time interactive management of PC hoisting. (2) This study verified the feasibility of using Dijkstra's algorithm to plan the hoisting path of PCs based on the established virtual model. (3) According to the system framework of virtual and real interactions, a method of real-time monitoring of the hoisting process of PCs is proposed to guide the management personnel in site construction. (4) Virtual model interactive management of PC hoisting management was realized to help site managers understand the location and related information of PCs. The method proposed in this study integrates DT, IoT, BIM and other technologies to improve the construction management level of prefabricated buildings, which is of great significance for future research.

3. DTm for PC Hoisting

According to the characteristics and requirements of PC construction and hoisting sites, this study establishes a DT five-dimensional model for prefabricated component lifting combined with the DT five-dimensional model proposed by Tao Fei et al. [27]. On this basis, the frame of a PC hoisting management system is proposed. As shown in Equation (1), H_{DT} represents the hoisting-oriented DTm; H_{PE} represents the hoisting of a physical entity, which mainly includes the component being hoisted, the surrounding physical environment affecting the hoisting and the actual changes in the component in the hoisting process, such as the position and attitude; H_{VE} represents the hoisting of virtual entities corresponding to the hoisting of physical entities, such as PCs, BIM models of building and surrounding site layout, etc.; H_{DD} represents the twin data system for storing and processing all kinds of data, such as building information model, data collected by sensor, etc.; H_{SP} represents the hoisting control service platform, which is used for various operations and presentations of the hoisting process visualization; and H_{CN} represents the relationships established among the four parts.

$$H_{DT} = (H_{PE}, H_{VE}, H_{DD}, H_{SP}, H_{CN}), \quad (1)$$

As shown in Figure 1, the hoisting service platform is used to display the information in the hoisting site environment and the hoisting process of PCs. Combined with the path-planning algorithm, the planned hoisting route of PCs is displayed. The service platform receives the virtual entity model, PC hoisting route data, PC position data and PC attitude data in the twin data service platform. H_{VE} is modeled manually according to H_{PE}. BIM modeling is conducted by hoisting surrounding environment (including building and site layout, etc.) and PCs and the data of H_{VE} are uploaded to H_{DD} for storage and processing. Combined with the cloud database, H_{VE} is uploaded to the cloud, which can be quickly inquired and viewed by the construction personnel. Meanwhile, the path optimization algorithm is used to optimize the PC construction path, determine the reasonable hoisting path and obtain the planning path data. During the hoisting process, the PCs are collected by a variety of sensor devices and uploaded in real time using LPWAN. H_{DD} collects and stores all the data in the hoisting scenario, mainly including the following four types: (1) H_{VE} data after modeling is completed; (2) multi-layer data collected by the sensor during hoisting; (3) through H_{VE} data combined with the path optimization algorithm, each PC hoisting planning path data; and (4) PC information data.

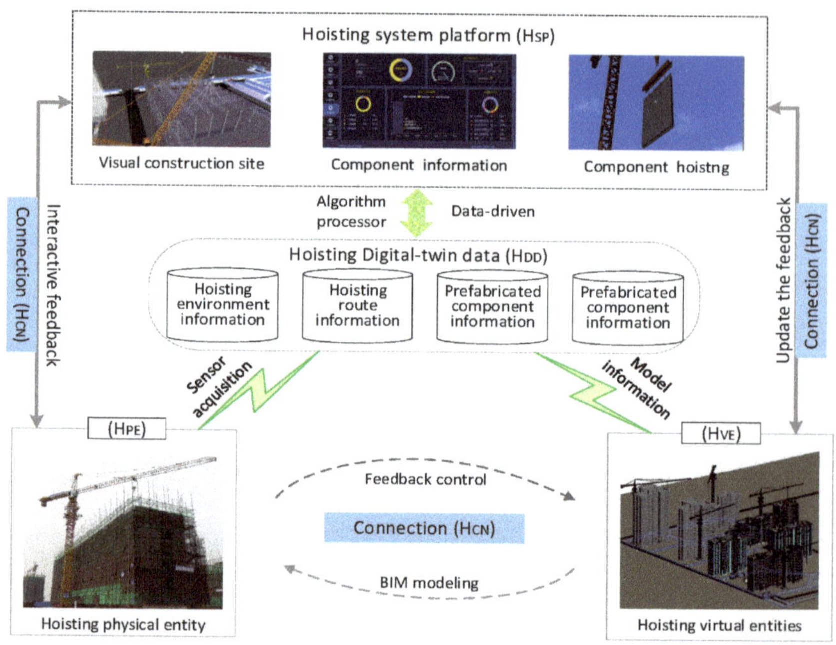

Figure 1. DT five-dimensional model for PC hoisting management.

Through the DT five-dimensional model for PC hoisting, the framework of the management system for PC hoisting is established in this study, as shown in Figure 2. Through H_{PE} in real object environment, H_{VE} is established for the visual display of PC hoisting. H_{VE} is processed and path planning for the hoisting route of PCs is conducted. The information of the PCs in H_{PE} in the hoisting process is collected through a variety of sensors. The collected data include position data, attitude data, stress–strain data, etc. The IoT is conducted to realize the real-time transmission of data in the hoisting process of PCs and the relationship between the variation in the hoisting process of PCs in H_{PE} and H_{DD} and

H_{SP} is established. H_{SP} can provide the visual display of PCs for tower crane operators, managers and project leaders. H_{SP} applies an intelligent algorithm to call and process the data in H_{DD}, so as to present the virtual construction site, PC position information and attitude information at the platform end. H_{SP} controls the swing of PCs in hoisting by setting the swing displacement threshold and the route offset threshold. When the threshold is exceeded, H_{SP} gives a danger alarm to reduce the risks in the hoisting process of PCs.

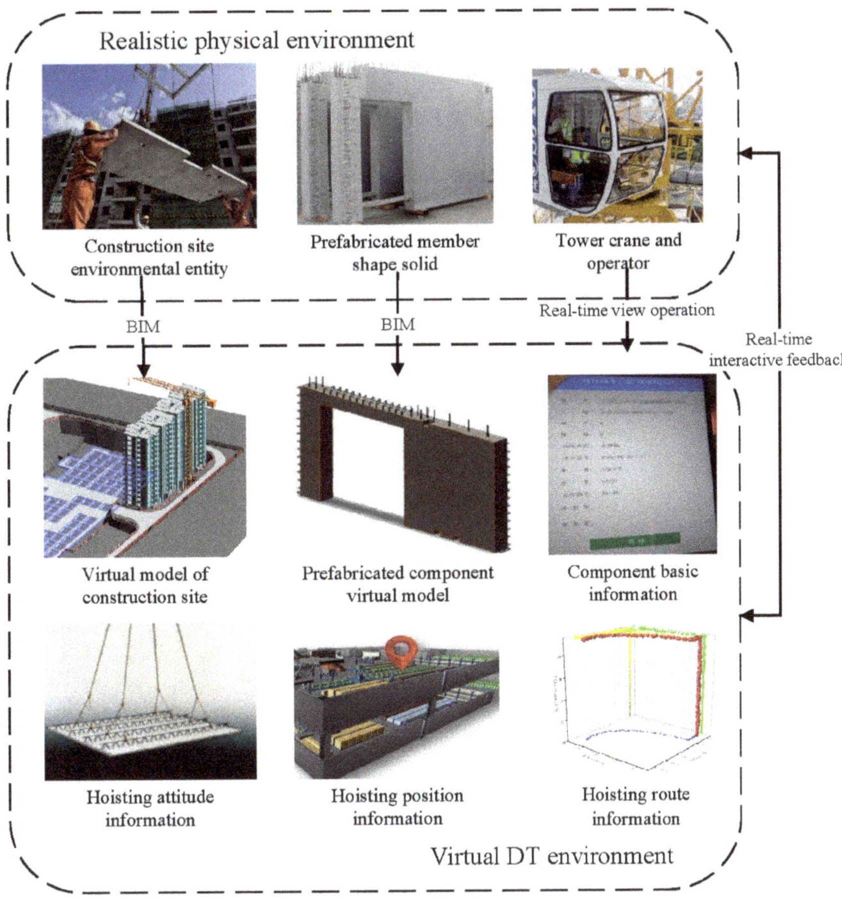

Figure 2. Framework of hoisting control method based on DTm.

4. DT-Driven Hoisting Management Framework Construction

Based on the concept of DT, this study establishes the connection between H_{PE} and H_{VE} through H_{DD} and H_{SP} and realizes the real-time interactive feedback between the real physical environment and the virtual digital environment. In the hoisting control management system established in this study, H_{PE} mainly includes construction site environmental entities, PC entities, tower cranes and operators. The construction site environmental entities and PC entities are modelled and digitized by virtual entities through BIM modelling. The virtual digital environment also includes the collection of a variety of data by a variety of sensor devices, including attitude information data and position information data, and uploads to H_{SP} through information transmission technology. By combining the H_{VE} data of PCs for data processing, visual presentation is carried out in H_{SP}. The position and attitude information displayed by the tower crane operator is used to assist decision making and operation, and real-time interactive feedback between H_{PE} and H_{VE} is realized. The information collection of PCs in the hoisting process adopts the Inertial Measurement Unit (IMU), GPS, RFID tag and other sensors. H_{PE} should be adjusted accordingly with the change in H_{PE}. If the building floor increases gradually, the hoisting virtual entity should be updated according to the actual change.

4.1. DT-BIM Model

Building information mainly comes from the BIM model, which highly integrates all relevant data in the building engineering project. Moreover, BIM, as a better information solution in the construction industry, has been widely applied. Based on the BIM model, this study establishes DT-BIM for hoisting. DT-BIM can realize the digital representation of H_{VE} on H_{PE} of construction sites, environmental entities and PC entities. Based on the BIM model in the design stage, the modelling of the surrounding environment and adjacent buildings should be conducted reasonably, so that the virtual entity can fully show the physical environment of the construction site, the construction field layout and the DT-BIM of the building. Meanwhile, reasonable modelling of the building model is conducted for all kinds of PCs that are hoisted on site and corresponding information is added into the model for unified information management of the PCs that are processed from the factory to the site construction and hoisting. The model of the PCs should be consistent with the entity of the PCs to fully display the information of the PCs in the hoisting process.

BIM model-type attributes are used to define common attributes of the same types of components, including section definition, geometric parameters and identification data information. The instance type is defined as each specific precast component model having unique attributes, including 3D graphics, dimension labelling, constraint mode, elevation and positioning, etc. In order to realize the clear presentation of the PC model in the BIM model, the PCs are named and coded in combination with the positioning and coding of components' names in the BIM model.

The BIM model is light-weighted to retain the 3D shape information of buildings and surrounding environment, so as to reduce the amount of model data and reduce the calculation pressure. According to the BIM model, combined with the layout location of the tower crane on the construction site, the location of the PCs, the location of the reinforcement processing zone and the surrounding building model, unified integration and lightweight processing are conducted, so that the construction hoisting site can be more clearly presented in the simplified BIM model. In the light-weight treatment of the BIM model, only relevant information affecting the hoisting of PCs in the construction site should be retained, such as the key location of the construction site as well as adjacent buildings and layouts that have an impact on hoisting, etc. Information that has no influence on hoisting, such as the specification of reinforcement, reinforcement and concrete of all parts of the building, is removed to reduce the information of the BIM model. The DT-BIM model is transmitted to the cloud server through the Industry Foundation Classes (IFC) interface; then, WebGL technology is used to mount it on the web page, so as to see the spatial environment layout of the construction and hoisting site of PCs on smart phones, tablets and other mobile devices.

In order to meet the requirements of the route optimization algorithm for highly abstracted construction site environments, this study proposes a plane model method which simplifies the DT-BIM model into a hoisting path-oriented model. The plane model facing the hoisting path is presented in the form of a spatial topology structure. The vertical hoisting plane model is established through the hoisting position of the PCs and the installation position of the PCs, which mainly includes the hoisting point of the PCs, the installation position of the PCs and the layout relationship of buildings and adjacent facilities, as shown in Figure 3.

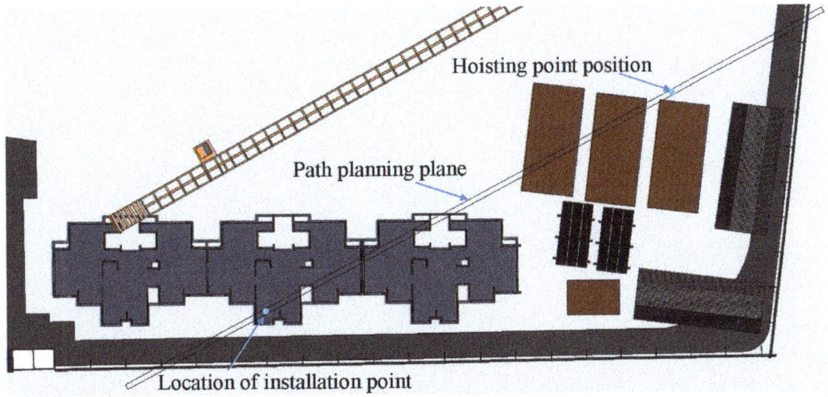

Figure 3. Hoisting location schematic diagram based on DT-BIM model.

4.2. Hoisting Route Planning Based on Dijkstra's Algorithm

According to the location of the PCs in the construction site, the path optimization algorithm is used to plan the hoisting path. According to the pre-developed building complex, the surrounding environment model and the principle of avoiding, the path is reasonably planned to establish the hoisting route from the storage location of the PCs to the installation point.

4.2.1. Prefabricated Component Hoisting Route Planning

This study takes the tower crane as an example to analyze the hoisting track of PCs and conduct hoisting route planning. In this study, the Dijkstra's algorithm is employed for improvement and the condition function of prohibition is added to the node topology to prevent the hoisting route from passing through the obstacle body. The hoisting DT-BIM model is processed and the plane model topology diagram facing the hoisting path is obtained through hoisting point and termination point. The target building and obstacle body area in the topology diagram are the hoisting prohibition area and the hoisting prohibition condition function is set, as shown in Equation (2). The forbidden condition function means that the distance between any two points in the restricted area of hosting is infinite, wherein the variable T represents the horizontal coordinate and the variable V represents the vertical coordinate. The forbidden conditional function is used to eliminate the case whereby the node connection (hoisting path) in the topology diagram intersects with the obstacle body region, preventing the hoisting object from contacting the obstacle body region. In the topological diagram, the simplified size of the obstacle body extends outward for a certain distance, representing as the obstacle region, and the conditional function of the region is established.

$$t_1 < T_x < t_2 \text{ and } v_1 < V_Z < v_2,\ d_{p_1 p_m} = \infty, \tag{2}$$

4.2.2. Application Principle of Improved Dijkstra's Algorithm

This section describes the principle of hoisting route planning of PCs in a construction site based on Dijkstra's algorithm. Dijkstra's algorithm adopts the pattern of a greedy algorithm, which is used to calculate the shortest path problem from one point to other points in a directed graph [44]. Its main feature is to extend the operation layer by layer from the starting point to the end point. The improved Dijkstra's algorithm can calculate the shortest path from the hoisting point to the installation point in a given topology based on the avoidance principle. The idea of the improved Dijkstra's algorithm is as follows:

(1) H = (S, U, E) is defined as a directed graph, where S is the set of nodes whose shortest path from source point to a node has solved and U is the set of nodes whose shortest path has not been solved.

(2) If there is a node P_n^1 for the source point P_m^1 and the edge d (P_n^1, P_m^1) that connects the two passes through the forbidden region (satisfies the forbidden condition function), then its path length is set to infinity. Meanwhile, if the slope of the side connecting source point P_n^2 to some node P_m^2 is less than 60 degrees, then the path length is set to infinity.

(3) The nodes in U are sorted according to the distance from the source point, the nodes with the minimum distance are transferred to set S and the nodes with the shortest path are recorded. Then, reorder the remaining nodes in U and select the node with the smallest distance.

(4) Repeat step 2. In the initial operation, there is only one point in S; find the shortest path each time and add it to set S, until all nodes are added to S and the algorithm ends.

In the above algorithm, the source points of the directed graph are updated according to the location of the PC and the directed graph and source points are updated according to the construction progress and the changes in the location of the PC.

4.3. *Method of Monitoring the Hoisting Process of PCs*

4.3.1. Fusion Algorithm Positioning in Hoisting Process

Based on acquisition layer sensor equipment such as RFID tag, IMU and GPS, it can collect detailed information of the components, position information and swing information in the process of assembly building construction in real time. The IMU is a device that measures an object's three-axis attitude angle (or angular velocity) and its acceleration. By means of three uniaxial accelerometers and three uniaxial gyroscopes of IMU, the oscillation information of the prefabricated component in three-dimensional space is obtained and the attitude of the object is calculated accordingly.

There are various representation methods for attitude monitoring, including three common ones, i.e., Euler angle, attitude matrix and quaternion. In this study, the attitude solution algorithm based on quaternion is adopted. When the quaternion represents the attitude, there are multiple ways to write it. When the Euler angle represents the attitude transformation, it represents three rotations around three axes. According to Euler's theorem, these three rotations can be equivalent to one rotation about one of the axes.

IMU was fused with GPS data and barometer data and the ETK (extended Kalman filter) algorithm was improved to monitor the hoisting trajectory of PCs. The algorithm architecture is shown in Figure 4. The improved fusion positioning algorithm architecture is shown in Figure 5. The fusion positioning algorithm fuses the data collected by IMU, GPS and barometer. The IMU collects the relative acceleration of the precast member in the hoisting process in real time, converts the coordinate to the absolute acceleration and obtains the three-dimensional displacement accumulation through two integrals. The collected angular rate is used to detect the stopping condition, information such as swing disturbance is zeroed and the quaternion is used to calculate the Euler Angle and the acceleration to monitor the swing. The above information collected by IMU forms a state matrix and Kalman filtering is carried out in combination with the position data collected by GPS and barometer, through which more accurate precast member position information can be obtained.

4.3.2. Information Transmission Based on LoRa Technology

According to the DTm for PC hoisting, this paper establishes the connection between H_{PE} and H_{DD} based on the IoT, which involves wireless information transmission technology, sensor technology, global positioning system and RFID technology, etc. In recent years, the IoT has developed rapidly. As one of the IoT methods, LPWAN has attracted more and more attention from enterprises and scholars. By analyzing and comparing the existing wireless transmission technologies, it was found that LoRa technology has the advantages of wider transmission range and lower energy consumption. According to the characteristics of complex construction processes of prefabricated buildings and the advantages of LoRa technology, such as flexible networking and low costs, compared with NB-IoT technology, LoRa technology is more suitable for information transmission in prefabricated building construction and hoisting sites.

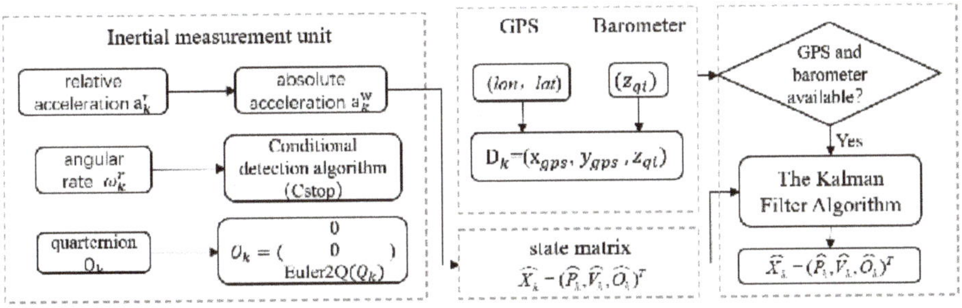

Figure 4. Fusion positioning algorithm architecture.

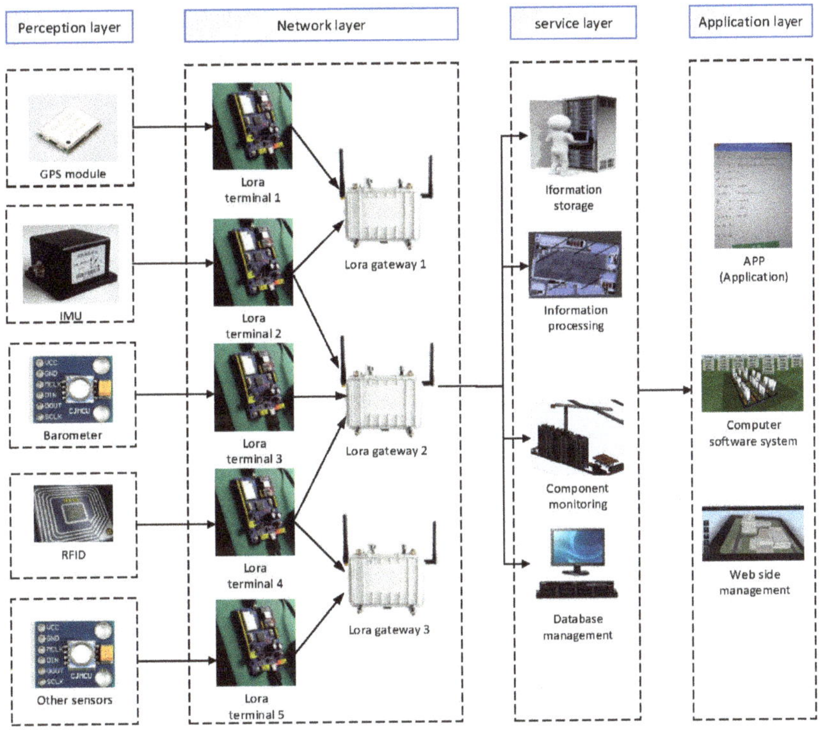

Figure 5. Hoisting monitoring method architecture based on LoRa WAN.

Based on the LoRa WAN network architecture, this study establishes the construction and hoisting site monitoring method architecture. LoRa technology is used for the wireless transmission of physical entity attitude information and position information during hoisting. In this study, the hoisting control terminal based on LoRa technology is used to transmit the information and data in the hoisting process wirelessly. The device has a variety of sensor interfaces, external IMU, GPS, barometer, etc. IMU collects attitude information, including acceleration, velocity, swing, etc., during the hoisting of PCs. IMU, GPS and barometer are used to collect the position information of PCs in the hoisting process through the fusion positioning algorithm. The collected information is transmitted in real time via the LoRa chip to the LoRa gateway, where it is uploaded to H_{DD}. The connection between H_{PE}, H_{DD} and H_{SP} is established through information collection, transmission and processing and the real-time feedback mechanism is established for visual display. As shown in Figure 5, the monitoring method architecture is mainly composed of perception layer, network layer, service layer and application layer. The perception layer refers to the information collection of the hoisting site through IMU, GPS, barometer, RFID tag and other sensors and information uploading through the LoRa terminal device. The network layer refers to the real-time and accurate transmission of the information collected by the perception layer through the LoRa network. The information collected by the perception layer is transmitted to the gateway through the LoRa WAN protocol and then to the service layer through the gateway. In the service layer, database technology is used to store and process the collected information, including component attribute information, hoisting monitoring information, etc. The information in the service layer can be displayed on apps and web pages through mobile phones, computers and other devices.

5. DT Application Case Study

Based on the DT theory and method, the framework of the control system for PC hoisting is established. In this system, the DT-BIM model of prefabricated building is embedded into the system, and the data collected by the sensor is also visually displayed at the H_{VE}. The DT-BIM model and the data collected by the sensor are stored in H_{DD}. According to these data and algorithms, component positioning, attitude monitoring, path-planning presentation and construction management functions are realized. In this study, the key technologies to realize these functions were experimentally studied and the above functions were preliminarily realized.

5.1. Application of Hoisting Route Planning Method

DT achieves accurate mapping of virtual and real objects that are similar in form, process and behavior by constructing virtual space objects corresponding to physical objects in physical space. By using digital twinning to achieve the accurate mapping of virtual and real conditions, we can have a more realistic grasp of the unpredictable situations in reality. The management method of twin hoisting based on twin PCs is shown in Figure 6. In this study, the real-time connection between the physical world and the virtual world for the hoisting of PCs was established through digital twinning and the dynamic interaction between the real situation and the virtual situation of construction hoisting process management was realized. At the same time, route planning and result prediction analyses of hoisting process were carried out through the interactive detection result analysis of processing architecture and algorithm. The mapping analysis of construction hoisting management is shown in the figure. Based on the concept of DT, this study started from model and data interaction, established the foundation of the virtual space model, perceiving and optimizing real-time information of physical entities, and promoted the technological integration and iterative integration of the BIM model, sensors, visualization software and IoT. In this study, the twin body mapping elements for prefabricated component hoisting were established, as shown in Figure 7. The driving factors of DT for the hoisting of PCs established in this study mainly include the following six aspects:

(1) BIM model—a pre-established digital carrier of virtual mapping based on real physical space objects;
(2) Sensor—actual hoisting process-related operation management and multi-dimensional and multi-level sensing tools;
(3) Data—including data collected by sensors, construction plan data and historical reference data (physical list, design specifications, engineering drawings, on-site feedback, etc.);
(4) Integration—the physical vehicle for the interaction between the physical and virtual digital worlds;

(5) Analysis/Forecasting—analyze data through artificial intelligence algorithms and visualization programs and provide analytical decision-making and forecasting solutions;
(6) DT—an accurate real-time digital model of the physical world.

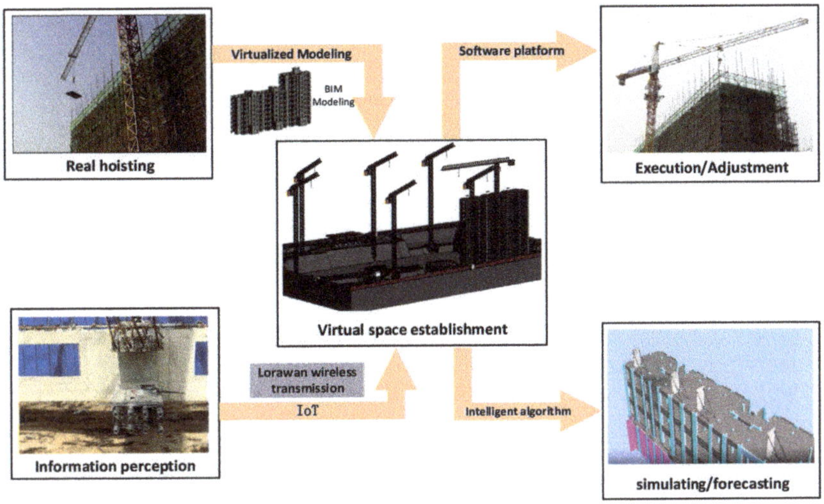

Figure 6. Twinning management method of hoisting of PCs.

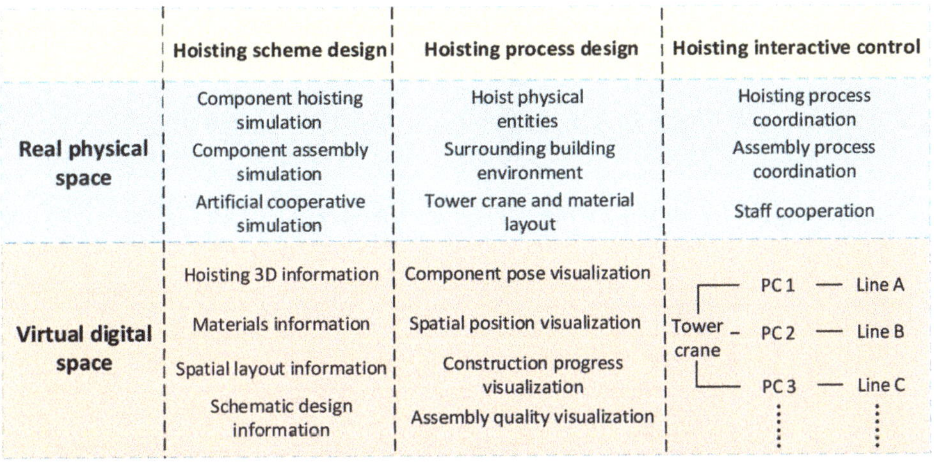

Figure 7. Mapping elements of virtual-real twin for hoisting of PCs.

This part takes the PC hoisting on the construction site of prefabricated buildings as an example, simplifies the DT-BIM (as shown in Figure 8) and preliminarily realizes the hoisting path planning with the improved Dijkstra's algorithm. The hoisting point position and the installation point position were determined in the lift-oriented virtual model and the plane model of the prefabricated floor hoisting route was obtained. As shown in Figure 9, the hoisting planning plane was divided into prohibited zones. In the figure, the Prohibited Zone 1, where the target building is located, and the Prohibited Zone 2, where the steel processing zone is located, affected the hoisting this time. The prohibited area is a certain length of the external contour extension of the obstacle as a safety area to avoid collisions between PCs and obstacles in the hoisting process.

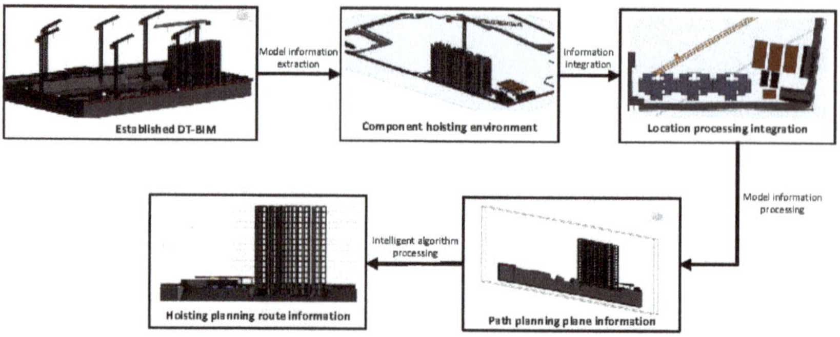

Figure 8. Hoisting planning plane obtained by model cutting.

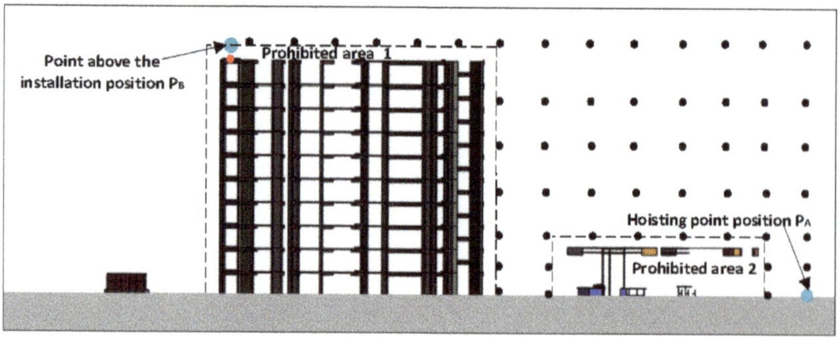

Figure 9. Hoisting topology diagram of PCs.

The hoisting route node set was generated by generating the hoisting route node passing nodes in the hoisting topology diagram, as shown in Equation (3).

$$P_L = \{P_A, P_1, P_2, P_3, \ldots \ldots, P_{n-1}, P_n, P_B\}, \quad (3)$$

The hoisting path was planned by using the improved Dijkstra's algorithm in Section 4.2. Firstly, the topology graph was represented by a topology matrix, in which the number represented the distance between its row node number and its column node number. The connected distance between nodes was expressed in the form of a matrix; then, the distance between nodes directly connected through the dangerous area was changed to infinity. This way, one can avoid collisions when using algorithms to calculate time paths. The established node topology matrix is shown in Figure 10. The data before and after algorithm optimization were compared and the results are shown in Table 1. It can be seen that the optimization effect was 27.369% when the safety angle of hoisting was not considered and 19.844% when the safety angle of hoisting was considered.

5.2. Data Acquisition and Transmission Unit

In this study, a sensor was installed on the PC for data acquisition during the hoisting process and real-time monitoring of the PC during the hoisting process was conducted to ensure the safety and visibility of the PC during the hoisting process. In this part, BIM technology, LoRa technology and inertial measurement unit were combined to realize the monitoring of PC in the hoisting process. In addition, taking prefabricated building construction in Tianjin as an example, data collection, uploading and webpage display were conducted.

	P_A	P_1	P_2	P_3	⋯⋯	P_{n-2}	P_{n-1}	P_n	P_B
P_A	0	4	8	14	⋯⋯	∞	∞	∞	∞
P_1	4	0	4	10	⋯⋯	∞	∞	∞	∞
P_2	8	4	0	6	⋯⋯	∞	∞	∞	∞
P_3	14	10	6	0	⋯⋯	∞	∞	∞	∞
⋮	⋮	⋮	⋮	⋮	⋱	⋮	⋮	⋮	⋮
P_{n-2}	∞	∞	∞	∞	⋯⋯	0	5	10	∞
P_{n-1}	∞	∞	∞	∞	⋯⋯	5	0	5	∞
P_n	∞	∞	∞	∞	⋯⋯	10	5	0	2
P_B	∞	∞	∞	∞	⋯⋯	12	7	2	0

Figure 10. Topology matrix of prefabricated component hoisting route planning.

Table 1. Effect analysis table of hoisting route planning for a prefabricated floor.

Hoisting Situation	Prefabricated Floor Hoisting Path	Length of Hoisting Route (m)	Reduced Length (m)	Optimized Effect (%)
1	$P_A \rightarrow P_6 \rightarrow P_{47} \rightarrow P_B$	113.35	/	/
2	$P_A \rightarrow P_{17} \rightarrow P_{39} \rightarrow P_{40} \rightarrow P_B$	90.857	22.493	19.844
3	$P_A \rightarrow P_9 \rightarrow P_{31} \rightarrow P_{46} \rightarrow P_{47} \rightarrow P_B$	82.33	31.02	27.367

Information acquisition in this case mainly includes data acquisition in the hoisting process of PCs by sensors such as attitude sensor, stress sensor and GPS sensor. Based on LoRa technology, the PC hoisting monitoring information collection and upload terminal and transmission gateway were developed. The physical structure is shown in Figure 11. The transmission terminal was composed of an MCU control module, an LoRa module and various sensor interfaces. Information of hoisting process was collected and upload-ed by the acquisition terminal mentioned above. Figure 10 shows the acquisition terminal installed on the PC and its composition. The information collected by the acquisition terminal was uploaded through the LoRa gateway. The LoRa gateway applied on the construction site is shown in Figure 11.

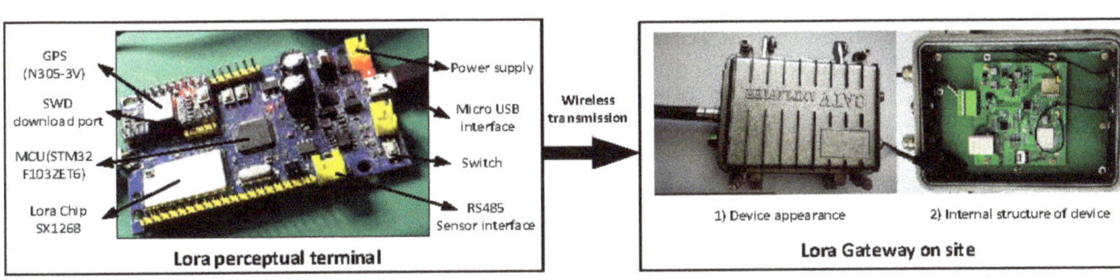

Figure 11. Physical view of LoRa information transmission terminal and gateway architecture.

In the case study, the BIM hoisting model established was exported in IFC format, which was parsed and read in the JavaScript environment for extension and light-weight processing, and the developed interface was used for transmission and uploading to H_{DD}. Additionally, sensor information was transmitted to the system and combined with the twin model for visual display, so as to realize the retrieving of model data on the system platform or software and the viewing of the overall environment affecting hoisting in the construction site. The access process of the web terminal system is shown in Figure 12. Through the functional analysis of the system platform, it mainly included eight functional modules, including staff management, system management, site safety management, information entry, DT database management and DT model management. The contents of each module are shown in Figure 13.

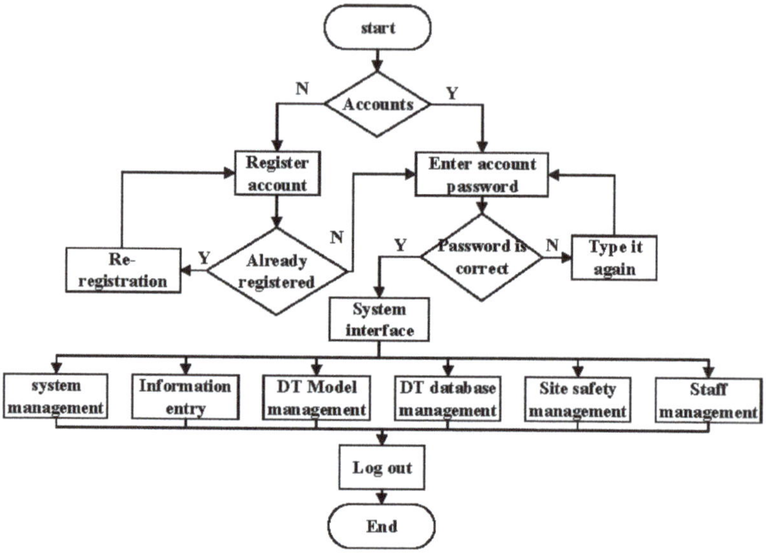

Figure 12. LoRa gateway installed on the construction site.

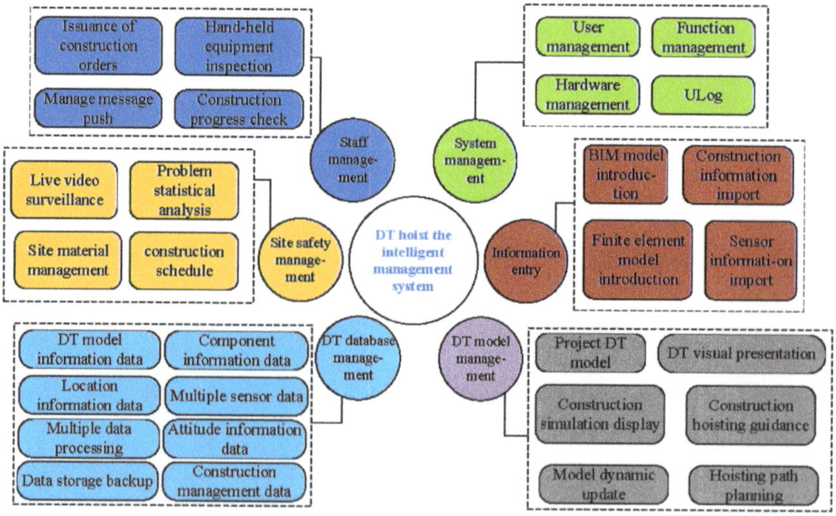

Figure 13. Analysis of system platform functional architecture.

5.3. *Prefabricated Component Hoisting Route Monitoring*

The precast floor numbered T1-11-2-3 in the case study established the hoisting route planning plane of the floor through the hoisting position and the installation position and determined the hoisting route of the precast floor through the path-planning algorithm of this study, as described in Section 4.2. The construction and hoisting of the prefabricated floor were conducted according to the path planned by the algorithm and hoisting control was conducted according to the prefabricated floor information transmitted during the hoisting process. After receiving the instruction from the construction site, the manager of the component yard transported the components out according to the information in the system and re-entered the working state through the control transmission module of hand-held devices for the real-time transmission of information. According to the information in the RFID spreader selection, floor ring inspection and other work, the correct hoisting needed to be checked. According to the information collected by the LoRa terminal during the hoisting process, the tower crane operator could view the real-time information of the components on the equipment and check the deflection, tilt angle and position information of the components during the hoisting process. The collected position information was processed and compared with the Kalman filtering effect, as shown in Figure 14. The position information directly measured by the sensor had a large error, especially the position of multiple projections. By comparing the position information of the three planes, it was found that the positioning accuracy was improved to a certain extent after the Kalman filtering correction of various sensors. By monitoring the location information of PCs, the tower crane was controlled timely to ensure that the components were lifted accurately to the installation point of prefabricated floor, thus improving the safety and visibility of the whole hoisting process.

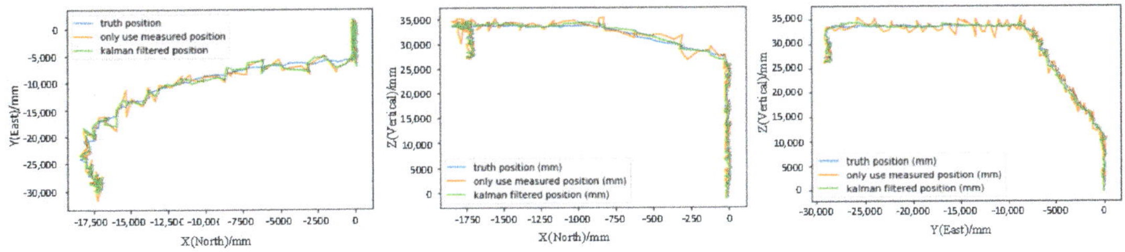

Figure 14. Hoisting route monitoring and comparison.

6. Discussion and Conclusions

Based on the concept of DT, this paper proposes the framework of a PC hoisting management system; furthermore, it intelligently controlled the hoisting process of PCs by using DT-BIM, IoT and sensor technologies. This study absorbed advanced technologies from other industries and applied them to the field of construction. An innovative solution is proposed for PC hoisting control, which has the following advantages.

(1) Aiming at the application method of assembly construction stage, the multi-technology integration mechanism of DT, IoT and BIM is explored. The twin feedback mechanism of virtual and real interaction provides a new idea for the realization of "real-time perception, intelligent analysis and intelligent decision" in the construction and hoisting management of PCs.

(2) In this study, DT-BIM hoisting planning model was established based on the DTm and hoisting path planning was carried out combined with the Dijkstra's algorithm. This paper uses virtual space data to deal with the problem of hoisting paths and verifies the effect of the optimized hoisting path. The application of DT model was well verified by this method.

(3) LoRa technology and multi-sensor fusion method were applied to the construction site monitoring of prefabricated buildings, realizing the twin interactive connection of prefabricated construction and opening up the DT interactive channel. This paper uses the fusion location algorithm to process the data and confirms the effect of the depth information fusion on the optimization control of PCs in the process of hoisting, which is helpful to promote the intelligent development of prefabricated building construction.

(4) PC hoisting construction intelligentization is a multi-disciplinary integrated engineering system. This research project adopted DT technology, IoT, BIM and the integration method driven by an intelligent algorithm to realize the virtual and real interaction of hoisting management, which greatly improved the intelligent management level of PC hoisting.

Through the introduction of DT, IoT, BIM technology and sensor technology, strengthening the intelligent management and control of the assembly construction process is feasible and can improve the visual and intelligent level of site construction, so as to ensure the accuracy and safety of construction. Meanwhile, with a background of intelligent construction reform and industrial upgrading of construction industry, this study actively explores research and application schemes of DT, IoT and other technologies in the field of intelligent construction of prefabricated buildings, providing a good reference for future research.

Author Contributions: Conceptualization, Z.L; methodology, Y.Z.; software, Y.Z.; validation, Y.Z., Y.Z. and C.C.; writing—original draft preparation, C.C.; writing—review and editing, Y.Z.; project administration, Z.L.; funding acquisition, Z.L. All authors have read and agreed to the published version of the manuscript.

Funding: This research study was supported by the Deep Learning-Based Lifting Safety Risk Prediction and Control Method of Assemblies Building (8201001) of Beijing Natural Science Foundation.

Institutional Review Board Statement: Not applicable.

Informed Consent Statement: Informed consent was obtained from all subjects involved in the study.

Data Availability Statement: The data used to support the findings of this study are available from the corresponding author upon request.

Acknowledgments: The authors would like to thank Beijing University of Technology for its support throughout the research project.

Conflicts of Interest: The authors declare no conflict of interest.

References

1. Arashpour, M.; Wakefield, R.; Blismas, N.; Minas, J. Optimization of process integration and multi-skilled resource utilization in off-site construction. *Autom. Constr.* **2015**, *50*, 72–80. [CrossRef]
2. Li, Z.; Shen, G.Q.; Alshawi, M. Measuring the impact of prefabrication on construction waste reduction: An empirical study in China. *Resour. Conserv. Recycl.* **2014**, *91*, 27–39. [CrossRef]
3. Liu, J.; Yi, Y.; Wang, X. Exploring factors influencing construction waste reduction: A structural equation modeling approach. *J. Clean. Prod.* **2020**, *276*, 123185. [CrossRef]
4. Zhong, R.; Peng, Y.; Xue, F.; Fang, J.; Zou, W.; Luo, H.; Thomas, N.S.; Lu, W.; Shen, G.Q.P.; Huang, G.Q. Prefabricated construction enabled by the Internet-of-Things. *Autom. Constr.* **2017**, *76*, 59–70. [CrossRef]
5. Zhao, L.; Liu, Z.; Jasper, M. Development of Intelligent Prefabs Using IoT Technology to Improve the Performance of Prefabricated Construction Projects. *Sensors* **2019**, *19*, 4131. [CrossRef]
6. Bortolini, R.; Formoso, C.T.; Viana, D.D. Site logistics planning and control for engineer-to-order prefabricated building systems using BIM 4D modeling. *Autom. Constr.* **2019**, *98*, 248–264. [CrossRef]
7. Ko, H.S.; Azambuja, M.; Lee, H.F. Cloud-based Materials Tracking System Prototype Integrated with Radio Frequency Identification Tagging Technology. *Autom. Constr.* **2016**, *63*, 144–154. [CrossRef]
8. Dave, B.; Kubler, S.; Kary, F.; Lauri, K. Opportunities for enhanced lean construction management using Internet of Things standards. *Autom. Constr.* **2015**, *61*, 86–97. [CrossRef]
9. Enrique, V.; Adán, A.; Carlos, C. Evolution of RFID Applications in Construction: A Literature Review. *Sensors* **2015**, *15*, 15988–16008.
10. Lee, S.K.; Kwon, S. A conceptual framework of prefabricated building construction management system using reverse engineering, bim, and wsn. *Adv. Constr. Build. Technol. Soc.* **2014**. Available online: https://www.iaarc-academy.com/download/CIB_IAARC_W119_CIC_2013_Proceedings.pdf#page=42 (accessed on 1 January 2022).
11. Wang, X.; Wang, S.; Song, X.; Han, Y. IoT-Based Intelligent Construction System for Prefabricated Buildings: Study of Operating Mechanism and Implementation in China. *Appl. Sci.* **2020**, *10*, 6311. [CrossRef]
12. Ma, Z.; Li, S. New Model of Project Management under the "Internet +" environment. *J. Tongji Univ. Nat. Sci.* **2018**, *46*, 135–139. (In Chinese).
13. Ma, Z.; Yang, Z. Bim-based Functional Demand analysis of Intelligent Residential Parts Production Operation Planning and Control System. *Civ. Archit. Eng. Inf. Technol.* **2015**, *7*, 1–7. (In Chinese)

14. Ma, Z.; Zhang, D.; Zhou, Q.; Liu, Z.; Yang, Z. Material management System of Subway Construction Site based on Mobile terminal and existing information System. *Constr. Technol.* **2012**, *41*, 5–9. (In Chinese)
15. Liu, S.; Liu, Z.; Wang, W.; Zhao, Y. Application of NB-IoT Technology in prefabricated Building Construction Management. *J. Civ. Eng. Manag.* **2019**, *36*, 178–184. (In Chinese)
16. Liu, Z.; Liu, S.; Wang, W.; Zhao, Y. An information solution for prefabricated building construction process based on low-power wide-area Internet of Things. *Constr. Technol.* **2018**, *47*, 117–122. (In Chinese)
17. Froese, T.M. The impact of emerging information technology on project management for construction. *Autom. Constr.* **2010**, *19*, 531–538. [CrossRef]
18. Wang, F.; He, F. Study of Hoist Perception System Based on IoT Technology. In Proceedings of the 2010 International Conference on Web Information Systems and Mining, Sanya, China, 23–24 October 2010; Volume 114, pp. 357–360.
19. Zhou, C.; Luo, H.; Fang, W.; Wei, R.; Ding, L. Cyber-physical-system-based safety monitoring for blind hoisting with the internet of things: A case study. *Autom. Constr.* **2018**, *97*, 138–150. [CrossRef]
20. Lee, G.; Kim, H.H.; Lee, C.J.; Ham, S.I.; Yun, S.H.; Cho, H.; Kim, B.K.; Kim, G.T.; Kim, K. A laser-technology-based hoisting-path tracking system for a robotic tower crane. *Autom. Constr.* **2009**, *18*, 865–874. [CrossRef]
21. Cai, P.; Cai, Y.; Chandrasekaran, I.; Zheng, J. Parallel genetic algorithm based automatic path planning for crane hoisting in complex environments. *Autom. Constr.* **2016**, *62*, 133–147. [CrossRef]
22. Zhang, J.; Shen, C.; Su, H.; Arafin, M.T.; Qu, G. Voltage Over-scaling-based Lightweight Authentication for IoT Security. *IEEE Trans. Comput.* **2021**, *99*, 1. [CrossRef]
23. Li, B.; Liang, R.; Zhou, W.; Yin, H.; Cai, K. LBS Meets Blockchain: An Efficient Method with Security Preserving Trust in SAGIN. *IEEE Internet Things J.* **2021**, *99*, 1.
24. Rosen, R.; Wichert, G.V.; Lo, G.; Bettenhausen, K.D. About The Importance of Autonomy and Digital Twins for the Future of Manufacturing. *IFAC Pap.* **2015**, *48*, 567–572. [CrossRef]
25. Grieves, M. *Digital Twin: Manufacturing Excellence through Virtual Factory Replication*; Dassault Systèmes: Paris, France, 2014; Available online: https://www.3ds.com/fileadmin/PRODUCTS-SERVICES/DELMIA/PDF/Whitepaper/DELMIA-APRISO-Digital-Twin-Whitepaper.pdf, (accessed on 23 March 2015).
26. Grieves, M. *Virtually Perfect: Driving Innovative and Lean Products through Product Lifecycle Management*; Space Coast Press: Cocoa Beach, FL, USA, 2011.
27. Negri, E.; Fumagalli, L.; Macchi, M. A Review of the Roles of Digital Twin in CPS-based Production Systems. *Procedia Manuf.* **2017**, *11*, 939–948. [CrossRef]
28. Tao, F.; Qi, Q.; Wang, L.; Nee, A.Y.C. DTs and cyber–physical systems toward smart manufacturing and industry 4.0: Correlation and comparison. *Engineering* **2019**, *5*, 653. [CrossRef]
29. Tao, F.; Liu, W.; Zhang, M.; Hu, T.; Qi, Q.; Zhang, H.; Sui, Y.; Wang, T.; Xu, H.; Huang, Z.; et al. Digital Twin Five dimensional Model and its Application in Ten Fields. *Comput. Integr. Manuf. Syst.* **2019**, *25*, 1–18. (In Chinese)
30. Wu, C.; Zhou, Y.; Pessôa, M.V.P.; Peng, Q.; Tan, R. Conceptual digital twin modeling based on an integrated five-dimensional framework and TRIZ function model. *J. Manuf. Syst.* **2020**, *58*, 79–93. [CrossRef]
31. Wang, P.; Luo, M. A digital twin-based big data virtual and real fusion learning reference framework supported by industrial internet towards smart manufacturing. *J. Manuf. Syst.* **2021**, *58*, 16–32. [CrossRef]
32. Qiu, C.; Zhou, S.; Liu, Z.; Gao, Q.; Tan, J. Digital assembly technology based on augmented reality and digital twins: A review. *Virtual Real. Intell. Hardw.* **2019**, *1*, 597–610. [CrossRef]
33. Bazaz, S.M.; Lohtander, M.; Varis, J. Availability of Manufacturing data resources in Digital Twin. In Proceedings of the 30th International Conference on Flexible Automation and Intelligent Manufacturing (FAIM2021), Athens, Greece, 15–18 June 2021.
34. Tao, F.; Zhang, M.; Cheng, J.; Qi, Q. Digital Twin Workshop—A New Mode of workshop Operation in the future. *Comput. Integr. Manuf. Syst.* **2017**, *23*, 1–9.
35. Tao, F.; Cheng, Y.; Cheng, J.; Zhang, M.; Xu, W.; Qi, Q. Digital Twin Workshop Information physical Fusion Theory and Technology. *Comput. Integr. Manuf. Syst.* **2017**, *23*, 1603–1611.
36. Liu, W.; Tao, F.; Cheng, J.; Zhang, L.; Yi, W. Digital Twin Satellites: Concepts, Key technologies and Applications. *Comput. Integr. Manuf. Syst.* **2020**, *26*, 565–588.
37. Meng, S.; Ye, Y.; Yang, Q. Digital twin and its aerospace applications. *J. Aeronaut.* **2020**, *41*, 1–12. (In Chinese)
38. CAICT—China Academy of Information and Communications Technology. What Does the Xiongan Digital Twin City and Smart City Look Like? Take a Look at Some Real-Life Example . . . . Available online: http://www.sohu.com/a/231025717_354877, (accessed on 8 May 2018).
39. CAICT—Chinese Academy of Information and Communications Technology. The Third Symposium on Digital Twin Cities Was Held in Beijing. Available online: http://www.caict.ac.cn/xwdt/ynxw/201806/t20180625_175807.htm (accessed on 25 June 2018).
40. Yua, J.; Songa, Y.; Tang, D.; Dai, J. A Digital Twin approach based on nonparametric Bayesian network for complex system health monitoring. *J. Manuf. Syst.* **2021**, *58*, 293–304. [CrossRef]
41. Liu, Z.; Shi, G.; Zhang, A.; Huang, C. Intelligent Tensioning Method for Prestressed Cables Based on Digital Twins and Artificial Intelligence. *Sensor* **2020**, *20*, 7006. [CrossRef]

42. Liu, Z.; Bai, W.; Du, X.; Zhang, A.; Jiang, A. Digital Twin-based Safety Evaluation of Prestressed Steel Structure. *Adv. Civ. Eng.* **2020**, *2020*, 8888876. [CrossRef]
43. Liu, Z.; Zhang, A.; Wang, W. A Framework for an Indoor Safety Management System Based on Digital Twin. *Sensor* **2020**, *20*, 5771. [CrossRef]
44. Liu, Z.; Zhang, A.; Wang, W.; Wang, J. Digital Twin Driven dynamic fire safety evacuation method for Winter Olympic Stadium. *J. Tongji Univ. Nat. Sci.* **2020**, *48*, 962–971. (In Chinese)

buildings

MDPI

Article

Research on Factors Influencing Intelligent Construction Development: An Empirical Study in China

Tao Li [†], Xiaoli Yan [*,†], Wenping Guo and Feifei Zhu

School of Management Studies, Shanghai University of Engineering Science, Shanghai 201620, China; litao0222@foxmail.com (T.L.); 13593373334@163.com (W.G.); z18605566905@163.com (F.Z.)
* Correspondence: yanxiaoli821@sues.edu.cn or yanxiaoli821@163.com
† These authors contributed equally to this work.

Abstract: Intelligent construction (IC) is an innovative development model of the construction industry in which construction is integrated with digital technologies against the backdrop of the new technological revolution. The development of IC involves many influencing factors which are actively promoting IC development. However, investigations focusing on identifying and examining the relationships among the factors necessary for IC development are limited. In contributing to bridging this gap, this paper investigated and analyzed influencing factors for IC development by developing structural equation modeling (SEM) based on 5 variables and 28 measures, including (1) identifying the factors and examining their influence on IC development in China and (2) clarifying the paths and key measures for successful IC development. The results showed that (1) the three variables of government, company, and technology had a direct and significant impact on the development of IC, (2) the three variables of industry, company, and technology actually formed a "closed-loop" within which they interact and promote each other, and (3) it was widely realized and accepted that IC development has bright prospects in China. Furthermore, four paths for IC development were obtained and the key measures of the five variables were further analyzed. This research contributes to the body of knowledge on IC by identifying the factors influencing IC development. The four paths and key measures were proposed to clarify the relationship between factors. Recommendations were put forward to promote IC development.

Keywords: intelligent construction; influencing factors; structural equation modeling

Citation: Li, T.; Yan, X.; Guo, W.; Zhu, F. Research on Factors Influencing Intelligent Construction Development: An Empirical Study in China. *Buildings* **2022**, *12*, 478. https://doi.org/10.3390/buildings12040478

Academic Editors: Hongling Guo, Jia-Rui Lin and Yantao Yu

Received: 26 February 2022
Accepted: 6 April 2022
Published: 12 April 2022

Publisher's Note: MDPI stays neutral with regard to jurisdictional claims in published maps and institutional affiliations.

Copyright: © 2022 by the authors. Licensee MDPI, Basel, Switzerland. This article is an open access article distributed under the terms and conditions of the Creative Commons Attribution (CC BY) license (https://creativecommons.org/licenses/by/4.0/).

1. Introduction

The global construction industry is booming, increasing building projects and demand for intellectual development [1]. Furthermore, given the increasingly frequent transfer of technologies such as building information modeling (BIM) and additive manufacturing, the intelligent development of the construction industry is inevitable [2].

At present, the expression "the process or product of construction using emerging digital technologies" primarily refers to "digital construction" [3], "smart construction" [4], or "construction 4.0" [5,6]. "digital construction", "smart construction", "intelligent construction", and "construction 4.0" have similar connotations; that is, the use of emerging technologies to achieve integrated collaboration of project approval decision making, planning and design, construction, and operation and maintenance services. Such integrated collaboration can significantly improve the efficiency and effectiveness of the construction industry. IC is an innovative development model of the construction industry [7]. Especially in China, IC is greatly advocated. In July 2020, the "Guiding Opinions on Promoting the Coordinated Development of Intelligent Construction and Building Industrialization" jointly issued by 13 Chinese government ministries and commissions proposed increasing the application of IC in all aspects of construction to form an IC industry.

In recent years, the application of IC technologies in construction has become increasingly extensive [8,9], and the research on IC has mainly focused on the application of BIM [10–13], intelligent equipment [14–16], information and communication technology [17], and additive manufacturing [18–20] in the construction industry.

It can be seen from previous studies that most of the research on IC development is concerned with IC technology, and investigations focusing on identifying and examining the relationships of the influencing factors necessary for IC development are limited. However, identifying the factors and their impact is crucial for IC development. Therefore, this paper investigated and analyzed factors for IC development by (1) taking IC development in China as an example and identifying the factors and examining their influence on IC development, and (2) clarifying the paths and key measures for IC development.

2. Theoretical Framework

The conceptual research model is constructed, hypotheses are provided, and the conceptual hypothesis model is presented in this section.

2.1. Conceptual Research Model

Explanatory factors and their relationships are discussed in classical theories of industrial development. A variety of analytical frameworks were proposed to analyze industry development. Bain (1956) proposed the structure-conduct-performance (SCP) analysis paradigm to explore the impact between national policies, technological upgrades, and corporate behavior. Porter (1990) presented four factors for determining industry development, namely production, demand conditions, the performance of related industries and supporting industries, and corporate strategy, structure, and competition in the industry, and claimed that two variables influenced these factors: opportunity and government. Industrial development theory studies the problems of technological innovation, industrial clusters, companies, and product service changes [21]. The theory assumes that the motivation of industrial development includes scientific and technological innovation and other factors related to production, policies, and markets.

In the case of the influencing factors of IC development, Ding [7] proposed that the government needs to actively support IC technology innovation to provide new opportunities, promote the transformation of construction companies to seize market opportunities, and encourage cross-border industries to extend the engineering market. Wang [22] believed that the current IC technology development is unusually rapid and technology upgrading assumes the role of a pioneer, promoting the high-quality development of the construction industry. Chen & Ding [23] also pointed out that the development of domain technology plays a key role in IC development. In July 2020, the "Guiding Opinions on Promoting the Coordinated Development of IC and Building Industrialization" pointed out that IC development involves the transformation and upgrading of technology, company, and industry.

From the above research review, this paper studied the following four factors that influence IC development: government, industry, company, and technology. Moreover, we built a conceptual model, as shown in Figure 1. Then, we clarified how these factors act on IC development by examining their relationships. Finally, we discussed the influencing paths and critical measures for IC development in China.

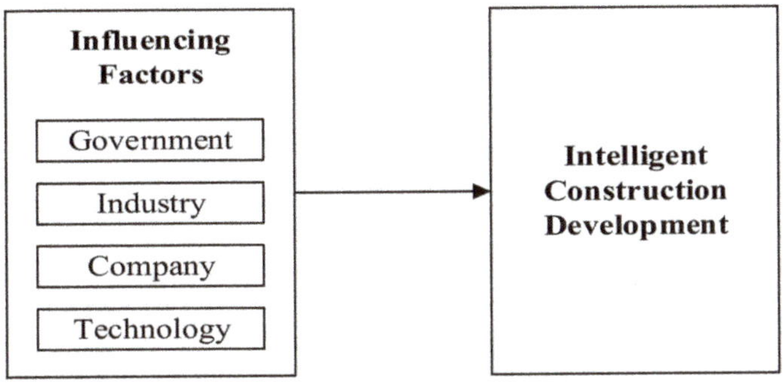

Figure 1. Conceptual research model.

2.2. Research Hypothesis

Based on a review of the research literature and expert interviews, five items including ten research hypotheses were proposed, then the conceptual hypothesis model was constructed.

2.2.1. The Relationship between Various Influencing Factors

(1) The relationship between the government and other factors

The government is an essential driver of IC development. IC development in China is driven by macro-development [24]. The role of the government affects the development of industry, company, and technology: government subsidies reduce economic risks to businesses and boost industry development [25], directly reducing the cost of research and development, encouraging companies to engage in research and development, and improving the degree of technological growth in the industry [26]. On the other hand, subsidy funds help reverse the disadvantage of cash flow shortages, reduce the risk of debt repayment, and provide more protection for the business development of companies [27]. Moreover, subsidies impact technological innovation output [28]. They make a significant contribution to technological advances [29], which also impact technological innovation behavior. Based on the above literature, we made the following assumptions:

Hypothesis 1 (H1). *Government has a positive impact on the industry;*

Hypothesis 2 (H2). *Government has a positive impact on the company;*

Hypothesis 3 (H3). *Government has a positive impact on technology;*

(2) The relationship between industry and other factors

The industry development trend has led to a mandatory structural adjustment on the supply of companies [30]. It has induced companies to carry out technological innovation [31]. According to the questionnaire survey, the industry development has had an important impact on individual companies. Based on the above literature, we made the following assumption:

Hypothesis 4 (H4). *The industry has a positive impact on business development;*

(3) The relationship between the company and other factors

Corporate strategy, innovation expectations, and research and development investment impact technology innovation [32–34]. Under the premise of pursuing their interests, companies have increased their investment in research, forming a virtuous circle that promotes technological progress [35]. For example, with the development of intelligent technology, architectural design companies need to study intelligent integrated systems to meet design requirements, solve difficulties, improve efficiency, and complete transformation and upgrading [36].

Based on the above literature, we made the following assumption:

Hypothesis 5 (H5). *The Company has a positive impact on technological development;*

(4) The relationship between technology and other factors

Schumpeter [37] argued that technology drove new market demand, incredibly innovative technology combined with marketing, which guides industry development. Many scholars have recognized this conclusion. With the current exponential growth of information technology, digital applications, networked applications, and intelligent integration innovation are the three driving forces of new technology [38]. Based on the above literature, we made the following assumption:

Hypothesis 6 (H6). *Technology has a positive impact on the industry.*

2.2.2. The Relationship between Various Influencing Factors and the IC Development

(1) The relationship between the government and IC development

IC development is a complex system. The government plays a key driving role in industrial development through overall planning and careful arrangements to create an effective environment for development [39]. The principle of market allocation and government guidance should be followed to realize IC development [40], which requires the government to actively cultivate new industries for IC and strengthen the transformation and promotion of scientific and technological achievements [41].

Based on the above literature, we made the following assumption:

Hypothesis 7 (H7). *The government has a positive impact on IC development.*

(2) The relationship between the industry and IC development

Industry development affects the production, development, and structural adjustment of the industry. IC development is not limited to the reform of the traditional construction industry, but other sectors are also gradually entering the field of the IC industry. Therefore, the rise and change of the industry is an important force driving the development of IC. Based on this, we made the following assumption:

Hypothesis 8 (H8). *The industry has a positive impact on the development of IC.*

(3) The relationship between the company and IC development

Companies make up the main body of the IC industry. At present, the intelligent transformation of the construction industry in China still relies on leading companies to drive change. The industry is gradually reaching a higher level of development through the competition and cooperation between companies. Mao et al. believed that traditional companies could use existing emerging technologies to achieve construction and management innovation [42]. While Xia [43] believed that construction companies, as the main body of IC development, should build a combination of production, research, and development of technological innovation systems, scientifically select the path of technological innovation, and carry out construction and exploration.

Based on the above literature, we made the following assumption:

Hypothesis 9 (H9). *The company has a positive impact on IC development.*

(4) The relationship between technology and IC development

The development of industry relies on the progress of technology. Therefore, the current integration of intelligent technology and construction technology is the key issue of IC development. Chen et al. [44] pointed out that recent IC technology development is unusually rapid and technology upgrading in the construction industry assumes the role of a pioneer, promoting the high-quality development of the construction industry. Chen & Ding [45] proposed that the development of domain technology as a hub connecting the underlying general technology with the upper business can play a key role in the development of IC.

Based on the above literature, we made the following assumption:

Hypothesis 10 (H10). *Technology has a positive impact on IC development.*

2.3. Hypothesis Model

We constructed a conceptual hypothesis model based on the previous assumptions to clarify the relationship between the influencing factors and IC development, as shown in Figure 2. Then we examined the model by conducting empirical research through a questionnaire and a structural equation model in the following sections.

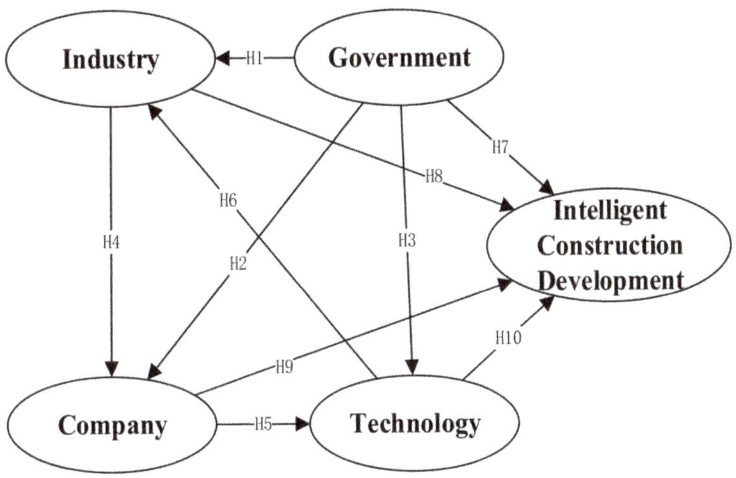

Figure 2. The conceptual model of the structural equation.

3. Research Methodology

We adopted a three-stage approach to achieve the set objectives shown in Figure 3. First, we proposed 28 measures through relevant literature and expert interviews to assess the five variables in the hypothesis model. Second, we conducted the questionnaires using a five-point Likert scale. Third, we analyzed the questionnaire data through the Harman single-factor test, measures statistical analysis, reliability and validity analysis, and structural equation modeling analysis.

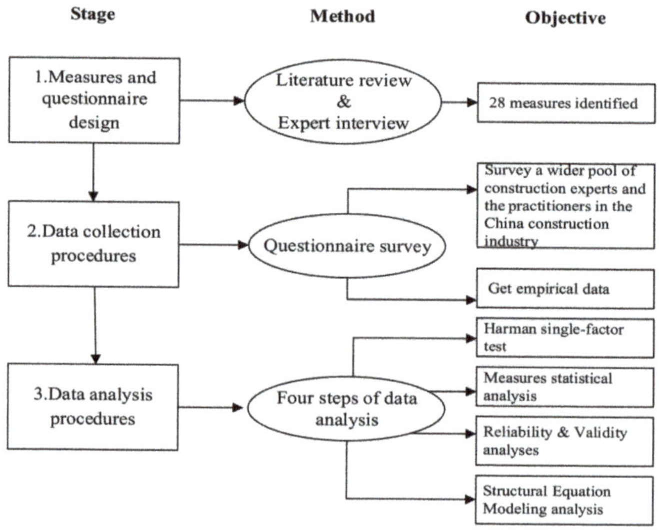

Figure 3. Three-stage research methodology.

3.1. Stage 1: Measures and Questionnaire Design

The influencing factors for IC development are latent. They are not directly measured and thus need to be specified by some measurement indicators. The measures were proposed in several ways, as listed in the following.

- The mature measurement scales are quoted as far as possible; and
- Most measures are designed based on literature reviews.

The appropriate revision and suggestions from a seven-expert group were incorporated to guarantee the rationality and completeness of the questionnaire measures. The seven-expert group included three practitioners in the construction industry, two government officials, and two scholars involved in IC. In the end, we proposed 28 measures to assess the five variables in the hypothesis model: government (four measures), industry (six measures), company (seven measures), technology (four measures), and trends of IC development (seven measures), as listed in Table 1.

Table 1. Influencing factors for IC development.

Code	Influencing Factors	Measures	Literature Reference
1	Government	The degree of government attention (G1)	[24,46]
2		The perfection and matching of regulations and standard systems (G2)	[47,48]
3		Government support policies and incentive policies (G3)	[49,50]
4		Demonstration project (G4)	[51,52]
5	Industry	Industry promotion/guidance measures (I1)	[53]
6		Industry training efforts (I2)	[54,55]
7		Market and consumer demand(I3)	[7,56]
8		The overall level of IC technology application capabilities (I4)	[36,57]
9		The number and type of IC companies (I5)	[58]
10		The degree of industry association (I6)	[59]
11	Company	The degree of attention of the companies (C1)	[53,59,60]
12		Company management system (C2)	[61,62]
13		The resource input of the companies (C3)	[63,64]
14		Company technology research and development capabilities (C4)	[65,66]
15		Vocational training for companies (C5)	[67,68]
16		Employee awareness and engagement with IC (C6)	[69,70]
17		Employee technical application capabilities (C7)	[71,72]

Table 1. *Cont.*

Code	Influencing Factors	Measures	Literature Reference
18	Technology	The maturity of the intelligent technology application system (T1)	[40]
19		Security issues for data and privacy (T2)	[73,74]
20		Hardware and software facilities (T3)	[40,75]
21		Technology convergence (T4)	[76,77]
22	Trends of IC Development	IC technology has been widely used, and the degree of intelligence has been greatly improved (TICD1)	[7,78]
23		Industry chain upstream, downstream extension, and market expand (TICD2)	[7,21]
24		The industry has achieved industrialization, service, and platform transformation (TICD3)	[7,79]
25		The industry can provide people-oriented, green sustainable products and services (TICD4)	[7]
26		Due to the concentration of industry, the degree of homogenization of companies is intensified (TICD5)	[41,80]
27		Company survival of the fittest, the industry as a whole to enhance innovation capacity (TICD6)	[81,82]
28		The construction industry traditional, and its future is not much different from that of the present (TICD7)	[83,84]

Then we designed the questionnaire based on the 28 measures of the five categories of influencing factors from the Porter Diamond Model Theory. The purpose of the questionnaire was to survey a wider pool of construction experts and practitioners in China's construction industry to obtain empirical data and determine the relationships of the identified factors. We used a five-point Likert scale [85,86] in the questionnaire: 1–5 respectively represented "strongly disagree", "disagree", "neutral", "agree", and "strongly agree", and all the measurement indices were declarative sentences.

3.2. *Stage 2: Data Collection Procedure*

We collected data by questionnaire. A total of 200 questionnaires were distributed online via a web-based platform. After recovery, 160 valid questionnaires were obtained and the effective recovery rate was 80%. The profile of the respondents is presented in Table 2.

Table 2. Respondents' characteristics.

Category	Characteristic	Frequency	Percentage (%)
Nationality	China	160	100
Organization	Construction companies	53	33.12
	Research/teaching institutions	81	50.63
	IC technology development company	4	2.50
	Government departments	6	3.75
	Other	16	10
Position level	Senior management	17	10.63
	Middle-level management	73	45.62
	Low-level physical operators	70	43.75
Working years	Less than 5 years	35	21.88
	5–10 years	38	23.75
	10–20 years	60	37.50
	More than 20 years	27	16.88
Level of understanding of IC	Greatly understand	10	6.25
	Understand	68	42.5
	Moderately understand	70	43.75
	Slightly understand	11	6.87
	Does not understand	1	0.88

3.3. Stage 3: Data Analysis Procedure

We analyzed the data in the four steps shown in Figure 4.

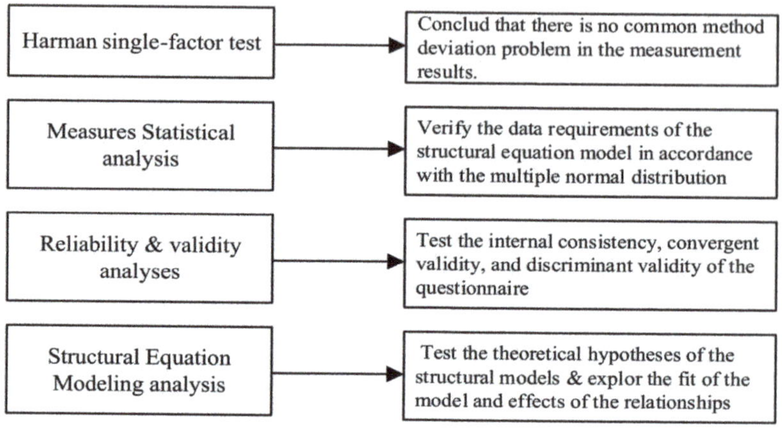

Figure 4. Four steps of data analysis.

First, all questionnaire questions were analyzed for gender factors through the Harman single-factor test. The variance interpretation rate of the first principal component obtained in the absence of rotation was 10.642%, which did not account for the majority. From the result of the composition matrix, the load of no question on the first main component exceeded 0.5. Therefore, it can be concluded that there was no common method deviation problem in the measurement results.

Second, we examined the correlation, collinearity, and normality of the variables. The thresholds of absolute skewness, absolute kurtosis, and variance inflation factor (VIF) were all less than or equal to 2, 7, and 5, respectively [87,88]. Finally, since the parameter estimation method in structural equation model analysis requires sample data to satisfy the multiple normal distributions, we tested the survey results for normal distribution. Table 3 shows the data that meet the requirements of the structural equation model under the multiple normal distributions.

Table 3. The result of measures statistical analysis.

Question	Maximum	Minimum	Standard Deviation	Skewness	Kurtosis	Kolmogorov–Smirnov D	p
1	5	1	0.758	1.958	6.167	0.298	*
2	5	1	0.798	1.706	4.886	0.276	*
3	5	1	0.881	1.830	3.954	0.332	*
4	5	1	0.886	1.117	1.864	0.255	*
5	5	1	0.737	1.654	6.255	0.303	*
6	5	1	0.774	0.877	0.851	0.268	*
7	5	1	0.723	0.779	0.587	0.247	*
8	5	1	0.782	1.438	4.354	0.279	*
9	5	1	0.800	1.586	4.590	0.311	*
10	5	1	0.839	0.885	1.790	0.287	*
11	5	1	0.853	1.566	3.609	0.260	*
12	5	1	0.746	0.107	−0.467	0.264	*
13	5	1	0.895	1.632	3.140	0.310	*
14	5	1	1.000	1.190	1.641	0.252	*
15	5	1	0.845	1.019	2.085	0.289	*
16	5	1	0.969	1.110	1.592	0.297	*
17	5	1	0.903	1.122	1.736	0.282	*
18	5	1	1.009	1.323	1.869	0.284	*

Table 3. *Cont.*

Question	Maximum	Minimum	Standard Deviation	Skewness	Kurtosis	Kolmogorov–Smirnov D	p
19	5	1	0.824	0.686	−0.003	0.242	**
20	5	1	0.849	1.071	2.134	0.244	*
21	5	1	0.857	1.215	2.366	0.245	**
22	5	1	0.961	0.757	0.451	0.250	*
23	5	1	0.926	0.643	−0.367	0.240	**
24	5	1	0.818	0.579	−0.114	0.251	**
25	5	1	0.958	0.974	0.904	0.268	*
26	5	1	0.866	0.661	1.055	0.253	*
27	5	1	0.895	0.777	−0.135	0.237	**
28	5	1	1.337	0.446	−1.119	0.290	**

Note: * means $p > 0.05$, ** means $p > 0.01$.

Third, we used the Kaiser-Meyer-Olkin (KMO) measure and Bartlett samples to test whether the data are suitable for factor analysis and the validity of the data. Experience showed that a KMO greater than 0.7 is suitable for factor analysis and KMO below 0.5 indicates unsuitable [89,90]. KMO and Barclay spherical significance tests of the study data were performed using SPSS and the results are shown in Tables 4 and 5, indicating that the research data were suitable for the principal component analysis.

Table 4. The result of Kaiser-Meyer-Olkin.

The Scope of the Inspection	KMO
The questionnaire as a whole	0.817
Government	0.719
Industry	0.742
Company	0.753
Technology	0.605
Trends of IC Development	0.851

Table 5. The result of Bartlett.

The Scope of the Inspection	Bartlett's Test of Sphericity Approx.		
	Chi-Square	df	Sig.
The questionnaire as a whole	1350.185	105	0.000
Government	34.767	10	0.000
Industry	214.632	30	0.000
Company	131.871	30	0.000
Technology	82.213	30	0.000
Trends of IC Development	380.407	30	0.000

Reliability is the overall consistency of a measure [91]. Cronbach's alpha test was performed to check the reliability of questions or items, which is much higher than the threshold of 0.70 [92], indicating internal consistency of the items and high data reliability. The Cronbach values of the five variables were higher than 0.7, as shown in Table 6, which denotes the high robustness and stability of the questionnaire.

Table 6. Results of measuring model validity.

Variables	Measures	CITC	Cronbach's α
Government	G1	0.659	0.840
	G2	0.751	
	G3	0.759	
	G4	0.548	

Table 6. *Cont.*

Variables	Measures	CITC	Cronbach's α
Industry	I1	0.686	0.883
	I2	0.616	
	I3	0.767	
	I4	0.731	
	I5	0.751	
	I6	0.710	
Company	C1	0.735	0.921
	C2	0.561	
	C3	0.808	
	C4	0.729	
	C5	0.835	
	C6	0.848	
	C7	0.771	
Technology	T1	0.792	0.892
	T2	0.745	
	T3	0.764	
	T4	0.762	
Trends of IC Development	TICD1	0.762	0.877
	TICD2	0.789	
	TICD3	0.798	
	TICD4	0.821	
	TICD5	0.783	
	TICD6	0.808	
	TICD7	0.207	

Fourth, as SEM is an appropriate technique for multivariate analysis that integrates factor analysis, path analysis, and multiple regression analysis [17,93], we used it to create the conceptual model and test the theoretical hypotheses of the structural models.

4. Research Results

We tested the aforementioned hypothesized model (Figure 2) with the SEM technique.

First, we used the standardized factor loading test for the measures, then we estimated the path coefficient by the maximum likelihood method and removed the path data indexes. Furthermore, we tested the model by factor analysis and repeated the above steps until the indexes were qualified. Finally, we got the verified model.

4.1. Data Results

First, through the calculation results, the 28 measures in which the loadings of the measurement variables and potential variables are more than 0.6 were retained. The revised indicators and the standardized load of each index are shown in Table 7.

Second, the maximum likelihood method was used to estimate the path coefficient, in which C.R. (i.e., *t*-test value) is the critical ratio. (1) in H1, T = 2.612, $p = 0.319 > 0.05$, indicating that the government has no direct relationship to industry development; (2) in H3, T = 0.124, $p = 0.409 > 0.05$, indicating that the government has no direct relationship to technology; (3) in H8, T = −1.012, $p = 137 > 0.05$, indicating that the industry has no direct relationship to the development of IC. Therefore, H1, H3, and H8 were removed, and the revised model test results are shown in Table 8, where all hypothesis tests passed.

The revised model was again tested for factor analysis. Each fitted indicator is shown in Table 9, and each indicator was within the standard range, so it could be considered that the model's overall fit is acceptable without the need for another correction.

Three main indices of the overall model fit were adopted in this paper to confirm the measurement model: the relative χ2 (χ2/degree of freedom), root-mean-square error of approximate (RMSEA), and comparative fit index (CFI). The upper thresholds of the relative χ^2/df and RMSEA were 3 and 0.08, respectively, while the lower threshold of CFI was 0.90 [94]. As shown in Table 9, χ^2/df of the model was less than 3; RMSEA was less than 0.08, indicating that the model was reasonable and not affected by the sample size; and CFI was greater than 0.9, meaning that the sample was stable.

Table 7. Results of the measures.

Variables	Measures	Standardized Factor Loading
Government	G1	0.925
	G2	0.814
	G3	0.878
	G4	0.904
Industry	I1	0.809
	I2	0.796
	I3	0.932
	I4	0.812
	I5	0.946
	I6	0.898
Company	C1	0.726
	C2	0.618
	C3	0.895
	C4	0.837
	C5	0.861
	C6	0.879
	C7	0.749
Technology	T1	0.863
	T2	0.814
	T3	0.941
	T4	0.938

Table 8. Results of the model test.

X	Path	Y	Hypothesis	T	*p*	Standardized Path Coefficients	Conclusion
Government	→	Company	H2	2.369	*	0.653	support
Industry	→	Company	H4	3.153	**	0.308	support
Company	→	Technology	H5	2.781	**	0.245	support
Technology	→	Industry	H6	6.572	***	0.735	support
Government	→	IC development	H7	9.454	***	1.180	support
Company	→	IC development	H9	0.336	*	0.144	support
Technology	→	IC development	H10	3.013	**	0.691	support

Note: *, **, and *** indicate that the statistics are at the levels of 0.05, 0.001, and 0.001, respectively, and the effect is significant.

Table 9. Calculation results of the fitting index of the modified model.

Index	Outcome
χ^2/df	2.662
GFI	0.847
RMR	0.057
CFI	0.935
NFI	0.657
NNFI	0.838
IFI	0.952

4.2. Verified Model

The hypothesis model was verified through SEM, as shown in Figure 5.

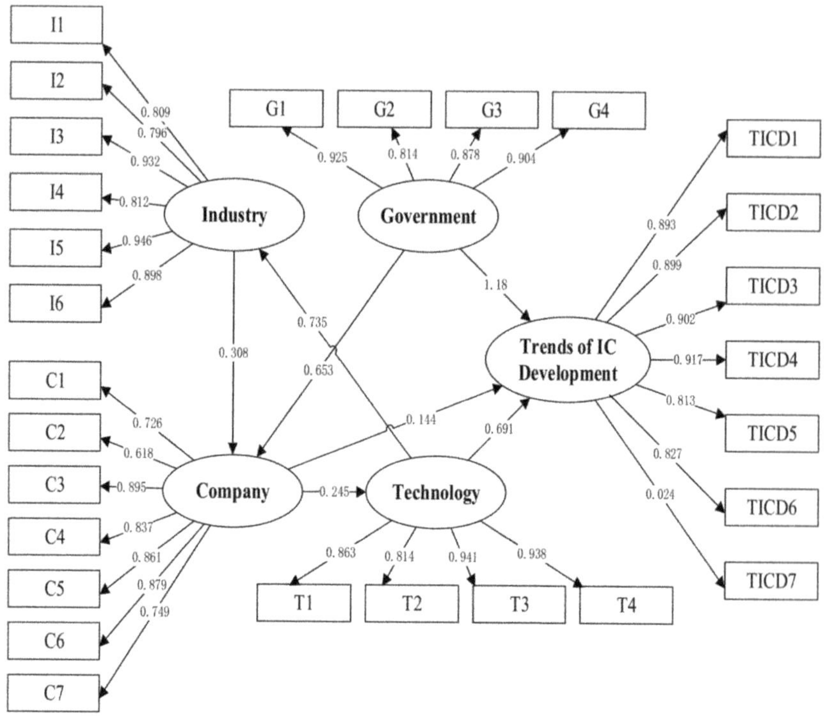

Figure 5. Influencers acting on the IC development mechanism model.

Taking into account these results of the model test, we summarized further conclusions as follows:

(1) The three variables of government, company and technology directly and significantly impacted IC development. These path coefficients were 1.18, 0.144, and 0.691, respectively. Although each path was significant, the coefficient load values were generally low, especially the impact of company, which means that the impact of company on IC was still at a low level.

(2) Industry did not directly affect the development of IC but rather indirectly affected the development of IC through company and technology. These three endogenous variables of industry, company, and technology form a "closed-loop" within which the three elements interact and promote each other.

(3) The standardized factor loading of TICD7 was 0.024 and less than the factor loadings of the trend of IC development, which shows that it was widely accepted that IC development has a bright outlook in China and IC can certainly change the relative backwardness of China's construction industry through the upgrading of technology, industry, and company.

5. Discussion

Based on the final model, we further summarized the paths among factors and identify key measures, hoping to provide a reference for the government and companies to formulate development strategies.

5.1. Path Analyses

(1) Government→Company→Technology→IC development. On the one hand, the introduction of policies forces companies to carry out IC development; on the other hand, preferential subsidies support companies in carrying out IC activities. As a result, companies vigorously carry out relevant practical activities, form a new industrial system, stimulate construction activities and the application of innovative technology, and improve the basic endowment of the development of the construction industry. The emergence of innovative technology also contributes to the development of the construction industry.
(2) Government→Company→Industries→IC development. The government's initiation and support has given birth to various companies with new business in the field of IC, increasing the complexity of corporate relations and the degree of competition within the industry. In contrast, this trend forces companies to change their management models and improve effectiveness to adapt to environmental changes. The adjustments and changes in corporate relationships reconstruct the industrial ecology, including the business model, business philosophy, market form, and industry management, creating a new steady state of IC industry development.
(3) Technology→Industry→Company→Technology→IC development. The iterative application of technology innovation affects the adjustment and development status of the industry's development. Therefore, industry development trends guide the direction of company development and companies adjust their development strategies, paying attention to investment in technological innovation and application and gradually forming a certain scale and level of economic benefits of IC and development of new industrial forms.
(4) Company→Technology→Industry→Company→IC development. Companies are the core of technological innovation and development. Intelligent technologies change the traditional working model of the construction industry which provides a realistic basis for IC development. Following the trend, more companies participate in the intelligent transformation.

5.2. Measures Analyses

The measures of the variables in the model were further analyzed. The standardized factor loading indicates the relationship between the variables and the measure; the higher the loading is, the closer the relationship with the corresponding variables. The indicator can be used as the basis for ordering the importance of the observed variables, and the order is shown in Table 10.

(1) Among the government measures, the degree of government attention was the most important. This result is in line with the current state of IC development in China. At present, most companies have recognized the necessity of upgrading and have a positive attitude toward development prospects. However, companies directly engaged in IC projects need to pay huge costs, so most have been in a wait-and-see state. In the above environment, the government was the most critical stakeholder. Survey data and the state of IC development in China showed that IC development depends on government policies. On the one hand, the government has a certain mandatory role; on the other hand, it provides preferential subsidies to encourage companies to carry out IC activities to protect companies motivated to undertake IC.
(2) Among the industry measures, the number and type of IC companies were of the utmost importance. In the infancy of IC development, most IC technology still belongs to the companies of developed countries. There is an urgent need for more companies to improve the ability to develop IC technology with independent intellectual property rights.
(3) Among the company measures, the resource input of companies was the most important. Most Chinese construction companies did not pay attention to IC in the past. Therefore, there currently exists a huge gap between actual development and the

vision of IC. In contributing to bridging this gap, a significant amount of resource input is needed to compensate for the backward development caused by traditional production methods so that it is possible to meet the current stage of IC development.

(4) Among the measures of technology, hardware and software facilities were the most important. Since most of the core technologies related to IC still belong to companies in developed countries, the application of IC technologies is limited. It is difficult for companies in China to find suitable hardware and software services companies to assist in construction projects.

Table 10. Order of measures.

Variables	Measures	Standardized Factor Loading	Order of Importance
Government	G1	0.925	1
	G2	0.814	4
	G3	0.878	3
	G4	0.904	2
Industry	I1	0.809	4
	I2	0.796	6
	I3	0.932	2
	I4	0.812	5
	I5	0.946	1
	I6	0.898	3
Company	C1	0.726	6
	C2	0.618	7
	C3	0.895	1
	C4	0.837	4
	C5	0.861	3
	C6	0.879	2
	C7	0.749	5
Technology	T1	0.863	3
	T2	0.814	4
	T3	0.941	1
	T4	0.938	2

5.3. Recommendations

Based on the above analysis, we propose the following strategies to enhance IC development:

(1) Understand the general path driving the development of IC. The process of influencing the development of IC has a certain regularity. Therefore, it is necessary to understand the influencing factors that employ the important role for its maximum utility and vigorously promote the development process of IC.

(2) Emphasize and play the role of policies. It is necessary to promulgate effective policies to ensure and encourage the willingness of relevant entities to practice IC, develop and improve the market, and establish a long-term force to promote IC development. In addition, the government should pay attention to the development and changes of IC to make corresponding policy adjustments to form a virtuous circle.

(3) Increase investment in research and development and overcome technical barriers. Different companies need to formulate appropriate integrated development plans and policy choices according to their business characteristics and foundation and carry out innovative research and development to promote the coordinated development of the IC industry.

6. Conclusions

IC is the key to adapting to the trend of intelligent development of the global construction industry, which involves multiple factors. This paper contributed to identifying and determining these factors through literature analysis and clarifying paths and key measures through SEM, which can help the government and companies better understand

IC development and provide a basis for the later introduction of policies and practice acceleration. Simultaneously, this paper offered a generalizable reference for other countries to develop IC.

In this paper, we analyzed the influence paths and key measures affecting IC development by SEM. This paper achieved the following results:

(1) We identified the following four variables that influence IC development: government, industry, company, and technology. Moreover, we built the conceptual model.
(2) Based on SEM method, we obtained four influence paths: (1) Government→company →technology→IC development; (2) Government→company→industries→IC development; (3) Technology→industry→company→technology→IC development; and (4) Company→technology→industry→company→IC development, which indicates that the government has a significant direct impact on the development of IC.
(3) We further analyzed the key measures of government, industry, company, and technology: the degree of government attention, the number and development capability of IC technology development companies, resource input, and hardware and software facilities.
(4) We proposed some recommendations to promote IC development.

This research contributed to the body of knowledge on IC by identifying the factors that influence IC development. The four paths and key measures were proposed to clarify the relationship between factors. Recommendations were put forward to promote IC development. Construction industries globally can leverage the factors influencing IC development in this research, which provides a valuable reference for further contextual investigations in their regions. Although the context of this study was China, the study findings can provide references for IC development in the construction industry globally, especially in those countries whose construction industries are in similar stages of development.

Author Contributions: Conceptualization, X.Y. and T.L.; writing—original draft preparation, X.Y. and T.L.; writing—review and editing, X.Y. and T.L.; investigation, W.G. and F.Z.; Data collection, W.G.; Data analysis, F.Z.; project administration, X.Y.; funding acquisition, X.Y. All authors have read and agreed to the published version of the manuscript.

Funding: This research was funded by the Project of 2020 Scientific and Innovative Action Plan of the Shanghai Science and Technology Commission: "Research on the Key Issues and Countermeasures of the Transformation of Traditional Industries Driven by Digital Technology-The Origin, Architecture and Realization of the Intelligent Construction Mode of the Construction Industry", grant number 20692101300.

Data Availability Statement: Some or all data, models, or code generated or used during the study are available from the corresponding author by request.

Acknowledgments: The authors are very thankful for the anonymous referees and editors whose suggestions and comments helped to improve the manuscript quality.

Conflicts of Interest: The authors declare that they have no conflict of interest. The funders had no role in the design of the study; in the collection, analyses, or interpretation of data; in the writing of the manuscript, or in the decision to publish the results.

References

1. Guo, P.; Tian, W.; Li, H.; Zhang, G.; Li, J. Global characteristics and trends of research on construction dust: Based on bibliometric and visualized analysis. *Environ. Sci. Pollut. Res.* **2020**, *27*, 37773–37789. [CrossRef] [PubMed]
2. Giel, B.K.; Issa, R. Return on investment analysis of using building information modeling in construction. *J. Comput. Civ. Eng.* **2013**, *27*, 511–521. [CrossRef]
3. Arunothayan, A.R.; Nematollahi, B.; Ranade, R.; Bong, S.H.; Sanjayan, J. Development of 3D-printable ultra-high performance fiber-reinforced concrete for digital construction. *Constr. Build. Mater.* **2020**, *257*, 119546. [CrossRef]
4. Štefanič, M.; Stankovski, V. A review of technologies and applications for smart construction. *Proc. Inst. Civ. Engineers.* **2018**, *172*, 83–87. [CrossRef]

5. Perrier, N.; Bled, A.; Bourgault, M.; Cousin, N.; Danjou, C.; Pellerin, R.; Roland, T. Construction 4.0: A survey of research trends. *J. Inf. Technol. Constr.* **2020**, *25*, 416–437. [CrossRef]
6. Sawhney, A.; Riley, M.; Irizarry, J.; Perez, T.C. A proposed framework for Construction 4.0 based on a review of Literature. In *Proceedings of the 56th Annual Associated Schools of Construction (ASC) International Conference_ASC, Liverpool, UK, 15–18 April 2020*; EasyChair: Lancaster, UK, 2020.
7. Ding, L.Y. *Digital Construction Introduction*; China Building Industry Press: Beijing, China, 2019.
8. Zhong, D.; Liu, X.; Cui, B.; Wu, B.; Liu, Y. Technology and application of real-time compaction quality monitoring for earth-rockfill dam construction in deep narrow valley. *Autom. Constr.* **2018**, *90*, 23–38. [CrossRef]
9. Zhang, D. Application of GIS+BIM technology in underground Pipe Gallery. *Low Temp. Archit. Technol.* **2019**, *41*, 121–123+133.
10. Al-Hammadi, M.A.; Tian, W. Challenges and barriers of building information modeling adoption in the Saudi Arabian construction industry. *Open Constr. Build. Technol. J.* **2020**, *14*, 98–110. [CrossRef]
11. Liao, L. Research on the application of BIM technology in the cost management of construction projects. *IOP Conf. Ser. Earth Environ. Sci.* **2021**, *783*, 012098. [CrossRef]
12. Zaid, N.U.B.M.; Hamzah, N.; Khoiry, M.A. Review building information modelling for infrastructure: Benefits for constructor. *J. Comput. Theor. Nanosci.* **2020**, *17*, 620–628. [CrossRef]
13. Jia, H.; Dong, S.; Fu, S. Application of BIM technology in intelligent construction and installation of prefabricated buildings. *Constr. Technol.* **2018**, *22*, 40–43.
14. Zhang, Q.; Liu, T.; Zhang, Z.; Huang, F.Z.; Li, Q.; An, Z. Unmanned rolling compaction system for rockfill materials. *Autom. Constr.* **2019**, *100*, 103–117. [CrossRef]
15. Kim, S.; Peavy, M.; Huang, P.C.; Kim, K. Development of BIM-integrated construction robot task planning and simulation system. *Autom. Constr.* **2021**, *127*, 103720. [CrossRef]
16. Pan, M.; Pan, W. Determinants of adoption of robotics in precast concrete production for buildings. *J. Manag. Eng.* **2019**, *35*, 05019007. [CrossRef]
17. Huang, Y.; Trinh, M.T.; Le, T. Critical factors affecting intention of use of augmented hearing protection technology in construction. *J. Constr. Eng. Manag.* **2021**, *147*, 04021088. [CrossRef]
18. Besklubova, S.; Skibniewski, M.J.; Zhang, X. Factors affecting 3D printing technology adaptation in construction. *J. Constr. Eng. Manag.* **2021**, *147*, 04021026. [CrossRef]
19. Craveiroa, F.; Duartec, J.P.; Bartoloa, H.; Bartolod, P.J. Additive manufacturing as an enabling technology for digital construction: A perspective on Construction 4.0. *Autom. Constr.* **2019**, *103*, 251–267. [CrossRef]
20. Ghaffar, S.H.; Corker, J.; Fan, M. Additive manufacturing technology and its implementation in construction as an eco-innovative solution. *Autom. Constr.* **2018**, *93*, 1–11. [CrossRef]
21. Yuan, X.; Chen, Y.W.; Fan, H.B.; He, W.H.; Ming, X.G. Collaborative construction industry integrated management service system framework based on Big Data. In *Proceedings of the 2019 IEEE International Conference on Industrial Engineering and Engineering Management (IEEM), Macao, China, 15–18 December 2019*; IEEE: Piscataway, NJ, USA, 2019; pp. 1521–1525. [CrossRef]
22. Wang, Y.; Lee, H.W.; Tang, W.; Whittington, J.; Qiang, M. Structural equation modeling for the determinants of international infrastructure investment: Evidence from Chinese contractors. *J. Manag. Eng.* **2021**, *37*, 04021033. [CrossRef]
23. Chen, K.; Ding, L.Y. Strategic thinking on technology development in key fields of intelligent construction in China. *Strateg. Study CAE* **2021**, *23*, 64–70.
24. Xiahou, X.; Yuan, J.; Liu, Y.; Tang, Y.; Li, Q. Exploring the driving factors of construction industrialization development in China. *Int. J. Environ. Res. Public Health* **2018**, *15*, 442. [CrossRef] [PubMed]
25. Simachev, Y.; Kuzyk, M.; Feygina, V. Public support for innovation in Russian firms: Looking for improvements in corporate performance quality. *Int. Adv. Econ. Res.* **2015**, *21*, 13–31. [CrossRef]
26. David, P.A.; Hall, B.H.; Toole, A.A. Is public R&D a complement or substitute for private R&D? A review of the econometric evidence. *Res. Policy* **2000**, *29*, 497–529.
27. Tzelepis, D.; Skuras, D. The effects of regional capital subsidies on firm performance: An empirical study. *J. Small Bus. Enterp. Dev.* **2004**, *11*, 121–129. [CrossRef]
28. Czarnitzki, D.; Toole, A.A. Business R&D and the interplay of R&D subsidies and product market uncertainty. *Rev. Ind. Organ.* **2007**, *31*, 169–181.
29. Yan, J.J.; Feng, J.Y. An empirical study on the impact of government innovation subsidy timing on enterprise technology leapfrogging. *Theory Pract. Financ. Econ.* **2021**, *42*, 98–105.
30. Li, X.A.; Shi, G.Q. On the influence of consumption demand change to industrial structure adjustment. *J. Hohai Univ. (Philos. Soc. Sci.)* **2002**, *4*, 34–36.
31. Wang, M.Y.; Li, Y.M.; Zhang, H.; Wang, H. Model construction and path analysis of green technology innovation driven by market orientation. *Sci. Technol. Prog. Policy* **2019**, *36*, 112–120.
32. Zhang, G.L.; Zhou, H.R.; Liao, J.Q. The impact of organizational structure on technological innovation from the perspective of knowledge transfer. *Sci. Sci. Manag. S. T.* **2009**, *30*, 78–84.
33. Miu, G.H.; Chen, W.M.; Tang, C.Y. A study on the influencing factors of talent aggregation in high-tech Enterprises—A case study of Commercial Aircraft Corporation of China LTD. *Sci. Technol. Manag. Res.* **2013**, *33*, 120–122+128.

34. Duan, X. Research on the effectiveness of enterprise R&D investment, technological innovation and resource investment. *Stat. Decis.* **2020**, *36*, 183–186.
35. Perkmann, M.; Walsh, K. The two faces of collaboration: Impacts of university-industry relations on public research. *Ind. Corp. Change* **2009**, *18*, 1033–1065. [CrossRef]
36. Fang, C.; Lai, Y.; Wen, Z. Research on key technologies of intelligent integrated system for architectural design enterprises. *IOP Conf. Ser. Earth Environ. Sci.* **2020**, *455*, 012209. [CrossRef]
37. Schumpeter, J. *Capitalism, Socialism and Democracy*; George Allen & Unwin Publishers: London, UK, 1942.
38. Zhou, J.; Li, P.; Zhou, Y.; Wang, B.; Zang, J.; Meng, L. Toward new-generation intelligent manufacturing. *Engineering* **2018**, *4*, 11–20. [CrossRef]
39. Luo, X.F.; Li, B.Z. The driving mechanism of new product demand on original innovation of large enterprises: An empirical study based on the comparison between domestic market and foreign market. *Sci. Technol. Prog. Policy* **2013**, *30*, 73–76.
40. Lu, C.; Liu, J.; Liu, Y.; Liu, Y. Intelligent construction technology of railway engineering in China. *Front. Eng. Manag.* **2019**, *6*, 503–516. [CrossRef]
41. Wang, Y.S.; Su, B.Y.; Zhang, Y.B.; Wu, T.Y. Study on the influence of industrial agglomeration on total factor Productivity of construction industry. *Build. Econ.* **2020**, *41*, 9–14.
42. Mao, C.; Zhou, Y. Analysis of core enterprise supply chain organization structure of intelligent construction industry. *Constr. Econ.* **2021**, *42*, 14–18.
43. Xia, X.G. Technological innovation practice of railway construction enterprises under the background of intelligent construction. *Constr. Econ.* **2020**, *41*, 43–47.
44. Chen, F.; Zhu, J.; Wang, W. Driving force of industrial technology innovation: Coevolution of multistage overseas M&A integration and knowledge network reconfiguration. *J. Bus. Ind. Mark.* **2021**, *36*, 1344–1357.
45. Zhang, J.; Long, Y.; Lv, S.; Xiang, Y. BIM-enabled modular and industrialized construction in China. *Procedia Eng.* **2016**, *145*, 1456–1461. [CrossRef]
46. Ogunrinde, O.; Nnaji, C.; Amirkhanian, A. Application of emerging technologies for highway construction quality management: A review. In Proceedings of the Construction Research Congress 2020: Project Management and Controls, Materials, and Contracts, Tempe, AZ, USA, 8–10 March 2020; pp. 1030–1039. [CrossRef]
47. Mao, C.; Zhang, L.M. Core industry selection of intelligent construction industry chain. *J. Eng. Manag.* **2021**, *35*, 1–6. [CrossRef]
48. Liu, Z.S.; Liu, S.N.; Zhao, Y.H.; Du, X.L. Development status and future trend of intelligent construction technology. *Constr. Technol.* **2019**, *50*, 772–779.
49. Hsiao, J.S.; Chen, J.C.P.; Shin-Jyun, T.L. Implementation of certified intelligent building in Taiwan. *IOP Conf. Ser. Mater. Sci. Eng.* **2019**, *652*, 012006. [CrossRef]
50. Shi, F.; Wang, Q.; Wang, Y. Research on Top-Level redesign of smart construction system based on case study. In Proceedings of the ICCREM 2019: Innovative Construction Project Management and Construction Industrialization, Banff, AB, Canada, 21–24 May 2019; pp. 117–124. [CrossRef]
51. Lin, M.; Wang, Q.; Wang, M.; Li, J. Exploration and practice of intelligent construction in island tunnel project of Hong Kong-Zhuhai-Macao Bridge. *Sci. Technol. Prog. Policy* **2018**, *5*, 81–85.
52. Binesmael, M.; Li, H.; Lark, R. Meta-standard for collaborative BIM standards: An analysis of UK BIM level 2 standards. *Work. Conf. Virtual Enterp.* **2018**, *534*, 661–668.
53. Chen, Y.; Yin, Y.; Browne, G.J.; Li, D. Adoption of building information modeling in Chinese construction industry: The technology-organization-environment framework. Engineering, construction and architectural management. *Eng. Constr. Archit. Manag.* **2019**, *26*, 1878–1898. [CrossRef]
54. Wang, G.M. Research on the implementation path of promoting the collaborative development of intelligent construction and new building industrialization. *Hous. Ind.* **2020**, *263*, 12–15.
55. Chen, B.; Wan, J.; Shu, L.; Li, P.; Mukherjee, M.; Yin, B. Smart factory of industry 4.0: Key technologies, application case, and challenges. *IEEE Access* **2018**, *6*, 6505–6519. [CrossRef]
56. Fitz, D.V.; Saleeb, N. Examining the quality and management of non-geometric building information modelling data at project hand-over. *Archit. Eng. Des. Manag.* **2019**, *15*, 297–310. [CrossRef]
57. Zhu, W. Intelligent construction and management of thermal power plant based on internet+ mode. *IOP Conf. Ser. Earth Environ. Sci.* **2019**, *252*, 032068. [CrossRef]
58. Shilan, L. Application research of computer information technology in intelligent building engineering management. In *Proceedings of the 2019 11th International Conference on Measuring Technology and Mechatronics Automation (ICMTMA),Qiqihar, China, 28–29 April 2019*; IEEE: Piscataway, NJ, USA, 2019; pp. 136–140.
59. He, R.; Li, M.; Gan, V.J.; Ma, J. BIM-enabled computerized design and digital fabrication of industrialized buildings: A case study. *J. Clean. Prod.* **2021**, *278*, 123505. [CrossRef]
60. Siebelink, S.; Voordijk, J.T.; Adriaanse, A. Developing and testing a tool to evaluate BIM maturity: Sectoral analysis in the Dutch construction industry. *J. Constr. Eng. Manag.* **2018**, *144*, 05018007. [CrossRef]
61. Mesároš, P.; Mandičák, T.; Behúnová, A. Use of BIM technology and impact on productivity in construction project management. *Wirel. Netw.* **2020**, *28*, 855–862. [CrossRef]

62. Bahrami, S.; Atkin, B.; Landin, A. Innovation diffusion through standardization: A study of building ventilation products. *J. Eng. Technol. Manag.* **2019**, *54*, 56–66. [CrossRef]
63. Bademosi, F.; Issa, R.R. Factors Influencing Adoption and Integration of Construction Robotics and Automation Technology in the US. *J. Constr. Eng. Manag.* **2021**, *147*, 04021075. [CrossRef]
64. Cesnik, J.; Zibert, M.; Lah, M.; Skalja, M. Required model content and information workflows enabling proficient BIM usage. *IOP Conf. Ser. Mater. Sci. Eng.* **2019**, *603*, 032074. [CrossRef]
65. Becerik-Gerber, B.; Jazizadeh, F.; Li, N.; Calis, G. Application areas and data requirements for BIM-enabled facilities management. *J. Constr. Eng. Manag.* **2012**, *138*, 431–442. [CrossRef]
66. Pishdad-Bozorgi, P.; Gao, X.; Eastman, C.; Self, A.P. Planning and developing facility management-enabled building information model (FM-enabled BIM). *Autom. Constr.* **2018**, *87*, 22–38. [CrossRef]
67. Villena, F.; García-Segura, T.; Pellicer, E. Drivers of innovation using BIM in architecture, engineering, and construction firms. In Proceedings of the Construction Research Congress 2020: Project Management and Controls, Materials, and Contracts, Tempe, AZ, USA, 8–10 March 2020; pp. 210–222. [CrossRef]
68. Yafeng, W. Research on the intelligent construction of prefabricated building and personnel training based on BIM5D. *J. Intell. Fuzzy Syst.* **2021**, *40*, 8033–8041.
69. Joshi, S.; Hamilton, M.; Warren, R.; Faucett, D.; Tian, W.; Wang, Y.; Ma, J. Implementing virtual reality technology for safety training in the precast/prestressed concrete industry. *Appl. Ergon.* **2021**, *90*, 103286. [CrossRef] [PubMed]
70. Qi, B.; Razkenari, M.; Li, J.; Costin, A.; Kibert, C.; Qian, S. Investigating US industry practitioners' perspectives towards the adoption of emerging technologies in industrialized construction. *Buildings* **2020**, *10*, 85. [CrossRef]
71. Ahuja, R.; Sawhney, A.; Arif, M. Developing organizational capabilities to deliver lean and green project outcomes using BIM. *Eng. Constr. Archit. Manag.* **2018**, *25*, 1255–1276. [CrossRef]
72. Nasir, A.R.; Bargstädt, H.J. An approach to develop video tutorials for construction tasks. *Procedia Eng.* **2017**, *196*, 1088–1097. [CrossRef]
73. Pradeep, A.S.E.; Yiu, T.W.; Zou, Y.; Amor, R. Blockchain-aided information exchange records for design liability control and improved security. *Autom. Constr.* **2021**, *126*, 103667. [CrossRef]
74. McNamara, A.; Sepasgozar, S.M. Barriers and drivers of Intelligent Contract implementation in construction. *Management* **2018**, *143*, 02517006.
75. McNamara, A.J.; Sepasgozar, S.M. Intelligent contract adoption in the construction industry: Concept development. *Autom. Constr.* **2021**, *122*, 103452. [CrossRef]
76. Wang, X.; Wang, S.; Song, X.; Han, Y. IoT-Based intelligent construction system for prefabricated buildings: Study of operating mechanism and implementation in China. *Appl. Sci.* **2020**, *10*, 6311. [CrossRef]
77. Kochovski, P.; Stankovski, V. Supporting smart construction with dependable edge computing infrastructures and applications. *Autom. Constr.* **2018**, *85*, 182–192. [CrossRef]
78. Vishnivetskaya, A.; Mikhailova, A. Employment of BIM technologies for residential quarters renovation: Global experience and prospects of implementation in Russia. *IOP Conf. Ser. Mater. Sci. Eng.* **2019**, *497*, 012020. [CrossRef]
79. Davtalab, O.; Kazemian, A.; Khoshnevis, B. Perspectives on a BIM-integrated software platform for robotic construction through Contour Crafting. *Autom. Constr.* **2018**, *89*, 13–23. [CrossRef]
80. Zhao, Z.Y.; Xu, K.; Zuo, J.; Tang, C. Developing the international construction contracting market: Enterprise niche approach. *J. Manag. Eng.* **2017**, *33*, 04016027. [CrossRef]
81. Osorio-Gomez, C.C.; Moreno-Falla, M.J.; Ospina-Alvarado, A.; Ponz-Tienda, J.L. Lean construction and BIM in the value chain of a construction company: A case study. In Proceedings of the Construction Research Congress 2020: Project Management and Controls, Materials, and Contracts, Tempe, AZ, USA, 8–10 March 2020; pp. 368–378. [CrossRef]
82. Tranchant, A.; Beladjine, D.; Beddiar, K. BIM in French smes: From innovation to necessity. *WIT Trans. Built Environ.* **2017**, *169*, 135–142.
83. Meng, Q.; Zhang, Y.; Li, Z.; Shi, W.; Wang, J.; Sun, Y.; Wang, X. A review of integrated applications of BIM and related technologies in whole building life cycle. *Eng. Constr. Archit. Manag.* **2020**, *27*, 1647–1677. [CrossRef]
84. Xu, X.; Wang, Y.; Tao, L. Comprehensive evaluation of sustainable development of regional construction industry in China. *J. Clean. Prod.* **2019**, *211*, 1078–1087. [CrossRef]
85. Dawes, J. Do data characteristics change according to the number of scale points used? An experiment using 5-point, 7-point and 10-point scales. *Int. J. Mark. Res.* **2008**, *50*, 61–104. [CrossRef]
86. Joshi, A.; Kale, S.; Chandel, S.; Pal, D.K. Likert scale: Explored and explained. *Br. J. Appl. Sci. Technol.* **2015**, *7*, 396. [CrossRef]
87. O'Brien, R.M. A caution regarding rules of thumb for variance inflation factors. *Qual. Quant.* **2007**, *41*, 673–690. [CrossRef]
88. Pektas, A.O. Determining the essential parameters of bed load and suspended sediment load. *Int. J. Glob. Warm.* **2015**, *8*, 335–359. [CrossRef]
89. Hadi, N.U.; Abdullah, N.; Sentosa, I. An easy approach to exploratory factor analysis: Marketing perspective. *J. Educ. Soc. Res.* **2016**, *6*, 215.
90. Williams, B.; Onsman, A.; Brown, T. Exploratory factor analysis: A five-step guide for novices. *Australas. J. Paramed.* **2010**, *8*, 1–13. [CrossRef]
91. Elhendawi, A.I.N. Methodology for BIM Implementation in KSA in AEC Industry. Master's Thesis, Edinburgh Napier University, Edinburgh, UK, 2018.

92. Nunnally, J.; Bernstein, I. *Psychometric Theory*; McGraw-Hill: New York, NY, USA, 1994.
93. Zhang, Q.; Yang, S.; Liao, P.C.; Chen, W. Influence mechanisms of factors on project management capability. *J. Manag. Eng.* **2020**, *36*, 04020045. [CrossRef]
94. Xiong, B.; Skitmore, M.; Xia, B. A critical review of structural equation modeling applications in construction research. *Autom. Constr.* **2015**, *49*, 59–70. [CrossRef]

buildings

MDPI

Article

Development of a BIM-Based Framework Using Reverberation Time (BFRT) as a Tool for Assessing and Improving Building Acoustic Environment

Antonio J. Aguilar [1,*], María L. de la Hoz-Torres [2], Mª Dolores Martínez-Aires [2] and Diego P. Ruiz [1]

1 Department of Applied Physics, University of Granada, Av. Severo Ochoa s/n, 18071 Granada, Spain; druiz@ugr.es

2 Department of Building Construction, University of Granada, Av. Severo Ochoa s/n, 18071 Granada, Spain; mlhoz@ugr.es (M.L.d.l.H.-T.); aires@ugr.es (M.D.M.-A.)

* Correspondence: antojes@ugr.es

Abstract: Both the building design and the construction process determine the indoor acoustic quality of enclosures. A suitable indoor acoustic environment is crucial for the productivity and well-being of users. For this purpose, Reverberation Time (RT) is often calculated or measured in situ. Recently, Building Information Modelling (BIM) has provided a new paradigm to face building projects. Nevertheless, little research has been conducted on the optimisation of indoor acoustics using BIM methodology. In this context, the objective of this work is to propose and develop a BIM-based framework for the analysis, evaluation and optimization of the RT. The proposed procedure allows designers to explore alternatives in order to achieve an adequate acoustic performance without any further needs of specific software. This proposal is devised to consider some important characteristics of the project, such as its location, applicable regulations, room uses, materials and costs. This framework calculates the solution set that meets the requirements, showing the set of optimal solutions according to the minimization of both the cost and the optimum absorbent surface area. BFRT contributes by offering a tool to support the decision making process of designers during the initial design phase in the field of acoustic conditioning of buildings.

Keywords: acoustic performance; building; Building Information Modelling (BIM); built environment; optimization algorithms; reverberation time

Citation: Aguilar, A.J.; de la Hoz-Torres, M.L.; Martínez-Aires, M.D.; Ruiz, D.P. Development of a BIM-Based Framework Using Reverberation Time (BFRT) as a Tool for Assessing and Improving Building Acoustic Environment. *Buildings* **2022**, *12*, 542. https://doi.org/10.3390/buildings12050542

Academic Editor: Jian Kang

Received: 1 April 2022
Accepted: 22 April 2022
Published: 24 April 2022

Publisher's Note: MDPI stays neutral with regard to jurisdictional claims in published maps and institutional affiliations.

Copyright: © 2022 by the authors. Licensee MDPI, Basel, Switzerland. This article is an open access article distributed under the terms and conditions of the Creative Commons Attribution (CC BY) license (https://creativecommons.org/licenses/by/4.0/).

1. Introduction

Between an 80 and 90% of the urban population spends most of its living time in interior spaces [1]. In this sense, the conditions of the indoor environment in buildings represent an important factor in the quality of life for occupants/users. According to the ISO 16814:2008 standard, indoor environment includes thermal, acoustic and lighting conditions, as well as indoor air quality (IAQ). All these elements taken altogether have been already identified and taken into account in different studies [2,3] as relevant aspects that will determine the environmental quality of interior spaces. Focusing specifically in the acoustic conditions, the acoustic behaviour inside rooms in a building is conditioned not only by external noise sources, but also by domestic sources and the characteristics of other adjacent spaces [4,5]. If such behaviour was not appropriate, the normal performance of human activities would be affected, and may even result in an increased risk of diseases related to exposure to noise [3].

In this sense, an appropriate acoustic behaviour for a given indoor space should be already guaranteed from the early design phase, working with some potential available objective parameters to allow evaluating the acoustic characteristics of a room [6–8]. Among these parameters this work focus in the reverberation time, denoted as RT, which is certainly one of the most important variables or physical factors considered by experts in the design of interior spaces [9]. In fact, the RT is used as the key parameter in the acoustic assessment of enclosures. RT is defined as the time required by the sound to "fade away" or decay in a given closed space. Specifically, it is the time taken from the sound pressure level to decrease 60 dB after ceasing the emission of a sound source in a room or closed space, or equivalently, the time needed for the acoustic intensity to decrease up to a million times its original value when the source is switched on. Its value depends on the constructive elements that make up the room and their fine finishes and coatings, which affect the overall sound absorption [10,11]. Its analysis is important because the presence of reverberant acoustic energy tends to mask the immediate recognition of any new incoming sound and makes it difficult for the speech intelligibility [11]. If the reverb is excessive, the speech intelligibility may be poor and/or the acoustic pressure be high, being able to adversely affect the performance of activities for which these spaces have been designed [12]. In fact, it is confirmed by experimental studies the strong empirical relationship between the characteristics of the RT of a room, its size and the amount of absorbent material of coatings [12]. So, the RT parameter will be optimized to control the main characteristics of the acoustic behaviour inside rooms in a building.

Furthermore, in spite of the great influence that the acoustic behaviour of buildings has on its occupants, it is often not taken into account from the early stages of the project (except in those buildings in which the acoustic requirements are essential, such as theatres and auditoriums). In general, the acoustic behaviour of spaces is analysed later, in an advanced stage in the construction projects when the geometry and configurations of the enclosures have been already set up. Therefore, if designers strive for achieving some minimum acoustic requirements, they realize that it becomes more complicated and expensive than if it had been handled during the design stage of the project [13,14]. What is more, the acoustic simulations are often carried out using specific software (i.e., Odeon, Catt-Acoustic, Ease, Soundplane, etc.) which are not usually integrated with the others used to design the building. In consequence, the use of these tools in later stages of conventional buildings projects involves an additional work that implies further costs in time and resources, but perhaps the most worrying issue is the fact that it not often concludes with an optimal result [13].

Therefore, the purpose of this research is to build an integrated framework in BIM-based software to generate a comprehensive scheme that allows the analysis of room acoustic performance from the early design stages to be included, in the same way as it was performed with other design disciplines in the construction sector (i.e., energy efficiency, sustainability, LCA, facilities, etc.) [15–19]. In this sense, this research assumes and will demonstrate that the use of BIM-based model into a reliable database with the acoustic characteristics of absorbent material of coatings will result in a reduction of the time invested in the definition of the project and, since the number of explored solutions is increased, the achieved acoustic performance is higher than the one obtained without the use of this framework.

This article is structured in five sections: Section 2 establishes the objectives and the methodology followed in this research. In this section it is outlined the problem of acoustic performance in buildings in the early stages of design of the project building using a BIM-based scheme. In Section 3, those parameters and tools that will be used in the subsequent BIM-based methodology development are defined and presented. The proposed framework based on BIM for the assessment of the acoustic performance of rooms in buildings is developed in Section 4, and in Section 5 the proposed scheme is evaluated on a study case of a building for educational uses. Finally, the main findings and conclusions of this research are drawn in Section 6.

2. Objectives and Research Methodology: Problem Identification and the Use of BIM as a Framework to Assess Acoustical Performance of Buildings

In addition to the comments given in the preceding section, it must be also emphasized that the processes of architectural design and construction are currently becoming increasingly specialized and complex tasks, not only due to the use of new technologies and materials, but also by the specific demands coming from the different regulations (acoustic, thermal, environmental, fire safety, etc.) [20]. In addition, it is quite common that multiple agents to be involved during the life cycle of the project and this fact causes an urgent need for a necessary communication, interaction and collaboration between the different agents from the very early stages of design. For all these reasons, Building Information Modelling (BIM) as a work methodology has attracted a great deal of attention by replacing the traditional methodology based on Computer Aided Design (CAD). The process of generation and data management related to the properties and characteristics of buildings (both geometric and non-geometric data), turns a BIM-based model into a reliable database. This database can be used throughout the whole life-cycle project allowing the effective exchange of information between the different agents involved.

Consequently, the application of the BIM methodology in construction projects offers an exceptional opportunity to assess the building performance from its initial phase [21]. BIM is so useful to test and visualize different scenarios during the process of design, construction and even maintenance; and therefore, it has the potential to improve the design process and to support designers and contractors in the decision-making process concerning the acoustic evaluation [13,22].

On the other hand, the use of BIM-based tools for the assessment of acoustic performance in buildings has been the focus of several studies. Among the most recent ones, it can be highlighted that conducted by Pauwels et al. [20] that proposes a method to analyze the acoustic performance of buildings by translating the design BIM data into ontology data and performing reasoning according to ontology-based rules. For the purposes of this research, it should be note that the BIM-based tool proposed by Wu [13] for the acoustical evaluation of simple rooms during the design stage of the building. The tool is comprised of four modules (BIM data extraction, analysis of frequencies, simulation of sound effects and auralization/visualization). In addition, for simulation purposes, Deng et al. [23] develops in a framework that integrates BIM and 3D-GIS for the assessment of traffic noise in both outdoor and indoor urban environments. This framework is based on four modules that allow calculating noise levels at the outside and the inside and generate output simulation results. As a tool for decision-making based on BIM, the proposal of Hammad et al. [24] allows comparing conventional construction methods and modular ones, considering different factors of sustainability (economic, social and environmental). Among the analysed factors are the study of environmental noise generated by the different processes during the construction process. Finally, Tan et al. [25] proposes an acoustic simulation approach supported by BIM to reduce the impact of noise on offshore platforms in maintenance work. BIM provides the information to configure and prepare the acoustic simulation, this is carried out in Comsol. In this framework, BIM is also used to integrate the obtained information with daily maintenance information. This developed tool can also be implemented in the early stages of design.

The above schemes propose the use of BIM in several acoustic problems but there is a need to develop a complete and fully integrated framework for the assessment of acoustical performance of buildings. For the aforementioned reasons, this work proposes a BIM framework for the assessment and optimisation of the RT in interior spaces. The objective is to propose and develop a framework that allows designers to explore design alternatives in order to achieve a suitable acoustic performance in the early stages of the design of buildings. This framework intends upon completion of the proposed steps to allow the assessment of RT in interior spaces in accordance to the legal regulations of each country or region, and for this commitment, the proposal incorporates an optimization algorithm for the selection of materials to ensure a suitable acoustic behaviour.

With the aim of accomplish the objectives outlined in the preceding paragraph, four consecutive phases were conducted in this research in the following order: (I) Identification of the problem, (II) Statement on Objectives, (III) Proposed Solution and (IV) Evaluation (see Figure 1). In the first phase, literature review allowed defining the context of the problem, i.e., the main characteristics of the room acoustic behaviour can be achieved by an adjustment of the RT parameter, and this acoustic behaviour should be addressed from the early stages of the building project, BIM methodology being a promising framework to accomplish it. With this starting-point, some evidence-practice gaps coming from RT-based analysis from the early stages of the building project made it possible to identify the needs, and objectives were defined on this basis. Therefore, on the basis of the information obtained from phases I and II, the objective of developing a framework that allows designers to achieve a suitable acoustic performance in initial stages of building design was configured. This framework is called "BIM-based framework for Reverberation Time" (BFRT) and it is developed in Section 4.

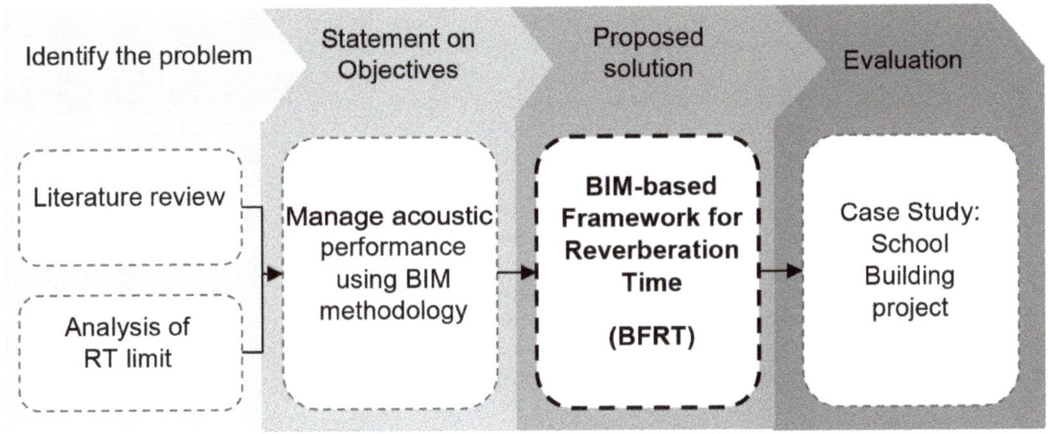

Figure 1. Research method.

Once the framework is defined and established, the feasibility of its implementation in real cases is tested and its results are evaluated in a study case. The project chosen corresponds to a building for educational purposes. In this study case, the application of the BFRT framework was made taking into account several regulatory requirements in European countries, and the obtained results were further compared each other to identify differences in possible options or solutions according the current different guidelines.

3. Parameters and Tools Used for the Development of the BIM-Based Framework

With the methodology scheme outlined in Figure 1, the following subsections establish the identification problem and the parameters and tools (Phase 2 in Figure 1) within a BIM-based methodology needed to develop the proposed framework scheme in Section 3.

3.1. *Using Visual Programming Language in BIM Methodology*

The use of the BIM-based methodology in the industry of Architecture, Engineering and Construction (AEC) has aroused a high impact for the last decade. One of the reasons for this growth is that the BIM methodology provides tools to comply with the Directive 2014/24/EU of the European Parliament and of the Council of 26 February 2014 on public procurement [23]. This Directive establishes in Article 22. Rules applicable to communication the following: "For public works contracts and design contests, Member States may require the use of specific electronic tools, such as of building information electronic modelling tools or similar."

On the other hand, the ability of BIM to support the decision-making process from the early stages of design has become in an effective tool for building performance modelling. In this sense, researchers have started to use BIM not only as a modelling tool, but also for what it was originally created, i.e., as a critical methodology and technology to achieve higher levels of performance and automatic simulations (such as predictive analysis of performance, sustainability performance [26], life cycle assessment (LCA) performance [27]).

This progress is accompanied by the appearance of tools based on Visual Programming Language (VPL) that make it easier for the designers, which are not usually programmers, extend the capabilities of BIM without the need for advanced knowledge in programming languages. There are different tools based on VPL for BIM (among the most known ones are Dynamo and Grasshopper). These tools enable us to expand the parametric functionalities of the BIM methodology and its use expands the options of iteration with the model, information extraction and development of tools.

There are many studies that have developed tools or frameworks based on BIM using tools with VPL, such as for example: multi-objective environmental optimization of buildings [28], Building Sustainability Assessment [29], evaluation of BIM-based LCA [30], Safety Analysis [31] or Design for Deconstruction [32].

In this work, VPL is used as a tool for the development of a BIM based-framework that allows designers to evaluate and optimize the acoustic behaviour of a room from an analysis of RT in interior spaces, all integrated into the own BIM software.

3.2. Reverberation Time (RT) for the Assessment of Acoustic Room Behaviour

The standard EN 12345-6:2003 [33] sets the calculation model for the estimation of the RT of enclosed spaces within buildings. Due to the strong dependence of the absorption on frequency, it is necessary to determine the RT for those most representative frequencies. In general, it is calculated as the average of the RT for 500, 1000 and 2000 Hz the frequencies. Equation (1) shows the classical Sabine formula for the calculation of the RT, taking as 345.6 m/s for the speed of sound in air [12]:

$$TR = 0.16\frac{V}{A} \tag{1}$$

where V is the volume of the room (m^3) and A is the whole room sound absorption (Equation (2)) given by [33]:

$$A = \sum_{i=1}^{n} \propto_{s,i} S_i + \sum_{j=1}^{o} A_{obj,j} + \sum_{k=1}^{p} \propto_{s,k} S_k + A_{air} \tag{2}$$

where:

$\propto_{s,i}$ is the coefficient of acoustic absorption of the i-th room surface.

S_i is the surface area (m^2) of the i-th room surface.

$A_{obj,j}$ is the equivalent sound absorption area of the j-th object (m^2).

$\propto_{s,k}$ is the coefficient of acoustic absorption of the k-th specific object configuration (for example rows of chairs, people sitting in line or children in a classroom with reflecting furniture).

S_k is the surface area covered by the k-th object configuration (m^2).

A_{air} is the equivalent sound absorption area of the air (m^2).

n is the number of absorbing surfaces in the room (excluding objects).

o is the number of absorbing objects in the room.

p is the number of the configuration of absorbing objects in the room.

The equivalent sound absorption area of the air is given by the Equation (3):

$$A_{air} = 4\,m\,V\,(1 - \Psi) \tag{3}$$

where

m is the sound attenuation coefficient of the air, in Neper per meter;

V is the volume of the empty closed space (m^3);

Ψ is the object fraction defined as the ratio between the sum of all the volumes of the objects and the volume of the empty space according to the ISO 12354-6:2004 standard (dimensionless).

To ensure a suitable acoustic behaviour, it is required to restrict the reverberating noise inside rooms. If the reverberation is excessive, the audibility may be poor and/or the acoustic pressure be high, which interferes with the appropriate performance of human activities for which these spaces were designed [8]. As a rule, to minimize the effects of an excessive reverberation is desirable to keep RT small. On the other hand, if the sized of the room is large and the sound source power is weak, it is advisable to achieve a higher RT to keep the sound audible at all points in the room. As can be observed, the choice of a suitable RT for a given enclosure depends on the final use of the room and the commitment/criterion to be reached in the design stages [9].

For this reason, in such a case in which rooms have an intended use for spoken word, current regulations of each country sets a maximum value for the RT, which should not be exceeded. In Table 1 it is showed those minimum requirements for the RT in different countries given the different uses of the spaces.

Table 1. RT minimum requirements in different countries.

Country	Type of Room	Requirement RT	Frequency Band	Comment
Spain [34]	classrooms and conference rooms	0.7 s	500–1000–2000 Hz	Unfurnished and unoccupied room. $V \leq 350$ m^3
	classrooms and conference rooms	0.5 s	500–1000–2000 Hz	Furnished room. $V \leq 350$ m^3
	Restaurants and canteens rooms	0.9 s	500–1000–2000 Hz	Unfurnished and unoccupied room
France [35]	Classrooms and polyvalent rooms	$0.4 \leq RT < 0.8$ s	500–1000–2000 Hz	Furnished and unoccupied room. $V \leq 250$ m^3
	Classrooms and polyvalent rooms	$0.6 \leq RT < 1.2$ s	500–1000–2000 Hz	Furnished and unoccupied room. $V > 250$ m^3
	Restaurant (School)	$0.4 \leq RT < 0.8$ s	500–1000–2000 Hz	Furnished and unoccupied room. $V \leq 250$ m^3
	Restaurant (School)	$0.6 \leq RT < 1.2$ s	500–1000–2000 Hz	Furnished and unoccupied room. $V > 250$ m^3. Special study required
	Sport	0.6 s	500–1000–2000 Hz	Furnished and unoccupied room. $V \leq 250$ m^3
Portugal [36]	Sport	$RT \leq 0.15\sqrt[3]{V}$	500–1000–2000 Hz	Furnished and unoccupied room.
	Sport	$RT \leq 0.12\sqrt[3]{V}$	500–1000–2000 Hz	Furnished and unoccupied room. With Public address
	Auditory, conference and polyvalent rooms	$RT \leq 0.12\sqrt[3]{V}$	500–1000–2000 Hz	Furnished and unoccupied room. if $V < 250$ m^3.
	Auditory, conference and polyvalent rooms	$RT \leq 0.32 + 0.17 \log V$	500–1000–2000 Hz	Furnished and unoccupied room. if $250 \leq V < 9000$ m^3.
	Auditory, conference and polyvalent rooms	$RT \leq 0.05\sqrt[3]{V}$	500–1000–2000 Hz	Furnished and unoccupied room. Furnished ≥ 9000 m^3.
Belgium [37]	classrooms and conference rooms	$0.35 \log(1.25V)$	500–1000–2000 Hz	Unfurnished and unoccupied room.
	Sport	$\log(V/50)$	500–1000–2000 Hz	Unfurnished and unoccupied room.
	Restaurant (School)	1.0 s	500–1000–2000 Hz	Unfurnished and unoccupied room.

Table 1. *Cont.*

Country	Type of Room	Requirement RT	Frequency Band	Comment
United Kingdom [38]	Clasrooms (primary school)	$RT \leq 0.6$ s[1] $RT \leq 0.8$ s[2]	500–1000–2000 Hz	Furnished and unoccupied room.
	Clasrooms (secondary school)	$RT \leq 0.8$ s[1] $RT \leq 1.0$ s[2]	500–1000–2000 Hz	Furnished and unoccupied room.
	Lecture rooms	$RT \leq 0.8$ s[1] $RT \leq 1.0$ s[2]	500–1000–2000 Hz	Furnished and unoccupied room. Fewer than 50 people
	Lecture rooms	$RT \leq 1.0$ s[1] $RT \leq 1.0$ s[2]	500–1000–2000 Hz	Furnished and unoccupied room. More than 50 people
	Gymnasium/activity studio	$RT \leq 1.5$ s[1] $RT \leq 2.0$ s[2]	500–1000–2000 Hz	Furnished and unoccupied room.

As can be observed in Table 1, in some country regulations there is some limit values for not occupied enclosures, furnished or unfurnished, or depending on the room volume. These considerations inherent in the regulations of each country will be taken into account in the proposed tool to define minimum and maximum values of RT.

Finally, in the design phase of the project these values will be used to define the range that allows us to check if the RT in a room will be acceptable or not, in accordance with its characteristics. In addition, the definition of an upper limit of RT will ensure the information coming from the sound source is intelligible inside the room. Furthermore, the establishment of a lower limit will ensure a suitable room acoustic performance.

4. Proposed BIM Framework for Acoustic RT-Based Design in Indoor Areas (BFRT)

The proposed BIM Framework based on RT (denoted as BFRT) is developed as an integrated tool for the assessment and optimisation of RT of interior spaces in buildings. The proposed tool generates a set of possible solutions from an extensive search for possible combinations of existing materials for the different surfaces of the room (wall, floor and ceiling) included in a database, so that the RT becomes suitable for the prescribed uses of the room. For this purpose, this work develops an optimization algorithm using combinations of finishing materials to propose alternatives of constructive design but always fulfilling the regulatory limits. The aim of this proposal is to help designers in the decision-making process in the initial stages of design.

The proposed framework is composed of 3 stages as shown in Figure 2. In the **Stage 1—BIM Modelling**, the building under consideration is designed and modelled using BIM, with special emphasis in including the information of constructive elements, materials, etc. In this stage, this modelling must ensure that all the necessary information (geometric and not geometric) of the project is available, accurate and reliable. After that, it is performed the zoning of the model and allocation of final uses for the building rooms. The **Stage 1** was developed using Autodesk Revit software [39], which is based on the building parametric information modelling.

In the **Stage 2—Data extraction and RT calculation**, the extraction of geometric data (dimensions of the enclosures, volume, areas of surfaces, etc.) and non-geometric data (item type, materials, etc.) from the BIM model is performed. This stage was developed using the Dynamo software [40] (as it is shown in Figure 3). At this stage, a connection with an external database of materials is made to gain access to absorption coefficients and price of the materials used in the model. From the data obtained from the BIM model and database, it is calculated the current RT and the targeted or desired RT (RT objective) is established, as well as the range of acceptance for each room according to their use and regulations in the specific country where the project is run.

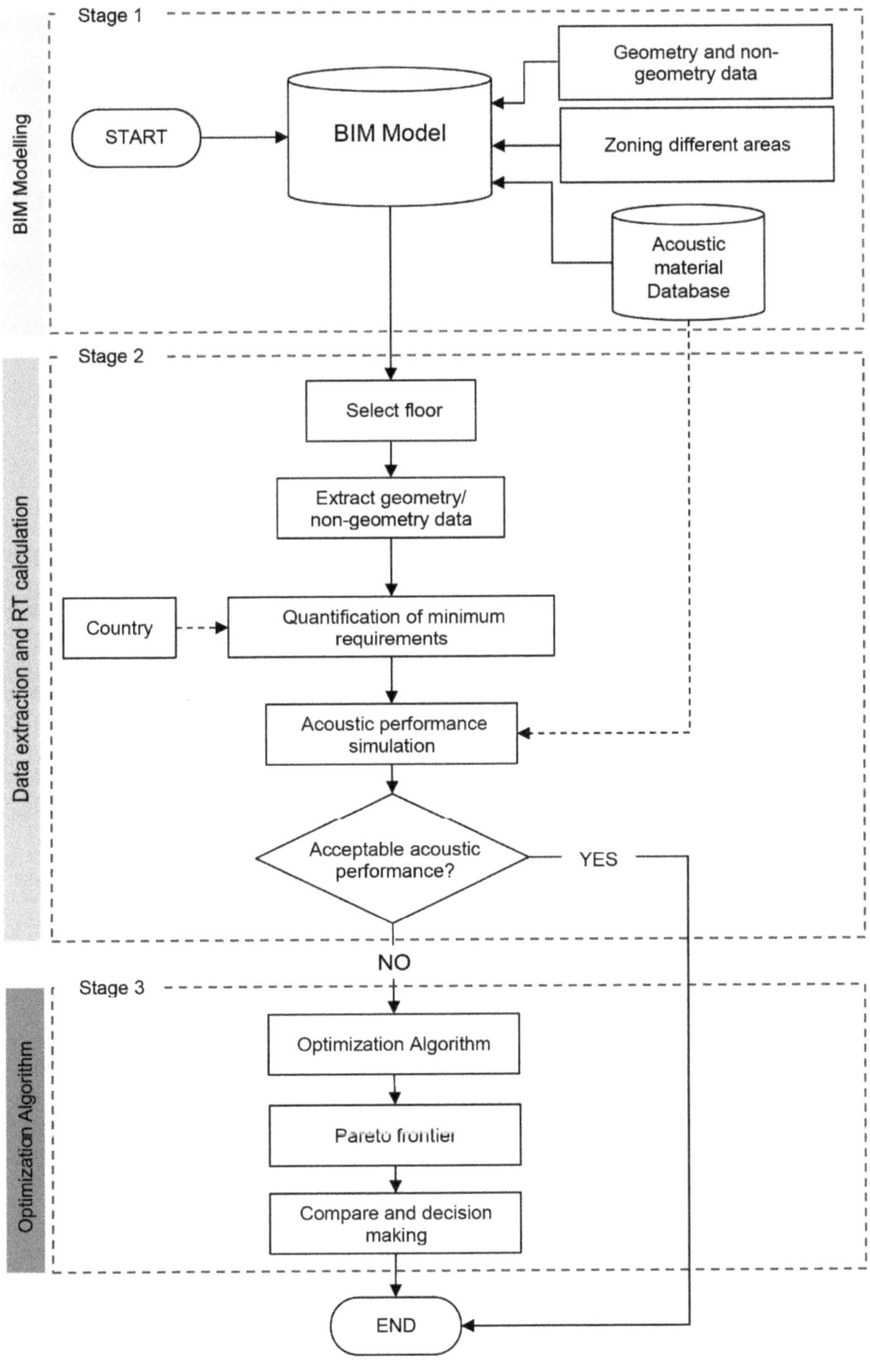

Figure 2. Proposed acoustic BIM framework.

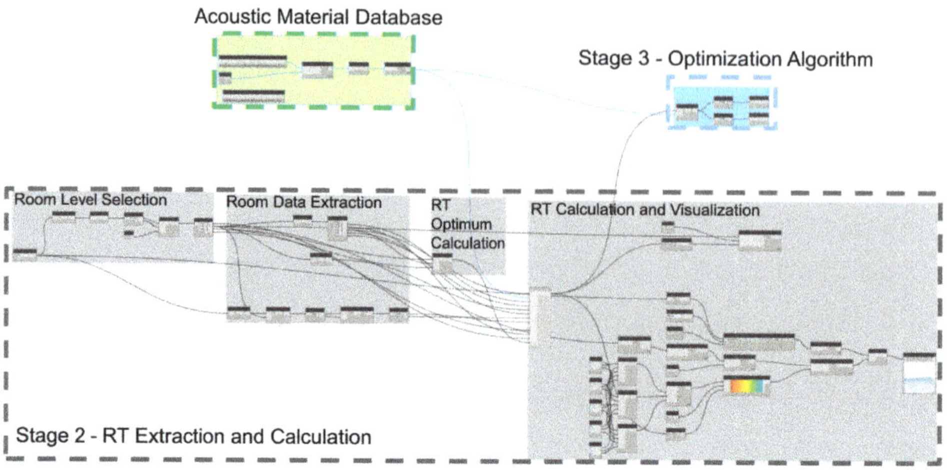

Figure 3. BFRT system in Dynamo (Stages 2 and 3).

Just in the case that the initial RT of the room (calculated from the original design) becomes unacceptable, it is undertaken the application of the optimization algorithm defined in Stage 3. If the initial RT were acceptable, the proposed procedure ends at this point. In any case, once the calculations are finished, the results are displayed in the Dynamo interface.

In those cases, in which the RT is not acceptable, the **Stage 3—Optimization Algorithm** starts. Now, these cases are evaluated using an algorithm that performs the search for the optimal solution of the RT using a branch and bound algorithm [41]. This stage is carried out using Dynamo software (Figure 3). The optimization is based on the search for all the appropriate combinations of existing materials in the database that can be used in the different surfaces of the room (wall, floor and ceiling) to accomplish the targeted RT. After performing this search, the algorithm shows a set of optimal solutions by the Pareto frontier. These stages are described in detail in the following Sections 4.1–4.3 of this work.

4.1. Stage 1—BIM Modelling

Prior to the design process of the building in BIM, it is necessary to define several multiple shared parameters (Figure 4) with that serve to save and communicate information about the BIM model components. The advantage of the shared parameters is that they can be used in other projects or ensembles of projects without the need to re-create them, as they are stored in different files separated from the main project files.

The above-mentioned shared parameters are required for the subsequent process of calculation and optimization of the acoustic room behaviour. In this case, it will be necessary to generate four shared parameters associated with the different components of the model. Table 2 shows the parameters used for the development of the BFRT.

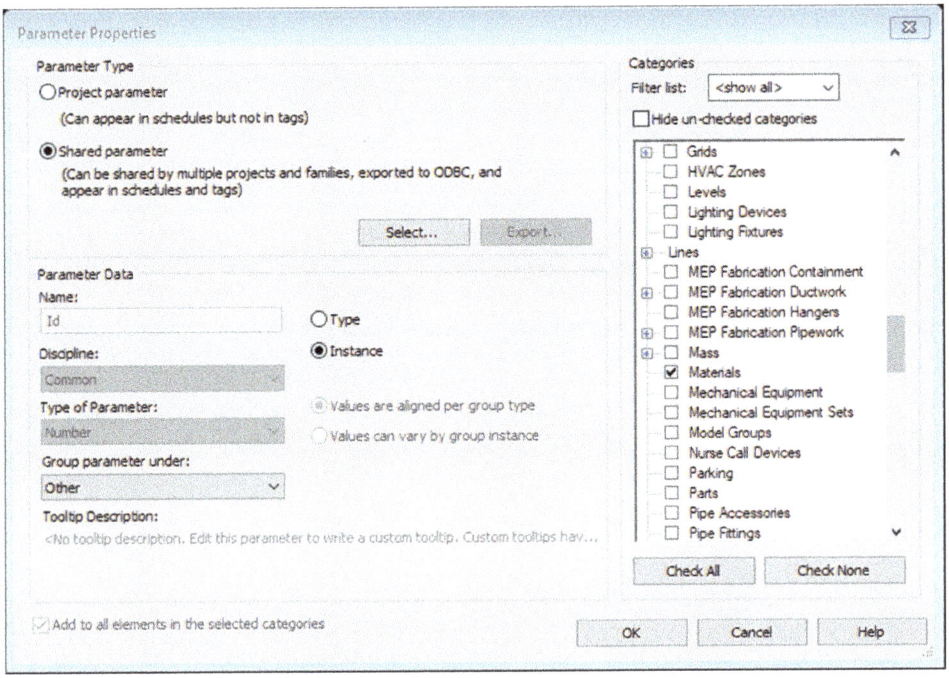

Figure 4. Example of creating shared parameters for its further use in BIM.

Table 2. Shared parameters of the BIM model.

Shared Parameter	Definition	Type Parameter	Caegory
RT	Reverberation time of room	Number	Room
Type room	Type of room	String	Room
Id	Identification number of material with Acoustic material Database	Number	Material/Door/Window
A_{furn}	Equivalent sound absorption area of furniture	Number	Room

Once the parameters are defined, it is accomplished the design and construction of the BIM model of the building (it would be also possible to use an already developed model and import the shared parameters, if it were the case). A minimum Level of Development (LOD) 300 is required for the analysis. Then, it should subsequently be defined the different constructive elements (walls, floors, ceilings, doors, windows, etc.) to be used in the rooms under analysis and the *Id* parameter is assigned to each material comprising the components. This parameter is used to properly relate the materials of the model with the external database of acoustic parameters of construction materials (AM database). This database must be defined and set up at this stage, and it is connected with the BFRT system. The AM database is composed of different fields (see Table 3), and it contains all the necessary information on materials for the making of calculations and the subsequent optimization. Its content and information have been obtained after a review of the currently materials available in the market and the information provided by the manufacturers. Users can easily modify the database manually, which may be broaden with new materials.

Table 3. Basic information scheme of the AM database for each material.

Element	Data-Type	Description
Name	String	Name or description of the construction material
Id	Number	Identification number of the acoustic material
α	Number	Absorption coefficient in the 125–250–500–1000–2000–4000 Hz frequency band.
Location	Number	Each finish material has a specific type of location (wall or/and ceiling or/and floor).
Cost	Number	Cost of material (€/m^2)

The next step is the zoning of the building. To accomplish this, it will be required the identification and definition of the use of each enclosure using the shared parameter Type room (Figure 5). In addition, in the event of the regulations of the country specify that the RT must be calculated considering that the room is furnished, the parameter A_{furn} must be fulfilled with the quantity corresponding to the amount of equivalent sound absorption area of the furniture that the designer should consider.

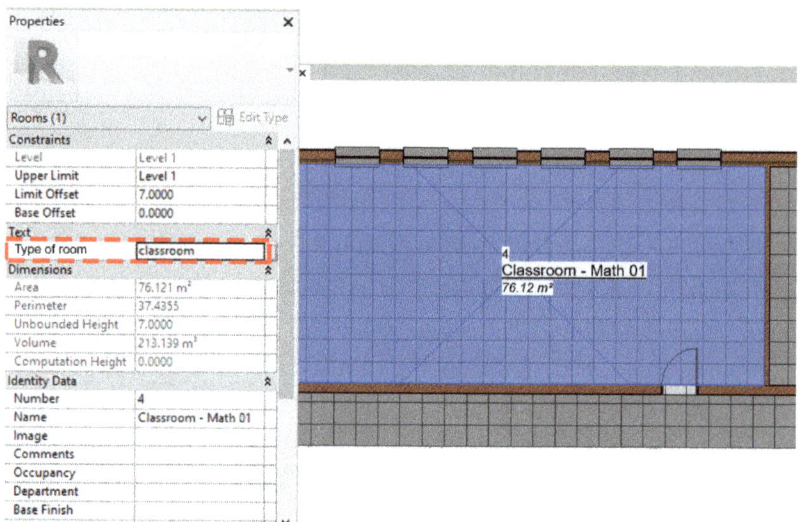

Figure 5. Example of Type room shared parameter setup.

4.2. Stage 2—Data Extraction and RT Calculation

This stage is composed of 4 groups of nodes developed using Dynamo, which will be defined in the following sections (Figure 3) i.e., Room Level Selection, Room Data Extraction, RT Optimum Calculation and RT Calculation and Visualization. Some of the used nodes are included in the basic library of Dynamo, while the more complex functions have been developed as nodes in a Python script to avoid the limitations from the basic nodes.

4.2.1. First Node Group—Stage 2: Room Level Selection

This first group is composed of 3 nodes. For RT assessment is necessary to gather some data from the BIM model. To accomplish this, firstly it will be necessary to choose the model floor for which the assessment is going to be performed. The levels node is used from the bookstore of Dynamo to select the corresponding plant model. Once the floor is selected, using the Room at level node it is performed a filtering process of the rooms classifying them by its floor (Figure 6). At this point, the Filter Room Regulation node makes a selection of rooms depending on the country where the project is located and the final intended use. In this node, only those rooms with a prescribed RT value addressed in regulations will only be selected for its subsequent evaluation.

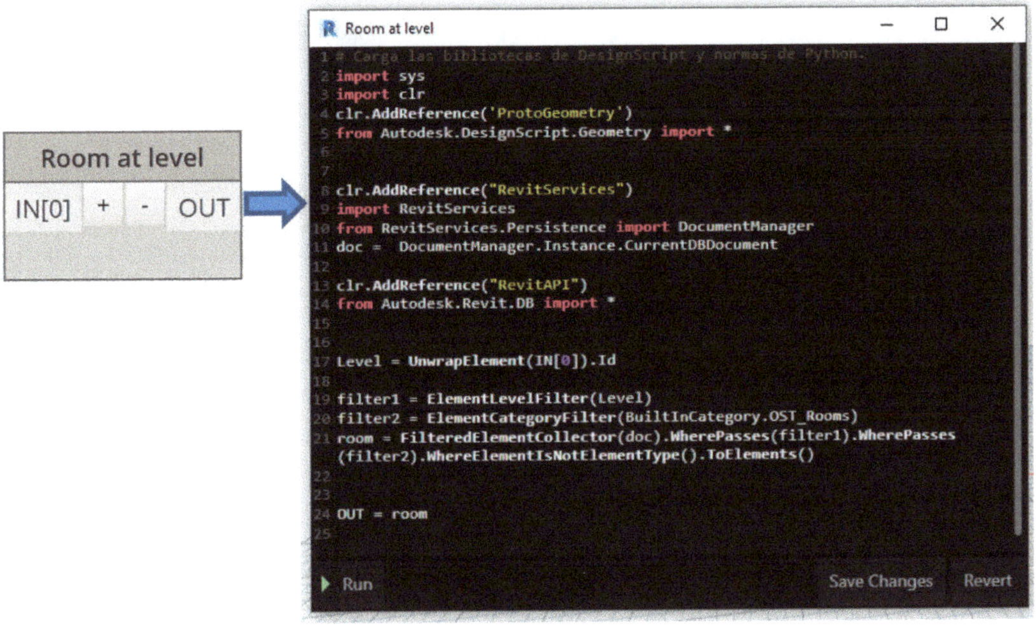

Figure 6. Room at Level Node.

4.2.2. Second Node Group—Stage 2: Room Data Extraction

The second group of nodes is comprised of 4 nodes. They extract information such as geometric data (dimensions, area, and volume) and non-geometric data (finishing materials, Id of the materials, element type, location, and the A_{furn} parameter) from the elements of the BIM model. Data Room and Element Room nodes extract the information related to the rooms selected by the First Node Group, while the Door/Window at Level and Door/Window Released nodes extract the information related to the doors and windows. To accomplish this, the Door/Window at room node filters out doors and windows of the model and then they associate them to the room where they are located. Finally, the Door/Window Released node obtains the necessary information for performing the calculation process.

4.2.3. Third Node Group—Stage 2: RT Optimum Calculation

Once data from the model have been extracted in the preceding sub-stages, it is calculated the minimum RT requirements demanded by regulations. For this task Regulation RT node has been developed. This node assigns the limit RT value depending on the country in which the project is going to be implemented and the room use, obtaining as output the maximum values that each of the rooms should meet. In a first approach of this framework, it has been implemented the limit or recommended values from the regulations of the following countries: Spain, Portugal, Belgium, Denmark and the United Kingdom. If new country regulations are required for RT assessment in other countries than the pre-set ones, the designers can add their own restrictions by editing de Regulation RT node. For this purpose, the code of the Regulation RT node must be edited, simply adding the new country, the types of room and the associated RT requirement values.

As noted in Section 3.2, it is necessary to set both the lower and upper limits to evaluate if the RT for a specified room can be acceptable or suitable in the design phase. In this way, if the RT belongs to the interval defined by those limits, it is ensured a correct acoustic behaviour of the room. The upper limit value (RT_{lim_sup}) of the acceptance interval corresponds to that established by the regulations as the value that should not be exceeded (see Table 1). In addition to this upper limit, it is necessary to establish the lower limit of the acceptance range to ensure a minimum suitable RT (RT_{lim_inf}) so that the speech intelligibility is reasonably good at all points in the room.

For this purpose, it has been set by default that the lower RT limit (RT_{lim_inf}) will be a value 20% lower than the upper limit RT_{lim_sup}. In this issue, the recommendation of DIN 18041 standard has been chosen, although the designer could set a different value or other requirement according to his own criterion (see Figure 7). In this sense, the value corresponding to the middle of the interval defined by both limits, RT_{lim_inf} and RT_{lim_sup}, is accordingly denoted as RT_{target}.

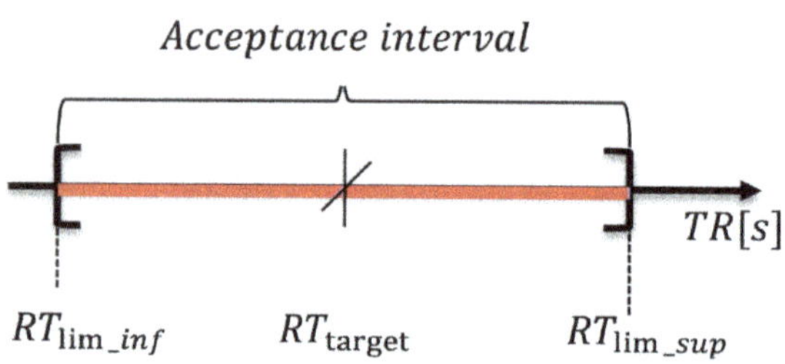

RT_{lim_sup}: Prescribed value from regulations

$$RT_{\mathrm{lim}\_inf} = 0.8 \cdot RT_{\mathrm{lim}\_sup}$$

$$RT_{\mathrm{target}} = \frac{RT_{\mathrm{lim}\_sup} + RT_{\mathrm{lim}\_inf}}{2}$$

Figure 7. Acceptance interval for the RT from the regulatory limits.

4.2.4. Fourth Node Group—Stage 2: RT Calculation and Visualization

Finally, the last group of nodes needs to perform the calculation of the RT from the data obtained through the previous node groups. The RT will be calculated for each room from the finishing materials defined in the initial design. The calculation is made using the absorption coefficients of materials provided by the AM database for the mid frequencies (500–1000–2000 Hz) and Equation (1) shown in Section 3.2. In those cases, in which regulations would require to consider the room furnished -that is to say, the equivalent sound absorption area of the furniture-, the A_{furn} parameter absorbent characteristic of each room will be added to the total absorption area.

Then, the RT value obtained for each room is checked upon it belongs to the acceptance interval established in the previous phase. At this moment, a preview of the rooms is displayed to the user in the Dynamo environment, so that those rooms marked in green stands for those ones with the value of the RT inside the acceptance interval, being marked in red otherwise. For those enclosures in which the RT value falls in the acceptance interval, the process finishes and this solution is taken as a valid one with the initially defined finishing materials. Lastly, the values of the RT obtained are exported to the model BIM to enrich and complete the information contained in the model database related to the acoustic behaviour of the building.

4.3. Stage 3—Optimization Algorithm

In this Stage 3 the optimization process is performed, in a Dynamo environment. This process is carried out over the rooms whose RT does not belong to the region of acceptance. The objective of this algorithm is twofold, firstly it tries to find out different design solutions to adapt the original RT to the prescribed targeted RT (see Figure 7) and secondly, it looks for minimizing the total cost, by the replacement of finishing materials once the AM database is connected. The information on materials included in the AM database is then used for proposal of new solutions in the optimization process. As a result, different solutions are shown, ordered as their average absorption coefficient increases.

Table 4 shows different types of action made for optimization, according to the replaced materials for the finishings of the room (wall, ceiling or floor) and their possible combinations.

Table 4. Types of optimizations according to the replaced material.

Type	Replaced Material
Type 1	Replace the wall
Type 2	Replace the ceiling
Type 3	Replace the floor
Type 4	Replace the wall-ceiling
Type 5	Replace the wall-floor
Type 6	Replace the ceiling-floor
Type 7	Replace the wall-ceiling-floor

To determine the possible potential solutions that would allow us to adapt the RT of the room, the implemented algorithm uses the branch and bound technique [41]. This technique is frequently used to solve optimization problems through the generation of a space of solutions defined in a tree (Figure 8). The use of this technique has the purpose of searching for a set of solutions that meet one criterion previously established, through a systematic path by the tree of solutions.

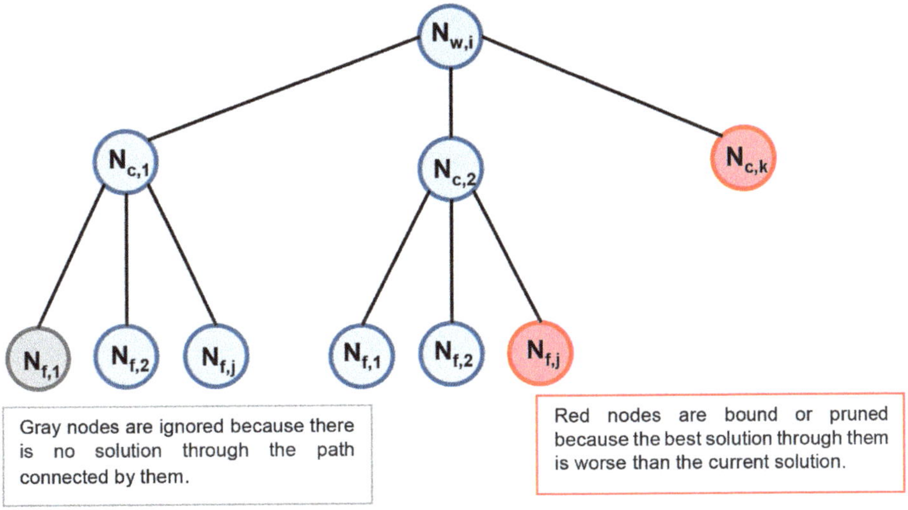

Figure 8. Branch and bound algorithm using to adapt the RT.

So, this procedure discards large subsets of unsuccessful candidates on the basis of the use of upper and lower limits by following different paths. The efficiency of this method depends mainly on the branching procedure of nodes and the strategy of bounding to remove those nodes that are not a feasible solution.

For this study, a FIFO (First In First Out) strategy of branching has been chosen, in which the path through the search space is made in the width of the tree. With regard to the bounding strategy, the range of acceptance interval of the RT is taken into account.

Thus, the procedure in the stage 3 remains as follows:

1. From the TR_{lim_sup} y TR_{lim_inf} limits of the acceptance interval, it is possible to calculate the minimum acoustic absorption surface ($A_{lim\ inf}$) y and the maximum

acoustic surface absorption ($A_{lim\ sup}$) that the room under analysis should have. The bounding strategy is set from the Equations (4) and (5),

$$A_{w,i} + A_{c,k} + A_{f,j} \geq A_{lim\ inf} \quad (4)$$

$$A_{w,i} + A_{c,k} + A_{f,j} \leq A_{lim\ sup} \quad (5)$$

where $A_{w,i}$ is the total absorption surface area corresponding to a wall coated with the *i*-th material; $A_{c,k}$ is the total absorption surface area corresponding to a ceiling coated with the *k*-th finishing material; and $A_{f,j}$ is the total absorption surface area corresponding to a floor with the *j*-th finishing coating material, i.e.:

$$A_{w,i} = \alpha_{w,i} \times S_w \quad (6)$$

$$A_{c,k} = \alpha_{c,k} \times S_c \quad (7)$$

$$A_{f,j} = \alpha_{f,j} \times S_f \quad (8)$$

In the above equations $\alpha_{w,i}$ is the average absorption coefficient of the wall coated with the *i*-th material; S_w it is the total surface area of the wall; $\alpha_{c,k}$ is the average absorption coefficient of the ceiling coated with the *k*-th material; S_c is the total surface area of the ceiling; $\alpha_{f,j}$ is the average absorption coefficient of the floor covered with the *j*-th material; and S_f it is the total surface area of the floor.

2. For each individual solution that meets the acceptance criterion, the objective functions are computed. These functions are two: the first one denoted as C_i is the cost of the investment (Equation (9)) and the second one is denoted as D_i (Figure 9) which is the absolute value of the difference between the total absorption surface area that provides such a solution with respect to the optimal absorbent surface area (A_{target}) of the enclosure (Equation (10)).

$$C_{ijk} = p_{w,i} \times S_{w,i} + p_{c,k} \times S_{c,k} + p_{f,j} \times S_{f,j} \quad (9)$$

$$D_{ijk} = \left| A_{target} - A_{ijk} \right| \ \ being \ \ A_{ijk} = A_{w,i} + A_{c,k} + A_{f,j} \quad (10)$$

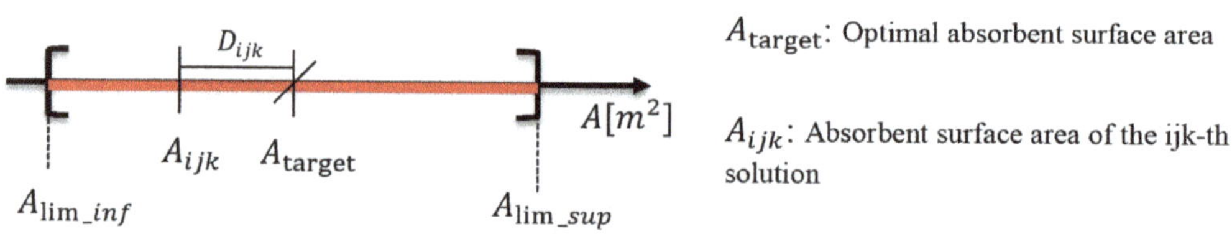

Figure 9. Calculation of D_{ijk} objective function.

In the above equation $p_{w,i}$ is the cost of the *i*-th finishing material covering the wall surface, $S_{w,i}$ being the total surface area of the walls coated with the *i*-th finishing material. $p_{c,k}$ is the cost of the *k*-th finishing material that coats the ceiling surface. $S_{c,k}$ is the total surface area of the ceiling coated with the *k*-th finishing material. $p_{f,j}$ is the cost of the *j*-th finishing material that covers the floor surface and $S_{w,j}$ is the total surface area of the floor covered with the *j*-th finishing material.

3. Once obtained the set of solutions for the studied problem, the optimum solutions are calculated using of the Pareto front or frontier. The criterion of optimization has been the minimization of the cost and the difference between the absorbent surface area of the solution and the optimum absorbent surface area. The Pareto front is the set of possible solutions of optimization that are not dominated; a non-dominated solution being a solution that is not dominated by any other solution. The optimal

Pareto solution will be that solution P_i such that there is no other solution P_j that will improve in a goal without becoming worse at least one of the other ones.

4. This algorithm ends by showing the solutions that belong to the Pareto fronts corresponding to each one of the 7 types of actions proposed in Table 3. Thus, the designer will be able to choose between the proposed solutions, as they all fulfil the criterion of a suitable RT.

5. Results: Case Study

In this section, the application of the proposed methodology is illustrated, using a study case. This example aims to show the type of solutions found and the potential of the proposed methodology to be applied in the design process.

5.1. Building for the Case Study

The proposed framework (BFRT) was applied to a building to be used for educational purposes. The building has a design area of 2500 m^2 with two floors (Figure 10), and is located in the city of Granada in Spain. The building comprises of different rooms for different uses (classrooms, laboratories, offices, conference rooms, library, dining, warehouses and facilities). Table 5 shows a summary of the materials used in the different rooms.

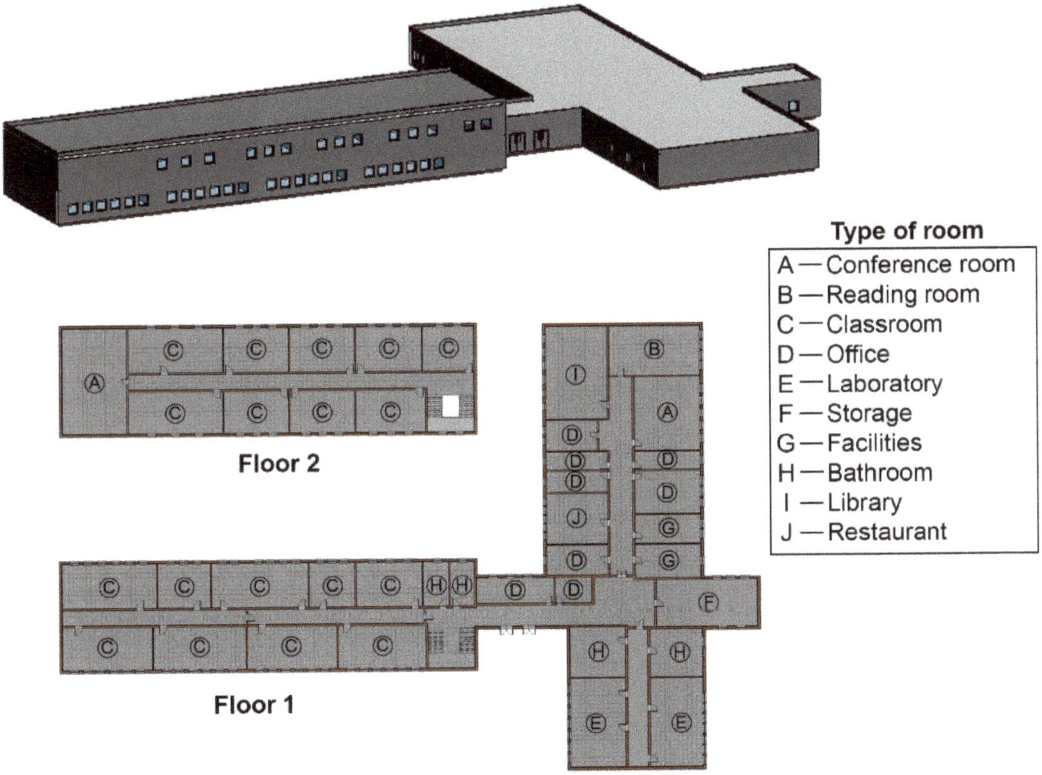

Figure 10. BIM 3D model of the educational building taken as a study case.

Table 5. Finishing materials according to the type of room.

Type of Room	Element	Finishing Material
Classroom/Reading room/Office/Laboratory	Wall Ceiling Floor	Plaster 15 mm gypsum board Ceramics
Storage/Facilities/Bathroom	Wall Ceiling Floor	Tile Ceiling Terrazzo
Library/Conference room	Wall Ceiling Floor	15 mm gypsum board Drop ceiling Parquet

The windows are composed of glazed surface area in the 90% of its surface, while the doors are made entirety of wood. The height of the rooms is 3.00 m. The 3D BIM model was created using Autodesk Revit 2021 and Dynamo 2.1.

5.2. Data Extraction and RT Calculation (Application of the Stage 2 of the Proposed Procedure)

Once the building is modelled, the package of nodes developed in Dynamo extracts the geometric and non-geometric data (related to its Node Group). Firstly, it performs a filtering of the rooms that have a regulatory requirement to fulfil. Given that this building is located in Spain, in accordance with the applicable regulations, only those rooms with a declared use of classroom, conference room and restaurant (see Table 1) must meet the prescribed limits. On the basis of these data, the RT is computed. Figure 11 shows the calculated values for the RT in the different selected rooms, as well as its use. Figure 12 shows the display of the RT fulfilment according to the Spanish regulation in the different floors of the BIM model. As shown in Figure 12 only the room with the use of dining room/cafeteria meets the criteria for acceptance of the RT.

A	B	C	D	E	F
Name	Level	Ud	Type of room	RT	Volume
A01	Level 1	Ud_1	classroom	0.76	120.26 m³
A02	Level 1	Ud_2	classroom	0.79	217.80 m³
A03	Level 1	Ud_3	classroom	0.75	103.59 m³
A04	Level 1	Ud_4	classroom	0.77	149.67 m³
A05	Level 1	Ud_5	classroom	0.78	194.87 m³
A06	Level 1	Ud_6	classroom	0.78	200.34 m³
A07	Level 1	Ud_7	classroom	0.78	196.52 m³
A08	Level 1	Ud_8	classroom	0.78	190.18 m³
A09	Level 1	Ud_9	classroom	0.87	213.14 m³
TEC	Level 1	Ud_12	classroom	0.88	249.21 m³
TECFP	Level 1	Ud_13	classroom	0.88	237.93 m³
Laboratory	Level 1	Ud_15	classroom	0.8	244.88 m³
Conference room 01	Level 1	Ud_20	conference	0.81	247.51 m³
Cafeteria	Level 1	Ud_23	restaurant	0.78	158.66 m³
Reading room	Level 1	Ud_30	classroom	0.8	229.43 m³
A10	Level 2	Ud_31	classroom	0.79	214.06 m³
A15	Level 2	Ud_32	classroom	0.79	205.55 m³
A16	Level 2	Ud_33	classroom	0.77	141.03 m³
A11	Level 2	Ud_34	classroom	0.77	146.87 m³
A12	Level 2	Ud_35	classroom	0.77	146.87 m³
A17	Level 2	Ud_36	classroom	0.77	141.03 m³
A18	Level 2	Ud_37	classroom	0.77	150.82 m³
A13	Level 2	Ud_38	classroom	0.77	149.67 m³
A14	Level 2	Ud_39	classroom	0.76	114.51 m³
Conference room 02	Level 2	Ud_40	conference	0.74	346.95 m³

Figure 11. Room schedule in the BIM model with RT calculated values.

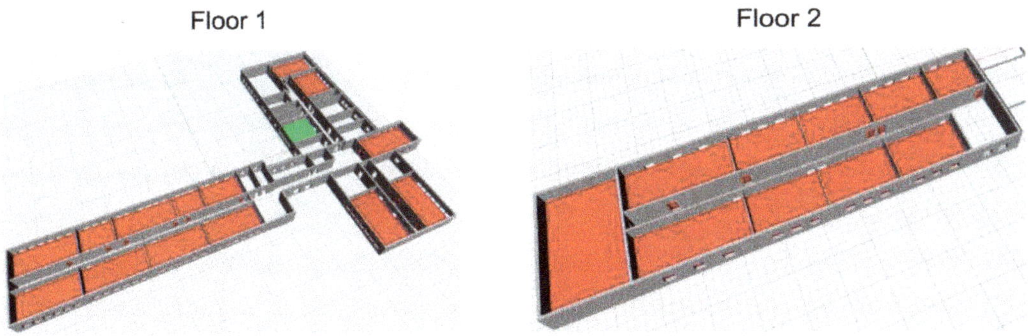

Figure 12. Visualization of the compliance with the RT limit values according to the type of room on the ground floor (project located in Spain).

5.3. Optimization (Application of Stage 3 of the Proposed Procedure)

Once it has been identified those rooms whose RT is not appropriate, a further analysis is performed applying the "Optimization Algorithm" node developed in Stage 3. The calculation is made automatically for all the on the same floor rooms that do not comply with the predefined criteria, in accordance with the acceptance interval based on the Spanish regulations.

An example of the results obtained through the optimization process is shown in Figure 13 for the "A" room whose use is classroom. The Branch and Bound technique is used to determine all the possible potential solutions. Every solution A_{ijk} providing a total absorption surface area contained in the interval $\left[A_{lim,inf}, A_{lim,sup}\right]$ is stored and classified according to the replaced material (i.e., Type 1, ... , Type 8). Based on this set of solutions classified for each typology, the optimum solutions are then calculated by making use of the Pareto frontier. In this sense, in this study case a database was used containing $i = 81$ wall materials, $k = 207$ ceiling coated materials and $k = 25$ floor materials. In fact, these numbers depend on the number of materials used by the design team. Accordingly, the size of the solution space is $n = 419,175$ cases. The algorithm has generated 46,437 feasible solutions. From these solutions and in different colour Figure 13 shows the Pareto fronts for each of the types of intervention defined in Table 4, based on the criteria of minimizing total cost and difference of absorption. In this figure, it can be also observed the results obtained for the different choices of elements for optimization: Wall (w), Ceiling (c) and Floor (f), as well as their diverse combinations. In type 1 (only wall) there are no solutions starting from the existing database. For the rest of elements (see Table 4) two sets of Pareto fronts can be identified.

According to the Pareto fronts obtained for the different elements, the different types can be grouped into two sets. The first set comprises the types 3, 2 and 6, in which all the solutions have a cost equal to or greater than 4000 €. The second set comprises the types 4, 5 and 7. In this set, the solutions provide values near to the optimal absorbent acoustic performance, but with a total cost less than the Pareto fronts in the first set.

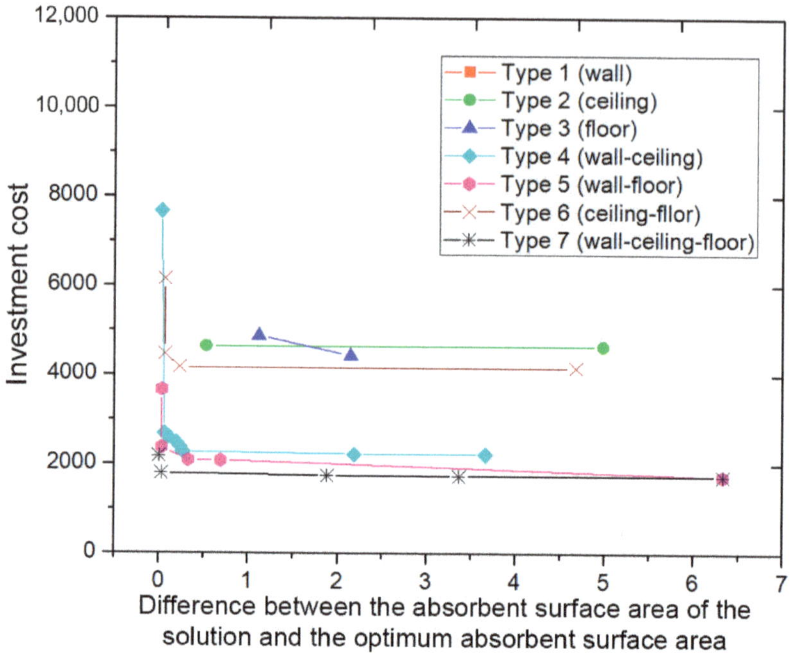

Figure 13. Results obtained from the Pareto frontiers calculation in Stage 3 for the optimization based on the acoustic absorption of materials and their investment cost. Red square—Pareto front type 1. Green dot—Pareto front type 2. Blue Triangle—Pareto front type 3. Cyan rhombus—Pareto front type 4. Purple hexagon—Pareto front type 5. Wine cross—Pareto front type 6. Black Cross—Pareto front type 7.

It is interesting to note that the use of the information provided by the stage 3 of the proposed procedure through the Pareto frontiers can be helpful to the designer/researcher in order to make a final decision by selecting a specific proposal. On the basis on these results, the designer could choose from several solutions depending on the preference for one criterion or another. Thus, the designer could prioritize whether to minimize the cost of the intervention, or minimize the difference between the optimal absorbent area and those provided by the tool, or the number of surfaces to adapt, or the material of the elements to be used in the project. The designer, depending on the level of requirement for acoustic comfort and other specific features or needs of the project, can thus implement this tool to make the final decision.

5.4. Solutions for the Study Case in Different Locations: Comparison of Results

As has been mentioned before, the regulatory requirements of RT vary by country (see Table 1). To assess the versatility of the proposed BFRT framework, the analysis of the same project of the case study has been performed, maintaining the initial configuration but changing the country of location of the building. For this, three different countries with different regulations were chosen and, accordingly, the results obtained are different depending on the selected country, since the regulatory limits established by each country are different. Figure 14 shows the results obtained following the implementation of the Stage 2 of the proposed BFRT to three different locations. It should be noted that the results differ from those previously obtained when the location was set in Spain (see Figure 12).

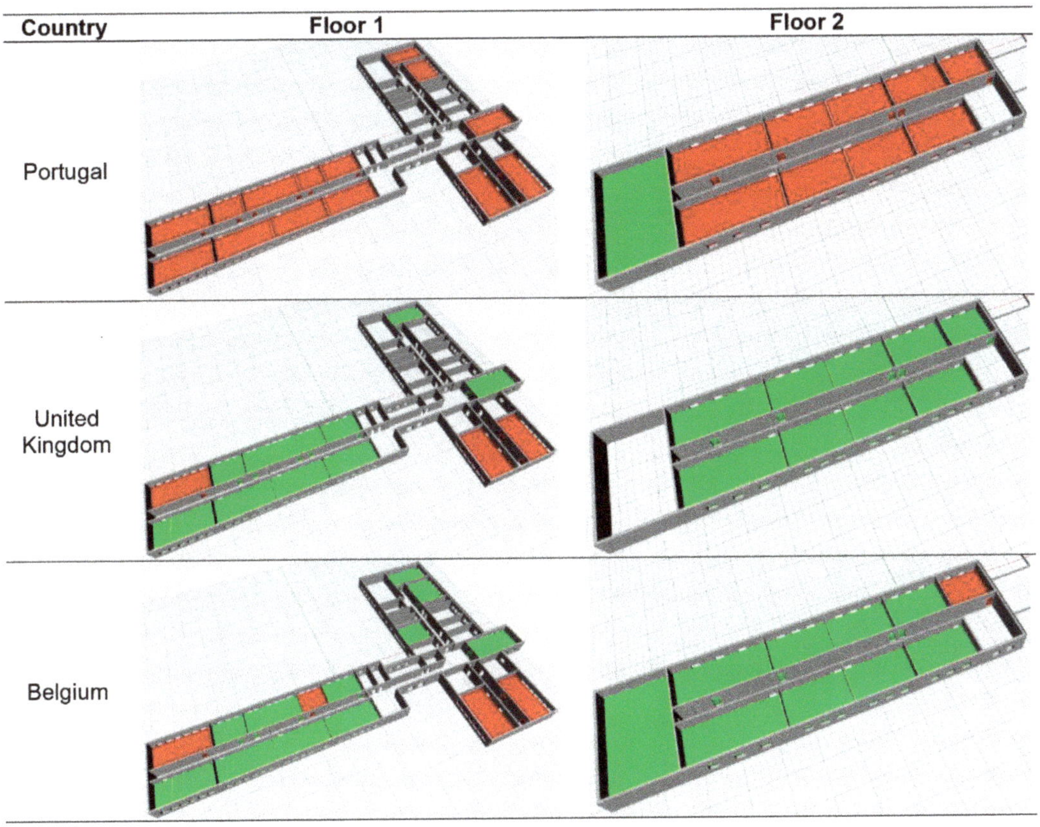

Figure 14. Compliance with the RT limit values by room type and floor for different locations of the study case.

Subsequently, it is interesting to make a comparison of the results obtained in Stage 3. To accomplish it, it has been selected the room of the study case ("A" room whose use is classroom). Figure 15 shows the Pareto fronts after the optimization process performed for this room.

For the case of the optimization of the room case study (classroom) located in Portugal, the algorithm generated 35,211 feasible solutions. If the classroom was located in United Kingdom, 25,642 solutions were obtained and 12,605 ones for the case of Belgium. The number of solutions obtained for each country is different because the RT limits are different for each country: the RT regulatory limits for Portugal is 0.76 s, being 0.8 s in the case of United Kingdom and 0.87 s in the case of Belgium. In this sense, the number of feasible solutions provided by the proposed optimization algorithm is going to depend on the RT regulatory limit established by each region or state, and of the initial finishing materials and the configuration of the rooms in the buildings.

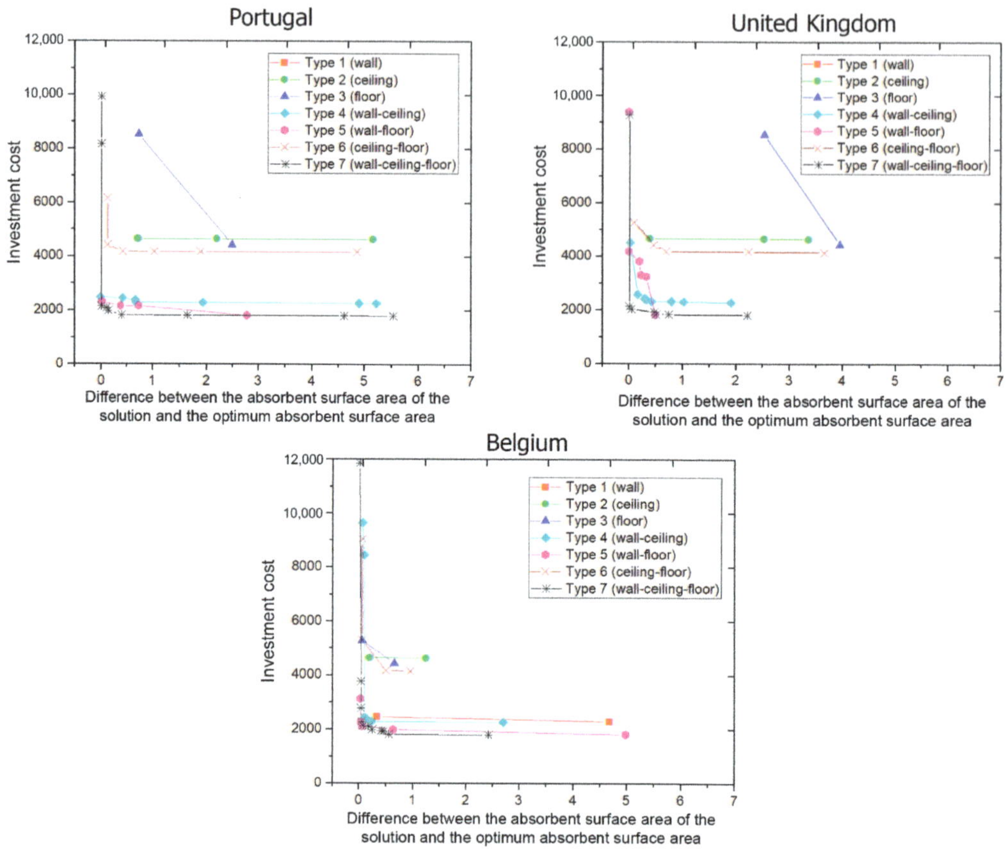

Figure 15. Results obtained from the Pareto frontiers calculation in Stage 3 for the optimisation based on the acoustic absorption of materials and their investment cost for the countries of Portugal, United Kingdom and Belgium. Red square—Pareto front type 1. Green dot—Pareto front type 2. Blue triangle—Pareto front type 3. Cyan rhombus—Pareto front type 4. Purple hexagon—Pareto front type 5. Wine cross—Pareto front type 6. Black cross—Pareto front type 7.

5.5. Discussion

The BFRT proposed framework implements a workflow based on BIM for the assessment and optimisation of RT in buildings. The BFRT framework allows to develop the acoustic analysis of spaces in the BIM design software itself, without having to resort to a specific acoustic software outside of the own BIM frame. This is an important advantage in contrast to other research approaches proposed in other studies which require the use of additional software, such as GIS [23] or Comsol [25].

Therefore, the scientific contribution of this research is the development of a framework for the integration of acoustic analysis in BIM-based software. This framework is developed in Section 4 (Proposed BIM framework for acoustic RT-based design in indoor areas: BFRT). The relevance of the proposed BFRT is that it contributes to solving the current problem of defining the design of spaces with a suitable acoustics according to their prescribed use with two important features: (1) it is carried out during the design stage and (2) it performs a systematic search for a large number of possibilities with a reduced time for analysis.

The BFRT allows the compliance with the limit values of RT depending on the country and its specific regulations to be analysed. The implemented procedure is based on the RT calculation, which is automatically computed for the different rooms of each floor of a building. In those cases where the value of the RT of the room does not belong to the acceptance interval, the optimization process provides solutions based on changing finishing materials of different surfaces of the walls, floors and ceilings. The design of the interior spaces is essential for a good acoustic conditioning. The selection of materials and composition of the constructive elements in the design phase allows us to anticipate the solution of arguments arising from a poor acoustic behaviour already in the initial design phases of building projects. Consequently, cost savings and better acoustic performance can be provided to building inhabitants compared to addressing the issue in subsequent phases.

The BFRT offers the possibility to evaluate the behaviour of the RT depending on the location of the project since it the limit values of the regulations can be included in the optimization process. In this first stage of development of this framework, the designer can supply other limits coming from specific regulations or requirements. This process requires editing the code of the Regulation RT node and a minimum knowledge of VPL and Python scripting in Dynamo is advisable. Results are displayed by using colours in the same interface of the BIM software, which greatly facilitates the visualization of the assessment in the same design interface.

In the proposed framework it has been chosen the optimization algorithm based on the technique of branching and bound since it is a flexible tool for calculating all possible solutions. It should be noted that the application of this procedure for optimization differentiates this research from other proposals that only evaluate the initial design solution and do not provide alternative design options [13,24]. Other multi-objective optimization tools could have been chosen at this stage, but results do not differ mainly from the proposed one, since the objective is that the algorithm does not calculate all possible scenarios for interventions in the rooms. In fact, the results obtained for the study case in the classroom, considering the location of the project in different countries (Spain, Portugal, United Kingdom and Belgium), shows that of the total number of combinations chosen by the branching and bound tool lies between 9% and 33% of the total (Spain: 33%, UK: 16%, Portugal: 23% and Belgium: 9%). So, the algorithm has discarded approximately a 67 to 91% of possible combinations without need to be computed, saving computational time. In this regard, the computation time needed by the proposed algorithm to obtain the results of the floor where the room study case is located (15 rooms) is 117 s and it obtained 639,647 workable solutions for all the different rooms. For the second floor (10 rooms), the computing time was 44 s obtaining 419,441 feasible solutions. The calculations were carried out with an Intel Core i7–9750H computer.

In summary, the main advantages and contributions offered by the application of BFRT in the field of acoustic engineering are: (1) it allows incorporating information related to the acoustic behavior of interior spaces and so the further enhancement of BIM model. In addition, the proposal allows for efficient connection with AM databases; (2) BFRT provides an automatic calculation of RT in all the rooms of the studied floor of a building (no need to re-enter data in other software); (3) Visualization of the fulfilment of the RT requirements in the design software allows the designer to work with a friendly tool for helping in decision-making process in the early design phase; (4) The proposed algorithm based on the technique of branching and bounding allows selecting the combination of finishing materials to obtain an optimal value of the RT without the need to evaluate all possible solutions. This implies a significant saving on computation time in the calculation process; and (5) the framework is flexible, i.e., it allows the user to add easily new RT limits according to the regulations of different countries in the code.

Finally, among the limitations presented by this study, it should be noted that the calculation of RT is based on Sabine's formula. Nevertheless, this formula has been implemented in the framework because national regulations in European countries state that it should be used to assess RT. In further research, complementary methods for the calculation of RT other acoustic parameters will be incorporated into the system.

6. Conclusions

This research develops a framework for the analysis of acoustic behaviour of rooms based on RT parameter (BFRT). Using both a BIM-based methodology and a graphical programming software (Dynamo) it has been developed a framework to support the decision-making process of designers during the early design phase in the field of acoustic conditioning of buildings. The proposed framework is embedded itself in the design software, so facilitating the evaluation of the RT without the need to use other specialized software. This is quite relevant since working time is saved and errors arising from manual data entry of a software to another are avoided. In addition, it allows an easy and automatic evaluation of the RT each time that a modification of the 3D BIM model is considered, showing a display of the results on the same interface design which is really comfortable for the designer.

The BFRT provides a framework for the integration of information on acoustic parameters within the building BIM model. The inclusion of parameters relating to the acoustic behaviour of the building allows additional features of the building to be taken into account and adds new information to the database so that it can be performed the analysis of the acoustic behaviour from the early stages of design in many ways. In this sense, the integration of the proposed BFRT into BIM design software simplifies the process, avoiding further rework and it reduces the time spent in RT assessment. Furthermore, it provides key information to designers for the decision-making process and improves the acoustic performance in buildings construction, which are key aspects in practical work. Finally, the automation of the assessment procedure encourages designers for optimisation of the building acoustic behaviour in their projects from the early stage of design, with the important fact that the acoustic data and parameters become integrated in the BIM model.

Finally, the management and consideration of the acoustic behaviour in the interior spaces from the initial stages ensures a further appropriate acoustic performance of the different rooms. This is an important issue since providing acoustic comfort and ensuring the correct performance of the activities that can be carried out according to its use without the need of subsequent costly and complicated actions in other phases of the project results in relevant time and economical savings and better final performances.

Author Contributions: Conceptualization, A.J.A. and M.L.d.l.H.-T.; methodology, A.J.A., M.L.d.l.H.-T., D.P.R. and M.D.M.-A.; software, A.J.A. and M.L.d.l.H.-T.; validation, A.J.A., M.L.d.l.H.-T., D.P.R. and M.D.M.-A.; formal analysis, A.J.A. and M.L.d.l.H.-T.; investigation, A.J.A. and M.L.d.l.H.-T.; resources, D.P.R. and M.D.M.-A.; writing—original draft preparation, A.J.A. and M.L.d.l.H.-T.; writing—review and editing, D.P.R. and M.D.M.-A.; supervision, D.P.R. and M.D.M.-A.; project administration, D.P.R. and M.D.M.-A.; funding acquisition, D.P.R. and M.D.M.-A. All authors have read and agreed to the published version of the manuscript.

Funding: This research was funded by the Consejo General de la Arquitectura Técnica (CGATE), the "Junta de Andalucía" (Spain) under project B-TEP-362-UGR18 and the State Research Agency (SRA) of Spain and European Regional Development Funds (ERDF) under project PID2019-108761RB-I00.

Institutional Review Board Statement: Not applicable.

Informed Consent Statement: Not applicable.

Data Availability Statement: Data are provided upon request to the corresponding author.

Acknowledgments: The first two authors wish to thank the support of the Ministerio de Ciencia, Innovación y Universidades of Spain under an FPU grant.

Conflicts of Interest: The authors declare no conflict of interest.

References

1. Sarbu, I.; Sebarchievici, C. Aspects of indoor environmental quality assessment in buildings. *Energy Build.* **2013**, *60*, 410–419. [CrossRef]
2. Pinho, P.; Pinto, M.; Almeida, R.M.; Lopes, S.; Lemos, L. Aspects concerning the acoustical performance of school buildings in Portugal. *Appl. Acoust.* **2016**, *106*, 129–134. [CrossRef]

3. Berglund, B.; Lindvall, T.; Schwela, D.H. Occupational and Environmental Health Team. In *Guidelines for Community Noise*; World Health Organization: Geneva, Switzerland, 1999; Available online: https://apps.who.int/iris/handle/10665/66217 (accessed on 1 March 2022).
4. Ryu, J.K.; Jeon, J.Y. Influence of noise sensitivity on annoyance of indoor and outdoor noises in residential buildings. *Appl. Acoust.* **2011**, *72*, 336–340. [CrossRef]
5. Jeon, J.Y.; Ryu, J.K.; Lee, P.J. A quantification model of overall dissatisfaction with indoor noise environment in residential buildings. *Appl. Acoust.* **2010**, *71*, 914–921. [CrossRef]
6. Pulkki, V.; Karjalainen, M. *Communication Acoustics: An Introduction to Speech, Audio and Psychoacoustics*; John Wiley & Sons: Hoboken, NJ, USA, 2015.
7. Vorländer, M. *Auralization: Fundamentals of Acoustics, Modelling, Simulation, Algorithms and Acoustic Virtual Reality*; Springer Science & Business Media: Berlin/Heidelberg, Germany, 2007.
8. Bradley, J.S. Review of objective room acoustics measures and future needs. *Appl. Acoust.* **2011**, *72*, 713–720. [CrossRef]
9. Lee, D.; van Dorp Schuitman, J.; Cabrera, D.; Qiu, X.; Burnett, I. Comparison of psychoacoustic-based reverberance parameters. *J. Acoust. Soc. Am.* **2017**, *142*, 1832–1840. [CrossRef]
10. Kinsler, L.E.; Coppens, A.B.; Frey, A.R.; Sanders, J.V.; Kinsler, L.E.; Coppens, A.R.; Frey, J.V. *Sanders, Fundamentals of Acoustics*; John Wiley & Sons: Hoboken, NJ, USA, 1992; ISBN 978-0-471-84789-2.
11. Young, R.W. Single-Number Criteria for Room Noise. *J. Acoust. Soc. Am.* **1964**, *36*, 289–295. [CrossRef]
12. Sabine, W.C.; Sabine, W.C. *Collected Papers on Acoustics*; Harvard University Press Cambridge: Cambridge, MA, USA, 1922; Available online: https://archive.org/details/collectedpaperso00sabiuoft (accessed on 10 March 2022).
13. Wu, C.; Clayton, M.C. BIM-based acoustic simulation Framework. In Proceedings of the 30th CIB W78 International Conference, Beijing, China, 9–12 October 2013; pp. 99–108. Available online: https://itc.scix.net/pdfs/w78-2013-paper-66.pdf (accessed on 10 March 2022).
14. Oral, G.K.; Yener, A.K.; Bayazit, N.T. Building envelope design with the objective to ensure thermal, visual and acoustic comfort conditions. *Build. Environ.* **2004**, *39*, 281–287. [CrossRef]
15. Jalaei, F.; Jalaei, F.; Mohammadi, S. An integrated BIM-LEED application to automate sustainable design assessment framework at the conceptual stage of building projects. *Sustain. Cities Soc.* **2020**, *53*, 101979. [CrossRef]
16. Mellado, F.; Lou, E.C. Building information modelling, lean and sustainability: An integration framework to promote performance improvements in the construction industry. *Sustain. Cities Soc.* **2020**, *61*, 102355. [CrossRef]
17. Safari, K.; AzariJafari, H. Challenges and opportunities for integrating BIM and LCA: Methodological choices and framework development. *Sustain. Cities Soc.* **2021**, *67*, 102728. [CrossRef]
18. Valinejadshoubi, M.; Moselhi, O.; Bagchi, A.; Salem, A. Development of an IoT and BIM-based automated alert system for thermal comfort monitoring in buildings. *Sustain. Cities Soc.* **2021**, *66*, 102602. [CrossRef]
19. Singh, P.; Sadhu, A. Multicomponent energy assessment of buildings using building information modeling. *Sustain. Cities Soc.* **2019**, *49*, 101603. [CrossRef]
20. Pauwels, P.; Van Deursen, D.; Verstraeten, R.; De Roo, J.; De Meyer, R.; Van de Walle, R.; Van Campenhout, J. A semantic rule checking environment for building performance checking. *Autom. Constr.* **2011**, *20*, 506–518. [CrossRef]
21. Carvalho, J.P.; Bragança, L.; Mateus, R. Optimising building sustainability assessment using BIM. *Autom. Constr.* **2019**, *102*, 170–182. [CrossRef]
22. Tan, Y.; Fang, Y.; Zhou, T.; Wang, Q.; Cheng, J. Improve Indoor acoustics performance by using building information modeling, ISARC. In Proceedings of the 34rd International Symposium on Automation and Robotics in Construction, Taipei, Taiwan, 28 June–1 July 2017; pp. 959–966. [CrossRef]
23. Deng, Y.; Cheng, J.C.; Anumba, C. A framework for 3D traffic noise mapping using data from BIM and GIS integration. *Struct. Infrastruct. Eng.* **2016**, *12*, 1267–1280. [CrossRef]
24. Hammad, A.W.; Akbarnezhad, A.; Wu, P.; Wang, X.; Haddad, A. Building information modelling-based framework to contrast conventional and modular construction methods through selected sustainability factors. *J. Clean. Prod.* **2019**, *228*, 1264–1281. [CrossRef]
25. Tan, Y.; Fang, Y.; Zhou, T.; Gan, V.J.; Cheng, J.C. BIM-supported 4D acoustics simulation approach to mitigating noise impact on maintenance workers on offshore oil and gas platforms. *Autom. Constr.* **2019**, *100*, 1–10. [CrossRef]
26. Van Eldik, M.A.; Vahdatikhaki, F.; dos Santos, J.M.O.; Visser, M.; Doree, A. BIM-based environmental impact assessment for infrastructure design projects. *Autom. Constr.* **2020**, *120*, 103379. [CrossRef]
27. Santos, R.; Costa, A.A.; Silvestre, J.D.; Pyl, L. Informetric analysis and review of literature on the role of BIM in sustainable construction. *Autom. Constr.* **2019**, *103*, 221–234. [CrossRef]
28. Kiss, B.; Szalay, Z. Modular approach to multi-objective environmental optimization of buildings. *Autom. Constr.* **2020**, *111*, 103044. [CrossRef]
29. Salimzadeh, N.; Vahdatikhaki, F.; Hammad, A. Parametric Modelling and Surface-specific Sensitivity Analysis of PV Module Layout on Building Skin Using BIM. *Energy Build.* **2020**, *216*, 109953. [CrossRef]
30. Hollberg, A.; Genova, G.; Habert, G. Evaluation of BIM-based LCA results for building design. *Autom. Constr.* **2020**, *109*, 102972. [CrossRef]

31. Cortés-Pérez, J.P.; Cortés-Pérez, A.; Prieto-Muriel, P. BIM-integrated management of occupational hazards in building construction and maintenance. *Autom. Constr.* **2020**, *113*, 103115. [CrossRef]
32. Basta, A.; Serror, M.H.; Marzouk, M. A BIM-based framework for quantitative assessment of steel structure deconstructability. *Autom. Constr.* **2020**, *111*, 103064. [CrossRef]
33. UNE-EN 12354-6:2003. Building Acoustics—Estimation of Acoustic Performance of Buildings from the Performance of Elements—Part 6: Sound Absorption in Enclosed Spaces. Available online: https://www.une.org/encuentra-tu-norma/busca-tu-norma/norma/?c=N0032294 (accessed on 10 March 2022).
34. Technical Building Code. Basic Document DB-HR Protection against the Noise, Ministerio de Fomento 2009. Available online: https://www.codigotecnico.org/DocumentosCTE/ProteccionRuido.html (accessed on 10 March 2022).
35. Conseil National du Bruit, Guide du CNB Réglementations Acoustiques des Bâtiments. 2017. Available online: https://www.bruit.fr/images/stories/pdf/guide-cnb-6-reglementations-acoustiques-batiments-novembre%202017.pdf (accessed on 10 March 2022).
36. Decreto-Lei n.° 96/2008. Diário da República n.° 110/2008, Série I de 2008-06-09. Ministério do Ambiente, do Ordenamento do Território e do Desenvolvimento Regional. Available online: https://dre.pt/home/-/dre/449682/details/maximized (accessed on 10 March 2022).
37. NBN S01-400-2:2012. Critères Acoustiques Pour les Immeubles Scolaires. 2012. Available online: https://environnement.brussels/sites/default/files/user_files/pres_20141204_bruit_ecoles_norme.pdf (accessed on 10 March 2022).
38. BB93: Acoustic Design of Schools—Performance Standards. Department for Education. Education Funding Agency. 2015. Manchester, England. 2015. Available online: https://www.gov.uk/government/publications/bb93-acoustic-design-of-schools-performance-standards (accessed on 10 March 2022).
39. Autodesk Revit 2020. Available online: https://www.autodesk.com/products/revit/overview (accessed on 10 March 2022).
40. Dynamo 2.1. Available online: https://dynamobim.org/ (accessed on 10 March 2022).
41. Land, A.; Doig, A. An automatic method of solving discrete programming problems. *Econometrica* **1960**, *28*, 497. [CrossRef]

buildings

MDPI

Article

BIM-Based Dynamic Construction Safety Rule Checking Using Ontology and Natural Language Processing

Qiyu Shen [1], Songfei Wu [1], Yichuan Deng [1,2,*], Hui Deng [1] and Jack C. P. Cheng [3]

1 School of Civil Engineering and Transportation, South China University of Technology, Guangzhou 510641, China; qushen@scut.edu.cn (Q.S.); scut_wsf@163.com (S.W.); hdeng@scut.edu.cn (H.D.)
2 State Key Laboratory of Subtropical Building Science, South China University of Technology, Guangzhou 510641, China
3 Department of Civil and Environmental Engineering, The Hong Kong University of Science and Technology, Hong Kong SAR, China; cejcheng@ust.hk
* Correspondence: ctycdeng@scut.edu.cn; Tel.: +86-020-87111030

Abstract: Real-time identification and prevention of safety risks in dynamic construction activities are demanded by construction safety managers to cope with the growing complexity of the construction site. Most of the studies on BIM-based construction safety inspection and prevention use data from the planning and design stage. Meanwhile, safety managers still need to spend a lot of time gathering reports about construction safety risks in certain periods or areas from inferred results in BIM. Therefore, this paper proposed an automatic safety risk identification and prevention mechanism for the construction process by integrating a safety rule library based on ontology technology and Natural Language Processing. An automatic inspection mechanism integrating BIM and safety rules is constructed, and a presentation mechanism of intelligent detection results based on Natural Language Processing is designed. The construction process safety rule checking system was developed, and the effectiveness of the system was verified by a case study. The outcome of this paper contributes to the development and application of ontology in construction safety research, and the NLP-based safety rule checking result presentation will benefit safety inspectors and construction managers in practice.

Keywords: construction process; safety compliance checking; ontology; BIM; NLP

Citation: Shen, Q.; Wu, S.; Deng, Y.; Deng, H.; Cheng, J.C.P. BIM-Based Dynamic Construction Safety Rule Checking Using Ontology and Natural Language Processing. *Buildings* **2022**, *12*, 564. https://doi.org/10.3390/buildings12050564

Academic Editors: Hongling Guo, Jia-Rui Lin and Yantao Yu

Received: 6 February 2022
Accepted: 27 April 2022
Published: 27 April 2022

Publisher's Note: MDPI stays neutral with regard to jurisdictional claims in published maps and institutional affiliations.

Copyright: © 2022 by the authors. Licensee MDPI, Basel, Switzerland. This article is an open access article distributed under the terms and conditions of the Creative Commons Attribution (CC BY) license (https://creativecommons.org/licenses/by/4.0/).

1. Introduction

As construction accidents continue to cause casualties and economic losses, the efficiency of safety management and the reduction of the occurrence of accidents remain an urgent problem to be solved. The current way of safety management heavily relies on safety managers' understanding of safety regulations and their experience. With the growth of scale and complexity of projects, the tasks of safety inspection and risk prevention are time-consuming and error prone. Analyzing the causes and formation mechanisms of safety accidents is the prerequisite for preventing safety risks [1]. At present, the accident causation theory is often used to analyze the causes of accidents. The theory of the cause of construction safety accidents refers to the analysis of typical accidents, thus as to extract the mechanism of accident occurrence and to establish the accident causation model. This can provide a scientific basis for the qualitative and quantitative analysis, prediction, and prevention of accidents, as well as the improvement of safety management [2]. The accident causation theory has experienced three stages, namely the early accident causation theory, the accident causation theory after World War II, and the modern system safety theory [3]. The early accident causation theory includes accident frequency tendency theory, Heinrich accident causation sequence theory [4], Frank Bode's modern accident causation theory [5], and the accident causation theory after World War II includes the trajectory cross theory and accidental release of energy theory. For building structures with increasingly complex structures and sizes, the sources of safety accidents are many and complex, including unsafe behaviors of construction workers, errors of safety management personnel, unsafe conditions of construction machinery and tools on-site, and unsafe conditions. Protective measures and environment. The modern system theory is a theory specifically aimed at complex accident systems. In order to comprehensively manage the hidden dangers and problems of the construction site, it is necessary to sort out the sources of various safety problems during the construction process in order to establish a comprehensive safety prevention system. This paper adopts the modern system safety theory, which believes that the most important factor that causes accidents is the various hazards in the production process. The theory also believes that the main work of safety management is to identify, evaluate, and control hazard sources.

Therefore, the automatic identification of construction safety risks and their corresponding solutions has gained more and more attention from researchers worldwide. However, current studies on the issue mainly focus on the use of BIM (Building Information Modelling) technology in safety inspection and risk prevention from the design stage. While the limited range of safety risks in construction are covered in current studies, such as openings or edges, it is believed that real-time identification and prevention of safety risks in dynamic construction activities will provide a more desirable solution to the ever-growing complexity of the construction site. In addition, the current research is unable to form a system of reusable safety knowledge in the form of standardized provisions for use in subsequent safety management projects. There still remains a gap in the establishment of reusable construction safety management knowledge and its application in the dynamic construction site.

In recent years, advanced information technologies such as BIM, ontology, NLP (Natural Language Processing) are introduced in the construction industry to improve production efficiency. Ontology, which can be interpreted as a knowledge description model, was originally derived from the field of philosophical concepts. A complete ontology consists of at least five parts: concept or class, relation, function, axiom, and instance. An instance is a concrete instance of a concept [6]. Ontology has been widely used in information retrieval, semantic web, knowledge management, digital library, and other fields. As an important knowledge representation model, ontology technology has been widely studied in the field of construction safety management, especially in information reasoning and extraction. Wang and Boukamp (2011) [7] established an ontology for analysis of potential job hazards, including construction activities and specific operation steps, and established a set of ontology reasoning mechanisms to realize the safety inspection of construction activities, with which the response rate of construction companies to construction activity risks was improved. To identify specific safety risks and automatically generate safety measures, Lu (2015) et al. [8] established five categories of CSCOntology (construction safety checking ontology), including the line of work, task, precursor information, hazard, and solution. The rules regarding high falls as promulgated by OSHA (Occupational Safety and Health Administration) were first translated into SWRL (Semantic Web Rule Language) rules that can be processed by computers, the JESS inference machine was then used to realize the inference for a given instance of ontology and generate corresponding solutions. However, the information needed for detection still needs to be created and entered manually, implying an insufficient degree of automation. Kim et al. (2016) developed a BIM-based scaffolding automatic planning system. The system used simulation engines and BIM to simulate daily construction progress and identify potential hazards according to safety specifications. Zhang (2015) et al. [9] proposed an automatic safety planning system combining BIM and ontology. This system established the mapping between BIM components and concepts in the ontology and then used SWRL rules in the ontology to carry out a safety inspection on automatically extracted masonry components in BIM and give a visual results report. In 2016, Ding et al. [10] applied ontology technology and semantic network technology in the safety risk management of deep foundation pit construction. There are several shortfalls in the current research of ontology applications in construction safety management. Firstly, the information for reasoning using ontology either comes from manual input or is extracted from a static BIM model in the phase, both of which cannot represent the dynamic nature of construction safety management. Secondly, a limited range of construction safety management is covered in current research, with emphasis on falls from height and border identification. Lastly, the results of safety checking are presented in the form of databases or tools for ontology reasoning, which imposes restrictions on construction managers who are not familiar with such tools.

Natural Language Processing is an important direction in the field of computer science and artificial intelligence, which helps computers understand the real meaning of human natural language. NLP is mainly used for information extraction and retrieval from documents written in natural language. At present, there have been increasing applications of NLP in construction safety management. One typical study was reported in Zhang and EI-Gohary (2015) [11], which used established semantic mapping rules and conflict resolution rules to convert some chapters of the International Building Code 2009 into logical sentences that can be used for safety code inspection. In this study, NLP was used for syntactic analysis, and ontology was used for semantic analysis. In order to automatically extract relevant information from the BIM model for automatic specification checking, Zhang and EI-Gohary (2016) [12] proposed the use of NLP to extract concepts related to consistency checking from the specification. The concepts were then mapped to the IFC (Industry Foundation Classes) hierarchical relationship to automatically match the information in the specification and the BIM model. In another study, Zhang and EI-Gohary (2017) [13] developed an NLP-based system to automatically extract and transform specification information and design information in BIM, which were then used in consistent reasoning. Lin (2015) [14] et al. proposed a method for intelligent retrieval and presentation of BIM cloud data based on NLP. The number of materials used on the construction site and the amount of progress completed can be presented in an intelligent way through inquiry.

In light of recent advancements in safety rule checking and result presentation, although progress was made recently towards intelligent construction safety management, there were some limitations that need to be resolved. Firstly, most of the studies were on BIM-based construction safety inspection and prevention use data from the planning and design stage. There were relatively few studies on real-time safety risk inspection and prevention of complex and changeable construction sites. Secondly, there were few studies using BIM as the information source for a safety inspection, which reported that a lot of manual intervention was still needed to name BIM model components in order to realize the mapping relationship between safety rules and BIM components. Lastly, while BIM has shown its potential in inferring construction safety risks and prevention measures, safety managers still need to spend a lot of time gathering reports about construction safety risk in certain periods or areas from these inferred results. There are few studies that focus on how to intelligently present construction safety reports thus that safety managers can quickly obtain safety issues for a real-time safety inspection.

This paper contributes to the body of knowledge of construction safety management by the establishment of a BIM-based construction process safety risk inspection system integrating ontology and Natural Language Processing. In this study, BIM is used to contain the dynamic construction process and serves as an efficient information retrieval hub. Combined with ontology and Natural Language Processing, the information in BIM is used to identify and prevent construction safety risks, thus as to achieve day-to-day dynamic and comprehensive safety management. Two research questions are addressed in this paper: (1) building an ontology-based computer-recognizable safety regulation knowledge base that helps to automatically identify and prevent safety risks in everyday construction activities; (2) comprehensive and easy-to-use ways of providing construction managers with real-time safety risk identification results and prevention measures. The work presented in this paper not only effectively reduces the workload of managers but also improves the accuracy of safety risk inspections and ultimately achieves the goal of reducing safety accidents and improving construction safety management.

2. Methodology

2.1. Methodology Overview

In this paper, a BIM-based dynamic construction safety rule checking framework using ontology and Natural Language Processing was proposed to overcome the shortcomings of current research. The methodology of the framework development included 3 major steps, namely the development of construction safety management ontology, the establishment of a safety rule library for the construction process and prosing a safety risk identification and retrieval mechanism. The framework can be illustrated in Figure 1.

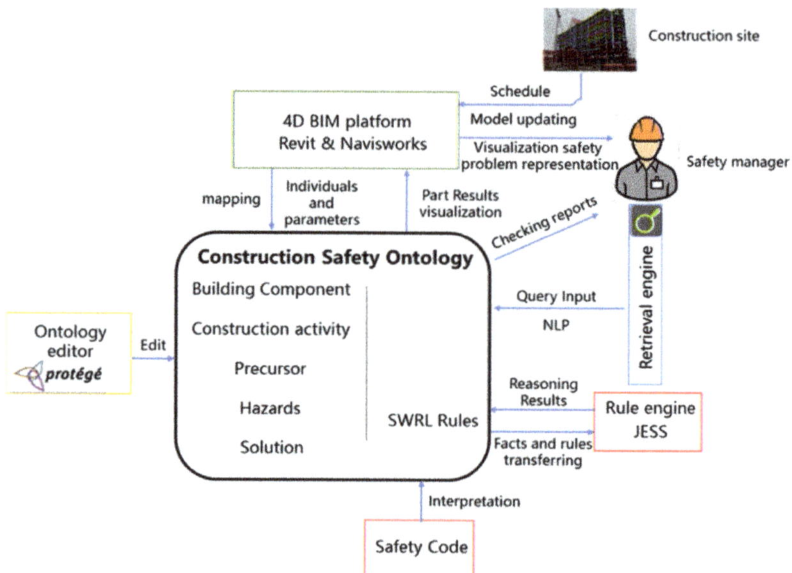

Figure 1. Illustration of the framework.

2.2. Development of Ontology

The construction safety management ontology integrates the relevant knowledge of the construction process, various types of safety accident knowledge related to the construction process, and the precursor information and corresponding safety solutions of the accident. The improved 7-step method was applied to complete the construction of the safety ontology. The standard 7 steps of ontology development were: (1) determine the domain and scope of the ontology; (2) consider reusing existing ontologies; (3) enumerate important terms in the ontology; (4) define the classes and the class hierarchy; (5) define the properties of classes—slots; (6) define the facets of the slots; and (7) create instances. The development of the safety ontology specifically includes 5 processes.

2.2.1. Determine the Scope of the Ontology

Since the establishment of the ontology is a complex and elaborate process, the first step in building the ontology is to determine the professional domain and scope of the established ontology, that is, to define the role of the ontology, the domain of the application, the user, and the maintenance of the ontology. In this paper, the developed ontology will focus on the civil construction process and renovation process of the building, as well as the corresponding safety issues and solutions. The development of the ontology will include the review of the concepts and interactions of the construction process, precursor, hazard, and solution based on relevant construction specifications and safety specifications.

2.2.2. Reusing the Existing Ontology

The existing ontology related to the scope of the study should be reused in order to reduce the difficulty of developing the ontology. Jiao [15] established a subway construction safety risk ontology knowledge database, and Zeng [16] established a knowledge database of subway construction safety accidents, both of which discussed the subway construction process from the perspective of accident causes, the fundamental aspects of the subway construction process, and the risk management ontology corresponding to the solution. Zhang [17] established a risk management ontology of construction engineering under a BIM environment to realize the semantic reasoning and retrieval of risk accidents. With reference to the structure and establishment process of these ontologies, this paper established an ontology that can effectively manage the safety risks of construction sites in view of the dynamic characteristics of the construction process in the construction field.

2.2.3. Hierarchy and Properties of Classes

This paper used a top-down approach to enumerate and define classes. The ontology classes constructed in this section mainly include 4 categories: construction process; precursor; hazard; and solution, each of which includes different sub-classes, hierarchy of these sub-classes, and the relationship between them (Table 1). In this paper, the WBS (Work Breakdown Structure) decomposition results included C1 (civil engineering), C2 (decoration engineering), and C3 (facility and pipeline). C1 (civil engineering) included 4 categories, and C2 (decorative engineering) included 7 categories.

Table 1. Category concept definition.

Name	Definition
Construction Process	The activities of construction
Precursor	Status or condition that may cause a safety incident
Hazard	Safety accidents caused by precursory information such as falling, collapse, and object striking
Solution	Preventing safety accidents or actions

The precursor information included 4 parts: structural member; material; equipment; and; environment, and these 4 precursor information classes were each divided into 3 different sub-classes. An example specification of the precursor information is shown in Table 2.

Table 2. Example of precursor definition.

Construction Process	Precursor	Location	Cause
C1.3 Formwork, scaffolding; C1.4 The stacking of masonry materials and construction machinery and tools in the masonry engineering; C2.1 Veneer engineering C2.2 Veneer engineering machinery stacking	Protective thickness at the top of the protective shed	Pedestrian access shelter	The thickness of the top of the pedestrian access shelter is not enough to resist falling objects from high places
	Geometry size of the protective shed	Pedestrian access shelter	The geometric dimensions of the pedestrian access protection shelter do not meet the requirements of the specification
	Safe net	Main building	The safety net around the main building is not erected or the erection quality/density does not meet the specification requirements
	Scaffolding position	Main building	The stacking position of the scaffolding is too close to the unprotected opening or the edge
	Tool stacking position	Building main body/scaffold board	The stacking position of tools is too close to the unprotected hole or the edge, or placed on the scaffold
C1.3 Tower crane installation and dismantling in the main structure project	Tower crane installation/removal	Tower crane	Tower crane installation and demolition have an intersection with construction pedestrians within a certain safe distance

The hazard component included 5 parts: fall hazard; struck; collapse; lifting injury; electric shock; fire; and explosion.

Solutions included staff protection system and safety control of material and equipment.

2.2.4. Definition of Class Attributes

The definition of class attributes in this paper included 2 aspects: the definition of the object property and the definition of data type property. The object attributes define the relationship between classes and classes, while the data attributes define the relationship between the classes and data.

Object property: the ontology established in this paper included 4 major categories, and the logical relationship between the 4 categories can be expressed by 3 relational terms. Some construction processes or precursors generated by unsafe construction state can be expressed by "Has_precursor"; the precursor may cause construction safety accidents (hazard), thus "Cause_hazard" can be used; and construction accidents (hazard) occur, which need corresponding countermeasures (solution) to prevent or solve, thus "Has_solution" can be used to express the relationship.

The formal expression of these attributes can be expressed by Domain U and Range U in the property's toolbar of Protégé 5.5.0 from Stanford University in the United States, an open-source ontology editing software. Each construction process may have several kinds of precursory information, but each specific precursory information instance can only belong to a construction process, and the setting of the attributes of objectivity can be seen in Table 3. The process of attribute construction in Protégé is shown in Figure 2.

Table 3. Object property definition.

Object Attribute	Domain	Range	Function Characteristics
Has_precursor	Construction Process	Precursor	Inverse Functional
Cause_hazard	Precursor	Hazard	Inverse Functional
Has_solution	Hazard	Solution	Inverse Functional

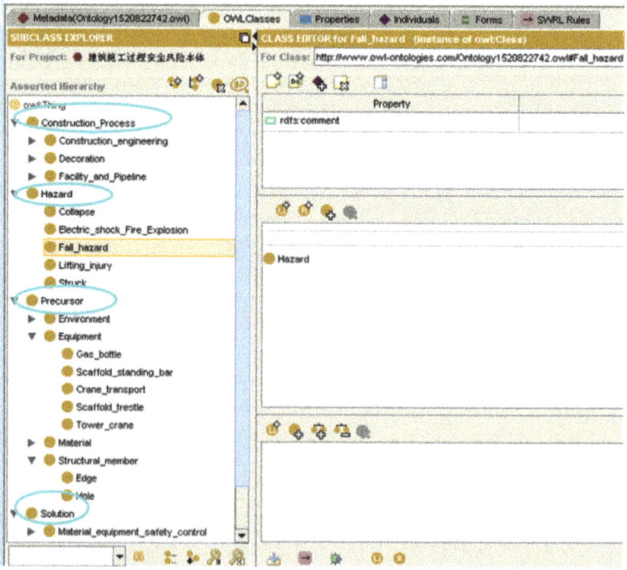

Figure 2. Object relationship setting.

The relationship between classes of the same level was defined above, and the relationship between different levels was an inclusion relation. According to the predefined definition of Protégé software from Stanford University in the United States, "is_a" represents the attribution relationship between the class and the inclusion class. For example, "Block_Masonry" of brick masonry projects belongs to a larger category of construction processes. Thus far, the definition of object type attributes of different classes at the same level was completed. Figure 3 presents an OntoGraf view of the class hierarchy.

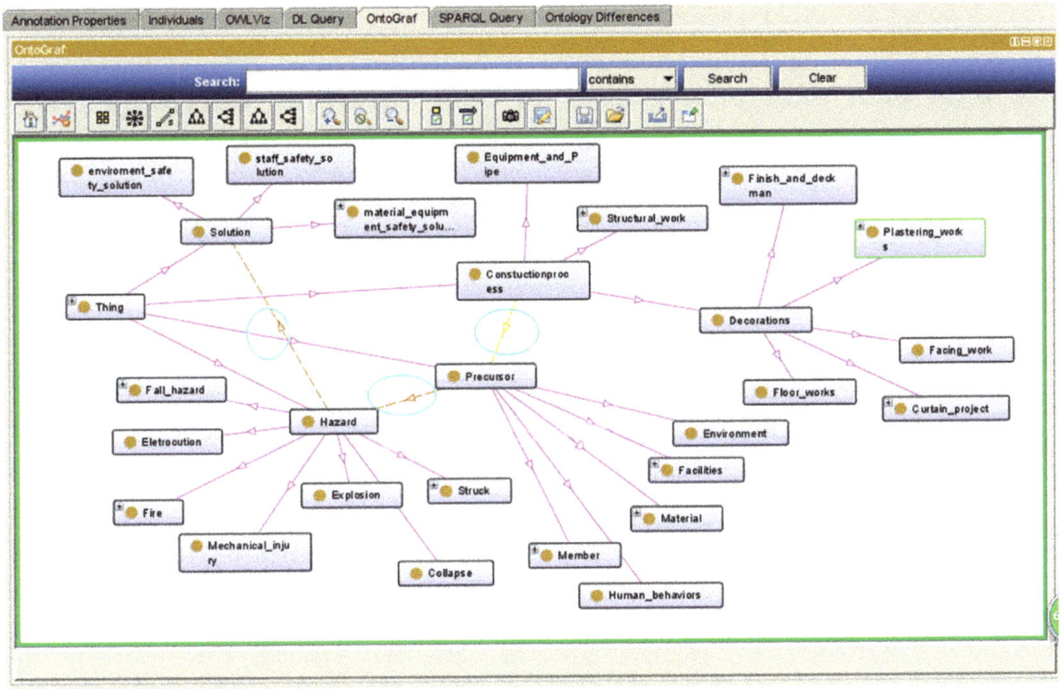

Figure 3. Class and class hierarchy relationship OntoGraf view.

Datatype Property: this paper defines data type attributes according to the characteristics of the 4 major classes, and the characteristics of the sub-classes contained in each class were relatively different; thus, for convenient construction, this paper summarizes the characteristics of each sub-class to the main class.

The datatype properties of the construction activities were summarized as follows: (1) the basic information includes the construction activity number, construction activity location, construction content, and start and end time; (2) the precursory information that may be included in the construction process, such as the components, materials, equipment, environment, specific attribute names, value types and function characteristics, as shown in Table 4.

Table 4. Construction activity data type attribute definition.

Data Type Attribute Names	Value Types	Function Characteristics
Construction_ID	int	Functional
Construction_Location	float	Functional
Construction_Task	string	Functional
Start_time	string	Functional
Ending_time	string	Functional
Construction_member	string	Functional
Construction_material	string	Functional
Construction_equipment	string	Functional
Construction_environment	string	Functional

The datatype properties of the precursory information were summarized as follows: the number of construction activities, the number of precursory information entries, and the description of precursory information, including the components, materials, equipment, and environment. Different types of precursory information have different characteristics; thus, different types of precursory information have different data attributes. For example, when considering the precursor information of the hole, the information regarding whether there was a cover (has_cover), whether there was a guardrail (has_guardrail), and the direction of the hole (has_direction) should be included. For each of the 4 types of values and functions, the characteristics are shown in Table 5.

Table 5. Precursor information data type attribute definition.

Data Type Attribute Names	Value Types	Function Characteristics
Precursor_ID	int	Functional
Belong_to_Constuction	int	Functional
Precursor_Description	string	Functional
Precursor_member	string	Functional
Precursor_material	string	Functional
Precursor_equipment	string	Functional
Precursor_environment	string	Functional

The datatype properties of the construction risk were summarized as follows: construction risk number, accident location, correspondence to the precursor information number, and accident occurrence time, as shown in Table 6.

Table 6. Construction risk data type attribute definition.

Data Type Attribute Names	Value Types	Function Characteristics
Risk_Number	int	Functional
Cor_to_Precursor	int	Functional
Risk_Description	string	Functional
Risk_Location	string	Functional
Risk_Task	string	Functional
Risk_Time	string	Functional

The datatype properties of the safety measures were summarized as follows: safety measure quantity, risk accident number, description of measures, safety measures for personnel, measures for materials, and measures for equipment, as shown in Table 7.

Table 7. Safety measure data type attribute definition.

Data Type Attribute Names	Value Types	Function Characteristics
Solution_ID	int	Functional
Risk_ID	int	Functional
Measure_Description	string	Functional
Solution_Person	string	Functional
Solution_Material	string	Functional
Solution_Equipment	string	Functional

With the object and datatype properties of the ontology class established above, the ontology was then stored in Protégé. Figure 4 shows the data type attributes of the construction process number (Construction_ID) in the construction process class.

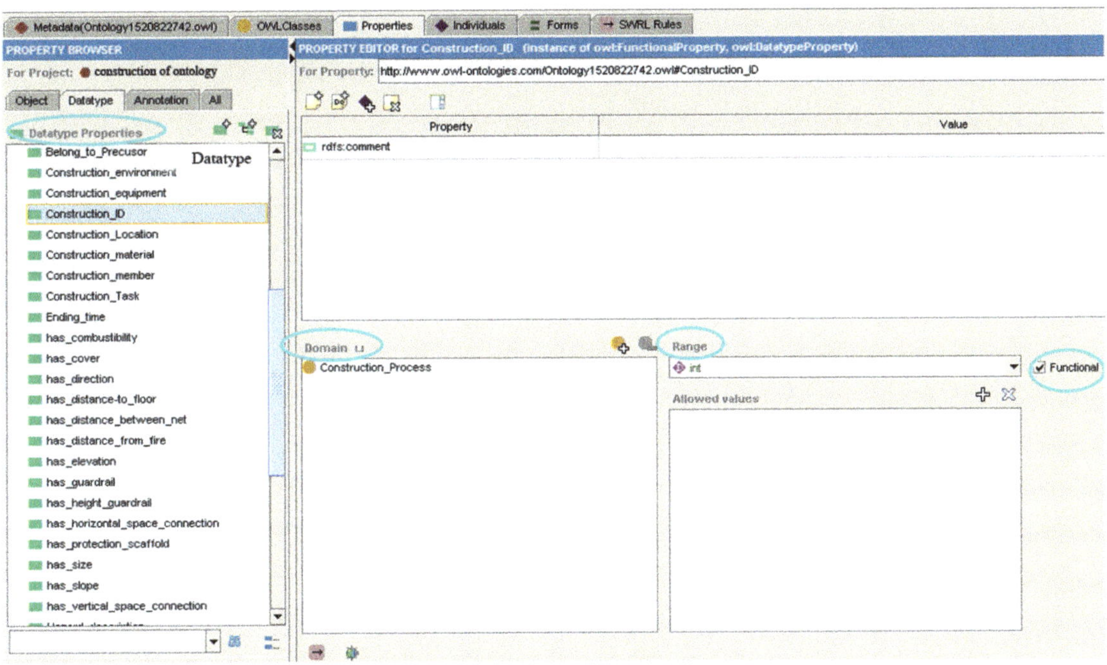

Figure 4. Various data type attribute definition interfaces.

2.3. Establishment of a Safety Rule Library for the Construction Process

2.3.1. Parameterization of Safety Rules

The safety rules of the construction process refer to the regular patterns or logical relationships that use certain pre-set parameters to determine whether certain precursors will lead to safety accidents in the construction process. The composition of safety rules generally includes 3 parts: the subject of the accident; the judgment conditions (parameters); and the safety risk. A general safety rule checking process starts by determining the subject of the accident through the inspected object and then continues to verify the inspected object through the distinguishing conditions to determine whether it has violated safety rules. The rule interpretation refers to the conversion of the normative terms expressed in natural language into a form that can be recognized by the computer, which forms the basis for automatic safety rule checking. Therefore, in order to fully identify the safety accidents in the construction process, the interpretation of each safety rule should consider the comprehensiveness of the discrimination conditions. The safety rule library formed by these safety rules in a unified form can become an important tool to help the safety managers at the construction site to identify potential safety hazards and promptly give safety precautions, avoiding complicated manual search and regulation checks.

The interpretation of safety rules needs to consider 2 key issues: the identifiability of the subject of the safety risk and the discrimination conditions. The premise of whether safety rules can be applied digitally is to find the "object/subject" that may cause accidents, and the key to the effective implementation of the judgment condition of the accident subject is the detail of the subject's attribute parameters. In addition, the safety rules were norm clauses compiled by experts of different professions and described in natural language, which makes the subject of the safety risk and the judgment conditions vague and abstract; moreover, for the same kind of safety risk, different norms and clauses have different focuses or overlap; thus the interpretation of safety rules needs to consider the expression and comprehensiveness of the clauses. Therefore, the process of interpreting safety rules was to clarify the subject of the accident. Moreover, for the same kind of risk subject, we should comprehensively collect the safety clauses describing the risk in the safety regulations with different expressions and then sort out the logical relationships among them according to the characteristics of the clauses to realize the operability or usability of safety rules.

In this paper, the establishment of safety rules can be preliminarily divided into 2 categories according to the different accident subjects and different information retrieval processes:

(1) The subjects of safety risks involve the components produced in the construction process, such as floors and balconies;

(2) The subjects of safety risks occur in dynamic construction operations or processes, such as welding operations.

A complete set of safety rule interpretations should include information about the subject, location, attributes, and parameters of the accident and solutions. In this paper, related safety standards in China [18–23] were combined with the ontology defined in Section 3.1 to establish the safety rules used in this study. The result of safety rule parameter establishment for a hole is shown in Table 8.

Table 8. Safety rule parameter establishment for a hole.

Accident Subject	Location	Attributes	Parameter Information	Treatment
Hole	Reserved hole (Floor, roof) platform	Landscape	$L < 25$ cm; No protection	Solid cover
		Landscape	$25 \leq L < 50$ cm; No protection	Cover
		Landscape	$50 \leq L < 150$ cm; No protection	Steel mesh
		Landscape	$L > 150$ cm; No protection	Protective fence, safety net
	Wall (Window)	Vertical	$H < 80$ cm and $V > 2$ m; Unprotected railing	1.2 m high temporary railing
	Elevator wellhead	Landscape	$W < 2 \times F$ and $W < 10$ m; No gate	Set up a safety net
	Other holes	Landscape	Same as reserved holes	Same as reserved holes

L: the length of the hole; H: the lower edge of the hole to the floor or the bottom; V: the side drop; W: the safety net spacing in the elevator wellhead; F: the floor height.

2.3.2. Building a Safety Rule Library Based on Ontology

(1) Semantic Web Rule Language (SWRL).

The semantic web rule language (SWRL) is a language that presents rules in a semantic way. SWRL integrates the ontology description language OWL Lite, OWL DL (description logic), and Datalog RuleML. SWRL is human-readable with the advantages of the rule editing function combined with the elements defined in the ontology to improve knowledge expression and reasoning [24]. The SWRL consists of 4 parts: Imp, Atom, Variable, and Building, as shown in Figure 5 [25].

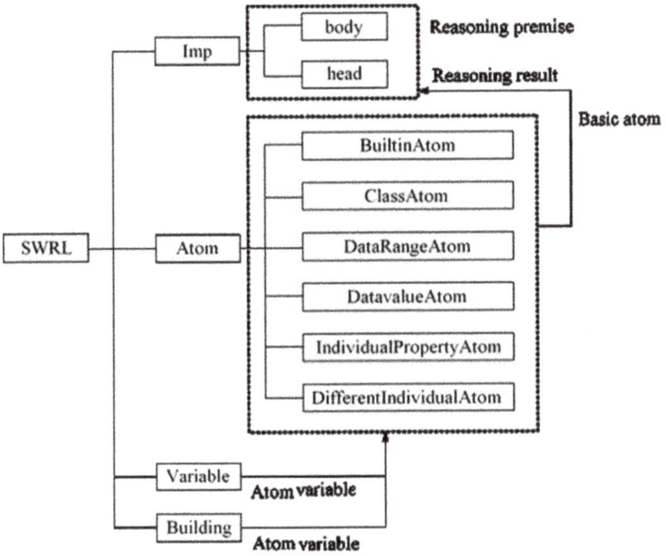

Figure 5. SWRL framework.

With the help of the developed safety management ontology classes and attributes, along with its rich arithmetic operations, the SWRL rules on safety management were established. To better connect the ontology and SWRL rules, the SWRLTab plug-in provided in the Protégé ontology editing software seamlessly connects the SWRL rule environment and the ontology editing environment, further improving the convenience of SWRL rule editing.

(2) Establishment of the SWRL safety rule database.

The ontology established in this paper included 4 categories: construction activities; precursor information; risk accidents; and solutions. The links formed by these 4 categories lay the premise for the establishment of SWRL rules. The elements (Atom) used in a complete SWRL rule for accident subjects, attributes, discriminant parameters, and prevention measures came from the 4 major classes of ontology, and the specific discriminant conditions or parameters were obtained in the interpretation of the safety rules. Taking the unprotected wall holes as an example, they were caused by the construction work brick masonry in the paving ash masonry and other construction steps for the outer window and other reserved holes. If the bottom edge of the hole was more than 2 m wide and the distance from the bottom of the floor was more than 80 cm without proper protection, accidents such as falling from a height may occur. The solution was to add a 1.2 m high protective railing around the hole in accordance with the specifications. The corresponding rule in SWRL database can be expressed as shown in Figure 6. The results of the complete set of safety rules for fall from height in the Protégé plugin SWRLTab developed in this study are shown in Figure 7. This paper also builds an SWRL safety rule base containing 5 major incident identifications in accordance with this process, as shown in Figure 8.

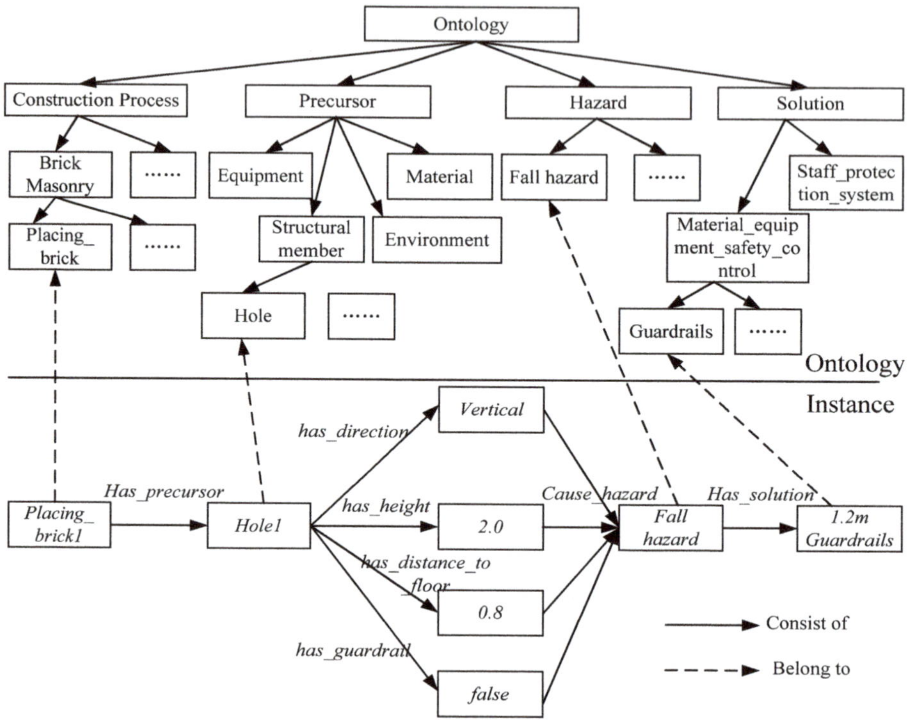

Figure 6. Vertical hole rule frame.

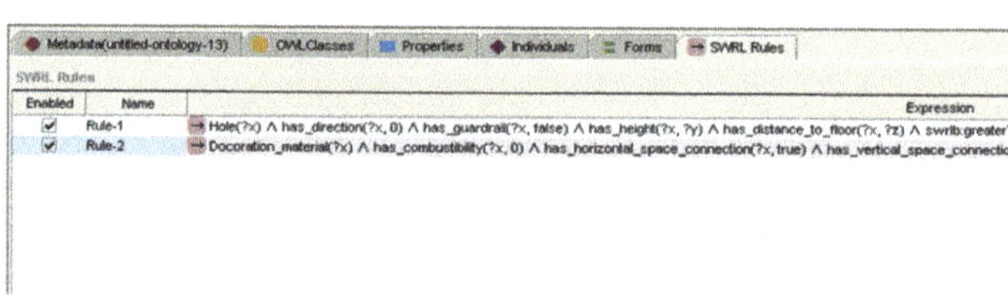

Figure 7. Vertical hole and cutting fire instance rules.

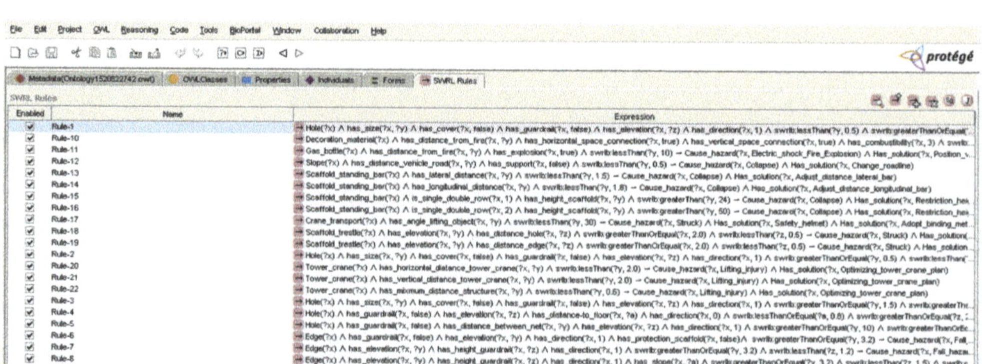

Figure 8. SWRL safety rule library for major safety risks.

2.4. Safety Risk Identification and Retrieval Mechanism

Safety rules were composed of accident subjects, judgment conditions, risks, and solutions, etc. The implementation of safety rules inspection needs to link the subjects of accidents in construction activities and extract discriminant parameters for verification. As a data source, the BIM model has its parameterized characteristics; thus that the information of each component and the relationship between each other can be directly attached to the component, which provides better convenience for the extraction of relevant information for a safety inspection. In addition, due to the characteristics of parameterization, modification or model building can be achieved directly by adjusting the parameters. This convenient way can save time and labor for safety managers on construction sites.

The safety risks in the construction process are dynamic and may only exist in a certain stage or process. For example, when the construction of the floor slab is completed, the unprotected periphery of the building forms edges that may cause the risk of falling from a height; and after the building of the walls eliminates the risk of the edges, the vertical opening of the windows on the masonry creates a new risk of falling from the height of the opening. Therefore, the process of safety inspection is not only the inspection of static safety risks but also includes more practical dynamic risk identification. BIM can realize the dynamic simulation of the construction process according to the given model and progress information (4D BIM) and serve as an intuitive process information hub. The dynamic risk inspection in the everyday construction process is only made possible by automatic identification and prevention of safety risks through the integration of BIM, ontology, and NLP. In light of the complexity of information related to safety risk, it was also necessary that safety managers can obtain access to the information in a convenient manner. Information related to safety management checked by the safety rules, such as accident subject/component attribute information, location information, time, over-limit parameters, etc., should be extracted as a safety problem report as required for the use of the safety managers.

2.4.1. Safety Risk Identification

The safety risk identification contains the following processes: first, use the BIM software tool to build the construction plan and schedule into a virtual construction process, and assign the necessary parameters and attribute information to it according to the characteristics of the construction components. Then, the ontology established a link with the detected object or construction process in BIM, automatically established an instance name in the safety ontology class according to the subject exposed to safety risks and extracted the required discrimination parameters to the ontology instance. Next, use the established SWRL rules to complete the inspection of safety risks and identify unsafe components and their parameters. Finally, all safety risks were summarized and output as an inspection report and displayed in the visual 3D model according to the inquiry made by construction safety managers. Considering that buildings are becoming more complex, and the content of the safety inspection report is complicated or cumbersome, safety managers still need to spend a lot of time finding safety management issues at different stages. For this reason, this paper uses the developed natural language search engine to achieve intelligence retrieval, which provides a convenient way for safety managers to deal with safety issues under their jurisdiction.

When the inspected object is a specific building component, the information of the component in the BIM model includes component name, attribute parameters, and other ancillary information, and the attribute parameters include direct attribute parameters such as elevation coordinate parameters, geometric information, material parameters, etc., and indirect parameters such as the topological relationship with the surrounding space. In building components such as stairs and elevator shafts, the names of building components have a corresponding relationship with the subject of the accident. This provides a preliminary way to link components with safety rules. After the connection was generated, multiple instance names were created by the scanning of the linked object, and the corresponding parameters or attributes were extracted to the instance to complete the instance creation. When the detected object was an abstract construction process, the link between the accident and the construction process was established through the name of the construction process, which was then used to detect related attributes or parameters of components or materials within a certain distance of the process, thus as to create an instance for safety rule checking. The process is shown in Figure 9.

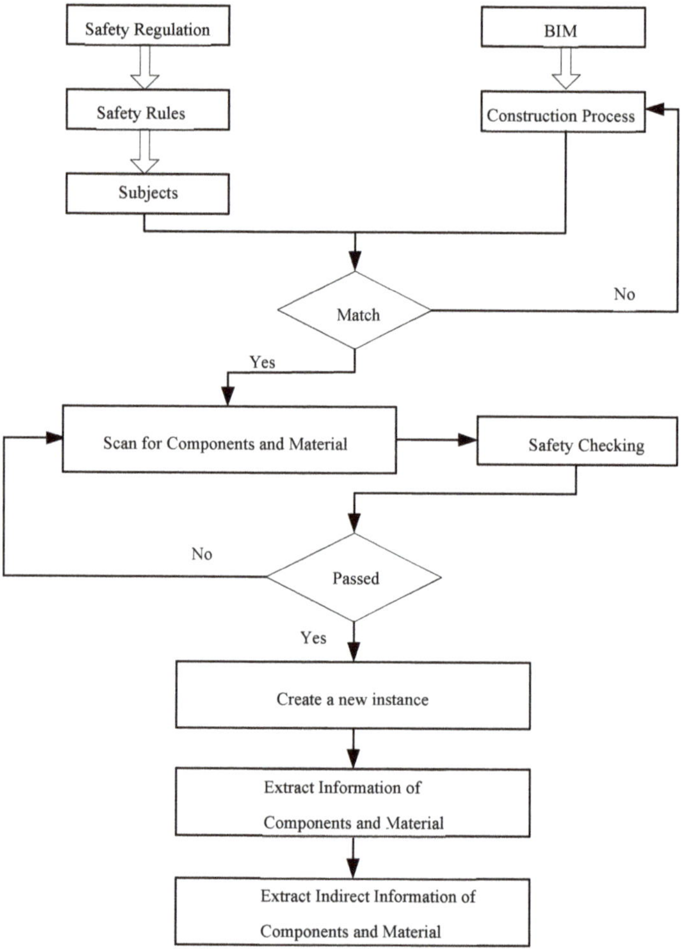

Figure 9. Linking building components and construction processes to safety rules.

When given specific domain rules and facts, the rule-based reasoning engine can be used to complete the corresponding reasoning tasks. However, the safety rules expressed by the SWRL rule language and the fact knowledge expressed in the form of ontology cannot be understood by JESS; thus, it was necessary to convert the SWRL rules and facts into the knowledge format supported by JESS to realize the reasoning. When the JESS inference engine is activated, the inference process begins. Pattern matching was used to determine what rules were executed and when to execute, after which the Agenda executes the commands in the activated rules. Then, the execution engine completes the execution of the rules. The processes are illustrated in Figure 10, which are then implemented in the case study.

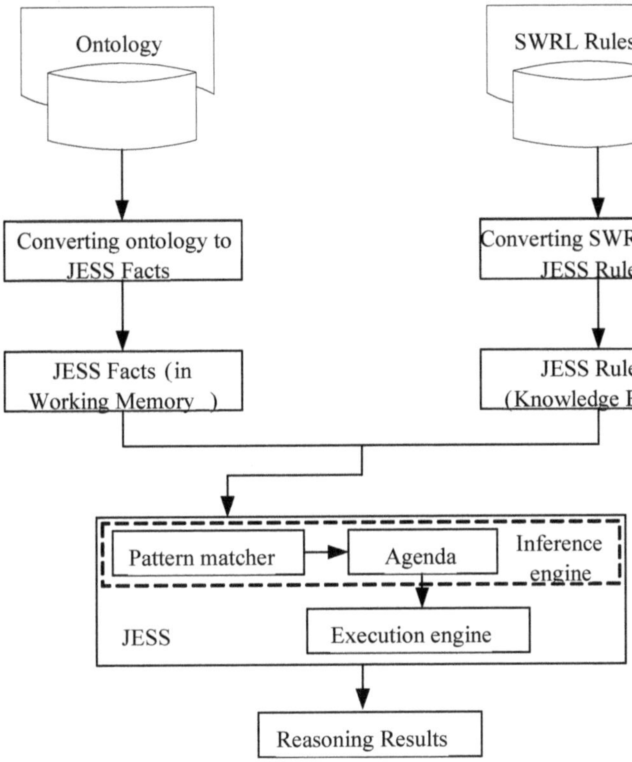

Figure 10. The reasoning processes.

2.4.2. Retrieval Mechanism

In contrast to traditional keyword-based retrieval technology, NLP-based retrieval technology was used to understand the semantics of the user's retrieval statement, which results in more accurate retrieval and allows safety managers to obtain construction safety information in a convenient and efficient manner. The retrieval mechanism designed in this paper included word segmentation, part-of-speech tagging, parsing, target keywords, and constraint sequence expansion.

(1) Word segmentation, in which the natural sentence from the user is divided into individual words after identifying each word and punctuation mark;

(2) Part of speech tagging, by marking the part of speech of each word based on statistical data or other methods, such as the Penn Treebank POS tag set;

(3) Parsing, designed to capture the relationship between words, forms a syntactic tree of queries through the Stanford parser. The NLP tool was used to obtain the relationship of different words in the sentence to form a syntactic structure tree. The results of the syntax analysis are shown in Figure 11, where IN, ADJ, NN, NNS, NP, and PP represent prepositions, adjectives, singular nouns or uncountable nouns, plural nouns, noun phrases, and prepositional phrases, respectively.

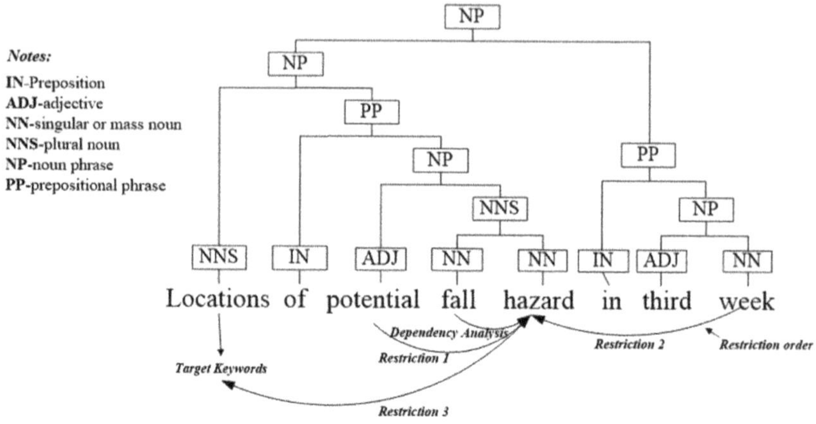

Figure 11. Syntactic structure tree.

The purpose of obtaining the syntactic tree was to find the words that actually represent the user's intent to search (also known as "target word") and the constraint relationship, while the keywords that contain the user's search purpose often exist in the noun phrase. Firstly, starting from the root node of the tree, create a line containing only nouns or noun phrases. The noun phrase closest to the root node was the target keyword, and the non-noun node (preposition phrase or adjective phrase with the same root node) was the constraint keyword of the target keyword. Secondly, for the subtree of the structure tree (the subtree root node is a noun or noun phrase), the leaf whose noun (noun phrase) was in the leaf node represents the target keyword of the subtree, and a similar case holds for the non-identical parent node. The noun node (prepositional phrase or adjective phrase) was the constraint word of the target keyword of the subtree. Then, a reverse analysis of the subtree to the root node was used to obtain a constraint sequence for the target keyword. According to the syntax tree analysis in the example sentence, the keywords "locations," "week," and the case of "fall" and "hazard" for the 2 keywords with the same root node were nouns. Finally, dependency grammar analysis was required to determine which word was a keyword and which word was a constraint word. After completing the analysis of this situation, the keywords of the whole sentence were "locations," "week," "hazard," and according to the order of constraints, "locations" was the keyword of the target (representing the purpose of the user). The user wants to obtain the location of the risk of falling from a height, and the risk of falling from a height was based on the third week of the project schedule. Therefore, the other keywords were all the constraints of this keyword and have a sequence (Restriction 1, Restriction 2).

(4) Keyword extension and mapping: since the query statement is expressed in natural language; the keywords or constraints of non-professional expression cannot be used for database retrieval. Therefore, this section designs the concept extension and mapping using ontology. This process consists of 3 tasks: (1) formalization of the extracted keywords or constraints; (2) query expansion using semantic similarity techniques to determine the final query term; and (3) mapping to the database.

Formal standardization of extracted keywords or constraints was used because the same term may have different expressions, such as abbreviations, spellings, or singular and plural numbers, such as "locations" becoming "location." Since the diversity of expressions leads to different expressions with the same meaning, the standardization of forms also includes conceptualization. For example, the concept of the use of the expression "location" in the ontology was "coordination," and the expression of risk was "hazard" instead of "risk." Therefore, formal standardization also includes the transformation of keywords or constraint sequence words obtained by search statements into standard concepts defined in the ontology, which were all integrated in the Protégé software.

The query expansion algorithm was used to improve the recall rate and retrieve the results that more strongly satisfy the user's needs. This paper determines the concept of extension based on the degree of approximation between ontology concepts. This process is called ontology mapping, which is the process of mapping to exchange information in a semantically sound manner. The target words obtained in the previous step were used to calculate semantic similarity of other words according to the concept hierarchy of the ontology (such as the parent class and subclass), which was then used as an expansion of query. This paper chose to use the Leacock–Chodorow formula [26] (Formula (1)) to characterize the degree of conceptual semantic similarity.

$$R(C,\ C_i) = -lg\frac{len(C,\ C_i)}{2Depth} \tag{1}$$

where $R(C,\ C_i)$ represents the similarity value; $len(C,\ C_i)$ represents the conceptual standard concept of the target keyword and the constraint keyword; C and the extended concept C_i represent the shortest distance in the construction safety ontology structure; *Depth* is the maximum depth of the construction safety ontology; and the coefficient 2 was used to ensure a positive value of similarity.

To remove the less similar concepts, this paper sets the similarity threshold to filter the concepts of lower similarity value, and the remaining expansion concepts and the initial standard concept were used as the final search keywords to complete matching with the Database. The final search keyword was used to match the database and sort the search results. Each target keyword and constraint sequence, extended keyword, and constraint sequence form a query vector and a record formed by the database record to calculate the similarity value, and the result was presented according to the similarity value. This paper used the Cosine space method to determine the size of the similarity value, as shown in Formula (2).

$$V = \frac{(X \cdot Y)}{\parallel X \parallel \parallel Y \parallel} \tag{2}$$

where

V: cosine of similar distance.

X: the feature vector formed by the target keyword or the extended keyword and its constraint sequence.

Y: the feature vector of the database data record.

2.4.3. Report of Retrieval Results

At present, the most common data presentation methods included 2-dimensional charts, network diagrams, documents, pictures, and animations, and this paper selected the document report + visual map in BIM to present the search results. The safety inspection report designed in this paper included the construction process, construction procedure, planning time, construction location (area, floor, etc., location map), type of accident, over-limit parameters, and recommended solutions. This information needs to be obtained from the database, and the information was presented according to the intention of the search phrase from the user.

3. Validation

3.1. Framework of the Proposed Safety Checking System

This paper chose the student dormitory building (C3 building) of the commuter college of a university as a case to demonstrate the framework. The student dormitory building (C3) covers an area of approximately 827.2 square meters. It extends seven floors above the ground and one floor below the ground. Each story height was 3.3 m, and its basement height was 3.6 m. The three-dimensional model of the dormitory building, which was generated in the Revit software (license number: 559-06926929), is shown in Figure 12 below.

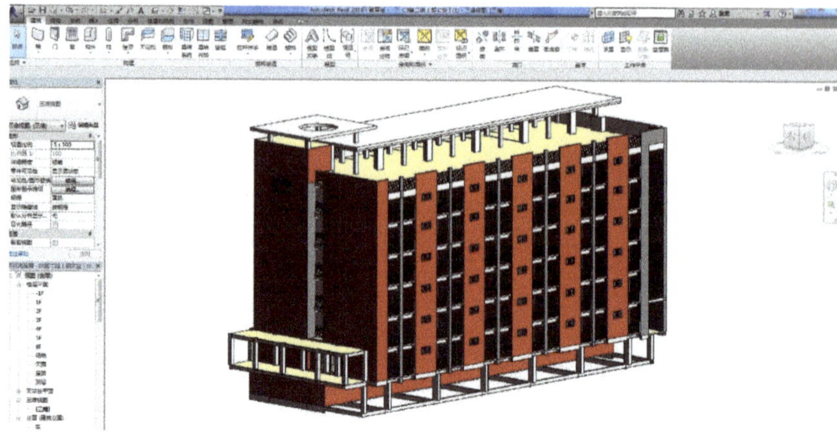

Figure 12. BIM model of a student dormitory building.

To realize the dynamic safety inspection process, this paper first used Microsoft Project (license number: W2JJW-4KYDP-2YMKW-FX36H-QYVD8) to develop a progress plan of simulated construction, as shown in Figure 13. The concrete pouring process was divided into structure construction and decoration construction. The schedule began on 1 July 2017, when concrete was poured into the basement beams and slabs. On 23 September 2017, the installation of the top doors and railings was completed. The total duration was 85 days.

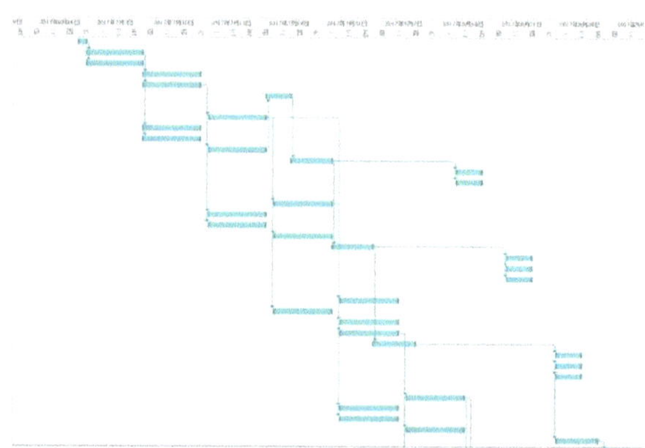

Figure 13. Construction schedule of the student dormitory building.

This paper then integrated the Revit information model of student dormitory building and MS project schedule into Navisworks software (license number: 559-06926929) and realized the dynamic simulation process by dividing the construction process and connecting time information of components in the construction process, as shown in Figures 14–17.

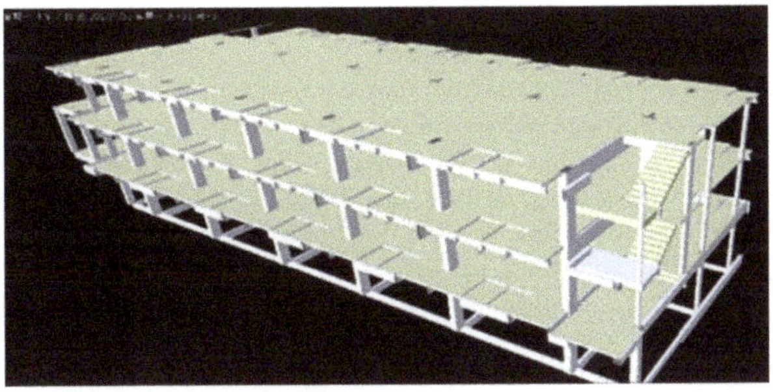

Figure 14. Project progress in week 4.

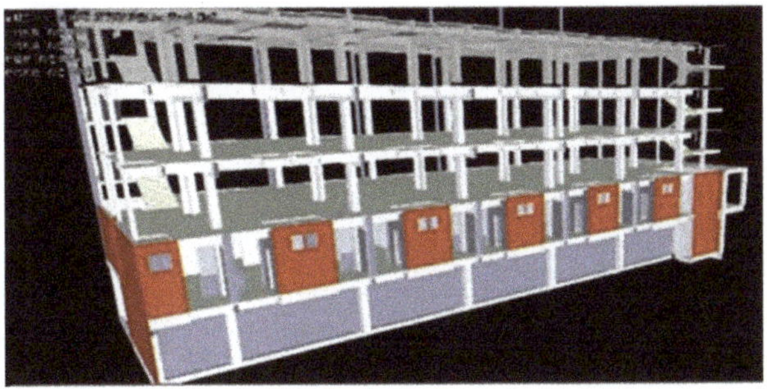

Figure 15. Project progress in week 7.

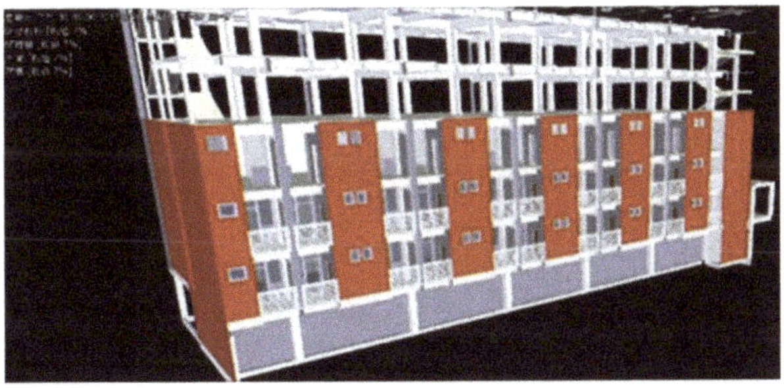

Figure 16. Progress of the project in week 8.

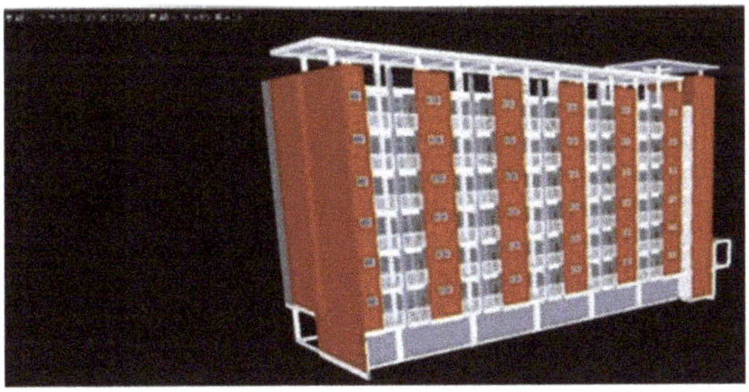

Figure 17. Progress of the project in week 13.

The following is an example of the process and results of the performed safety inspection on holes. According to the construction schedule, the time period for the top floor to complete the floor and stairs was 10 weeks (26 August to 2 September 2017), while the maintenance wall of the top floor took 12 weeks (15 September to 19 September), and the opening formed by the stairs and slabs had not undergone corresponding protection measures during the period of 10 to 12 weeks (Figure 18). This condition may lead to an accident involving falling from a high place.

The automatic identification process was performed by scanning the construction process files containing time information by the developed framework, which established the connection with the top-level components according to the detection rules of the openings and extracted the relevant parameters. The results from the reasoning are shown in Figure 19. Under the implementation of a reasoning engine, the safety parameters can be identified automatically, and the risks and solutions can be identified (Figure 20).

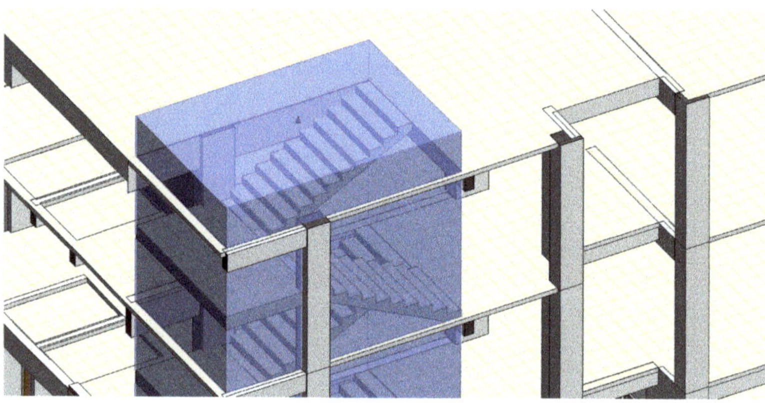

Figure 18. Top floor stair opening.

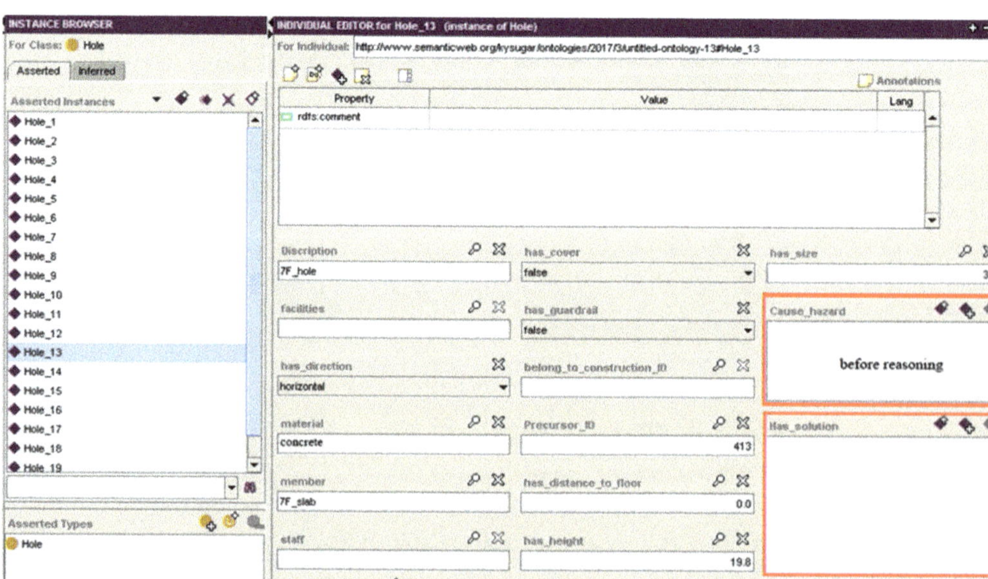

Figure 19. Extraction of the top floor stair opening information to the ontology instance.

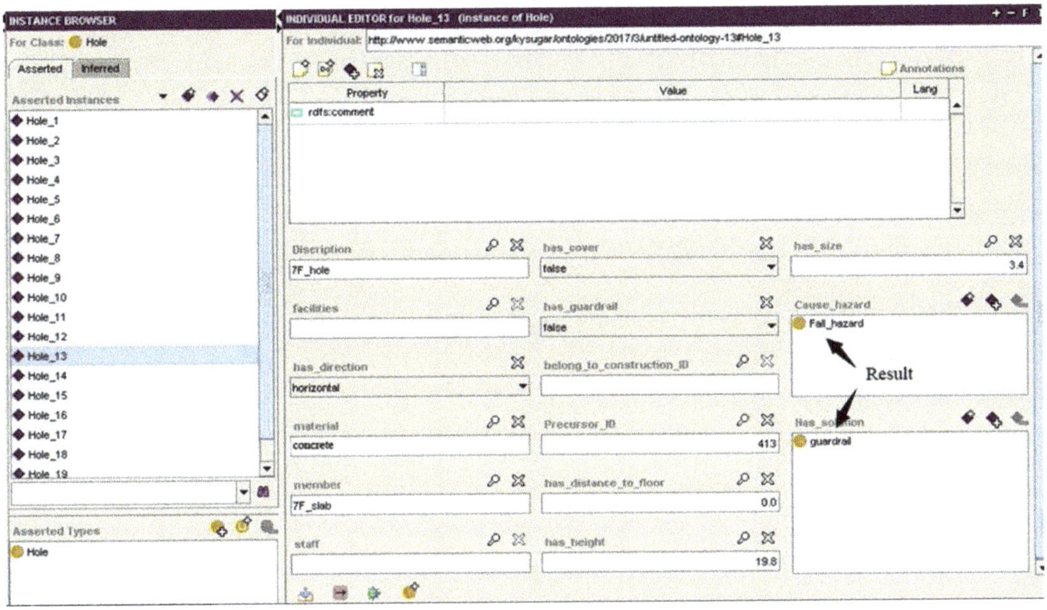

Figure 20. Risk reasoning results of the top floor stair opening.

The NLP-based query system to present the identification results of related risks according to the user's requirements was then illustrated. The login interface and the query interface are shown in Figures 21 and 22, respectively. This paper takes the query statement "I need to know about any potential hazards on the third floor" as an example to show the query results (Figures 23 and 24) and query reports (Figure 25). The project manager of the case study stated that the developed plugin was user-friendly, and the results could help improve the construction safety management since they were rational according to their experiences and requirements. Similar to many other industries, people tend to use a system that can make their queries easily without previous knowledge of semantic technologies [27].

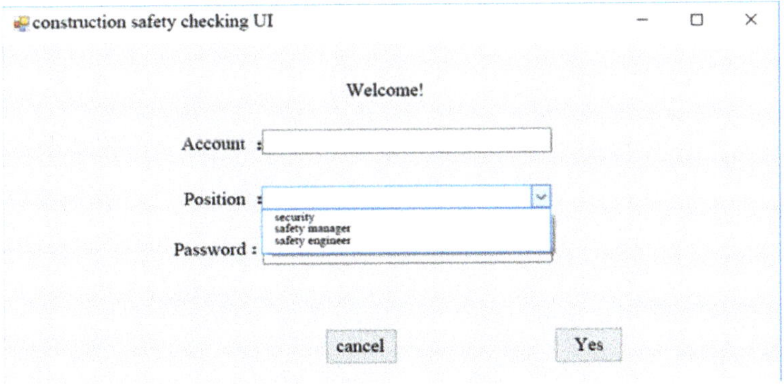

Figure 21. Landing interface of the developed safety risk inspection system.

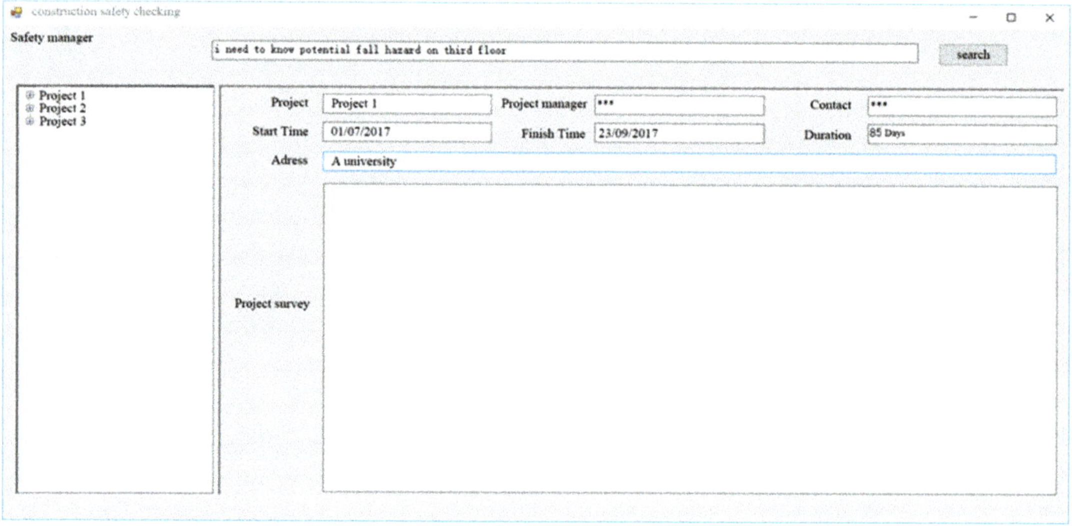

Figure 22. Landing interface of the safety risk inspection system in the construction process.

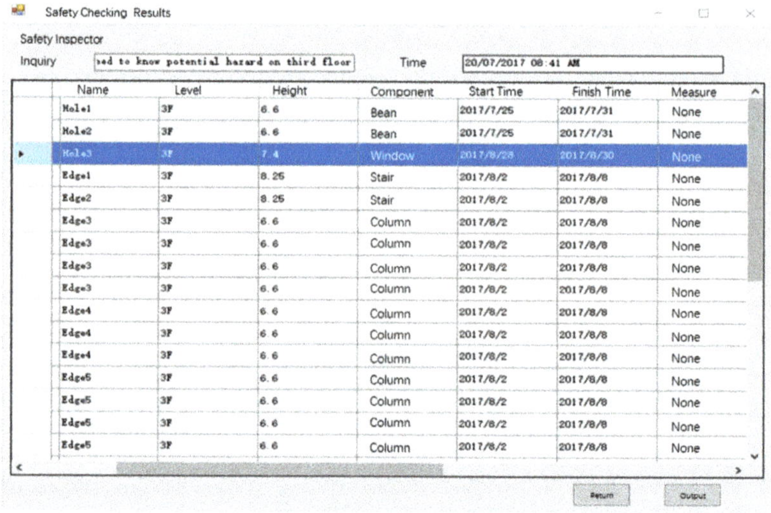

Figure 23. Safety risk query result for potential risks on the third floor.

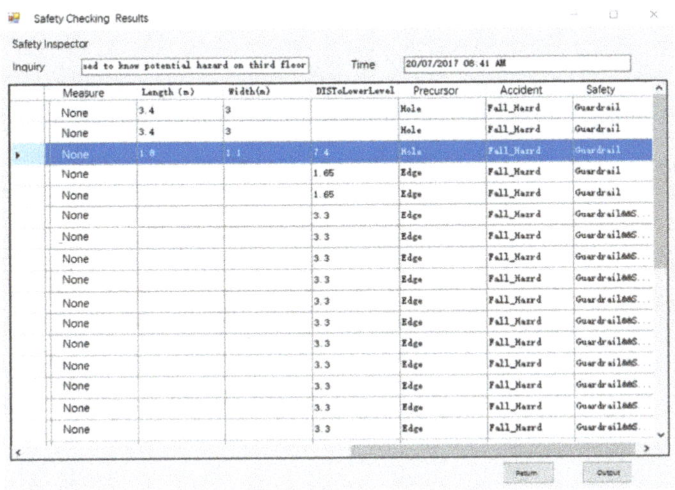

Figure 24. Safety risk query result and the suggested safety measures.

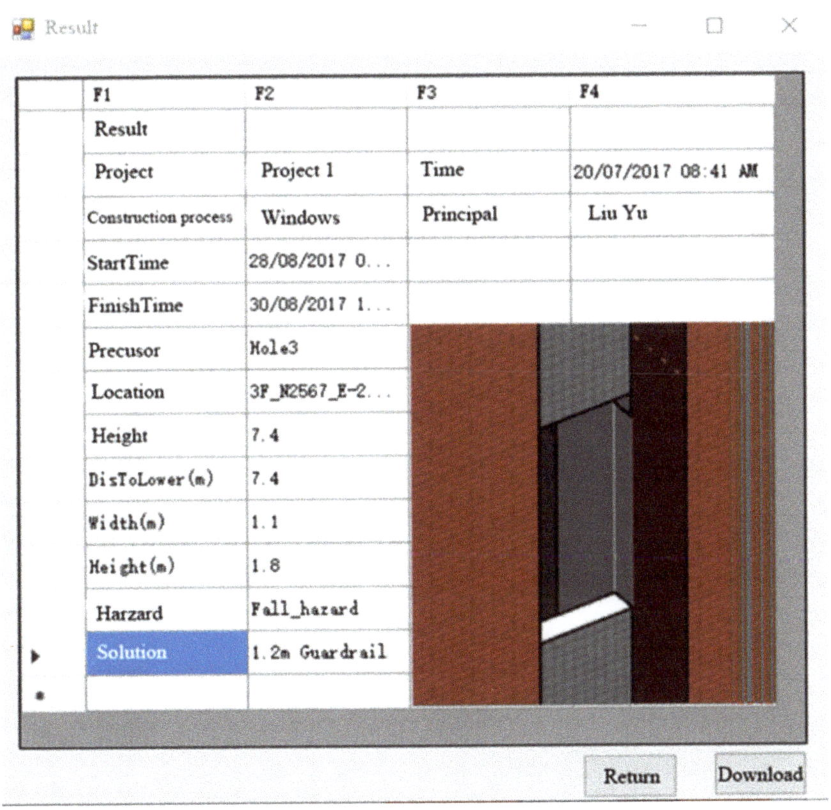

Figure 25. Query report.

3.2. Experiment Results

In the experiment, the developed system presents the safety inspection results of the third floor. It was shown that the system successfully identified the safety risks in the construction process of the third floor, such as the stairway opening and window opening, as well as the multiple adjacent edges. The reasoning accuracy by virtue of the ontology was close to 100%. For the evaluation of retrieval effectiveness, this paper adopted two indicators: recall rate and precision rate. Suppose A was the correct number of results retrieved by a certain retrieval sentence, B was the number of incorrect results retrieved by a certain retrieval sentence, and C was relevant results that were not retrieved by a certain search sentence; thus, the precision rate can be expressed by A/(A + B), and the recall rate can be expressed by A/(A + C).

However, these two indicators have a mutual restriction relationship; that was, if the recall rate was required to be high, the search result would contain more wrong results; thus, the accuracy rate would be reduced, and if the accuracy rate was high, the search result will miss a certain correct result. For this reason, an F-value metric that takes the two into account was introduced to consider the retrieval level of retrieval technology. The Formula (3) is as follows.

$$F_\beta = \left(1 + \beta^2\right) \cdot \frac{Pre \cdot Re}{\beta^2 \cdot Pre + Re} \tag{3}$$

where *Pre* stands for accuracy;

Re stands for recall rate; and

β Represents the weight of recall relative to accuracy;

This article randomly tested about 50 natural language search sentences and made statistics on accuracy and recall. The results are shown in Figure 26. According to the chart, the accuracy rate of the retrieval system was higher than 60% when the recall rate was lower than 50%.

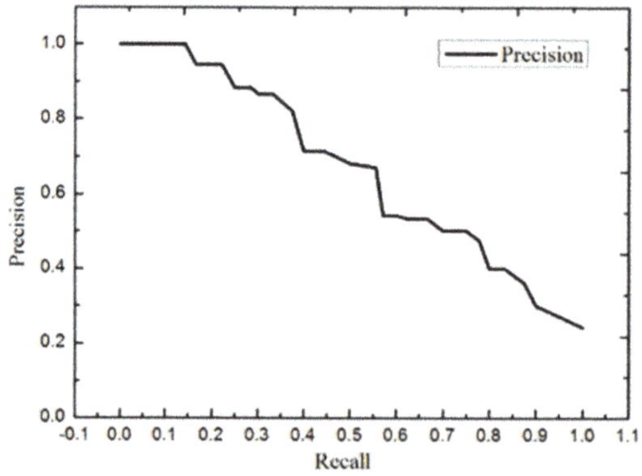

Figure 26. Retrieval recalls and precision curves.

4. Conclusions and Future Work

This paper built an automatic mechanism for identifying and preventing safety risks during the construction process by integrating the safety rules library and BIM technology with advantages in visualization and parameterization. Moreover, an intelligent presentation method based on the Natural Language Processing technology was proposed to intelligently present identified results from a developed system.

In the building of the ontology, the hierarchical construction activities of residential buildings were broken down by using WBS. Moreover, then the precursory information of four categories, including structural components, materials, equipment, and environment, were established by combining construction activities and types of safety accidents. Then, domain ontology was built depending on the seven-step method, which consists of construction activities, precursor, hazard, and solution. Next, on the basis of the established precursor and relevant codes, the parameterized forms for five kinds of accident safety rules were summarized and converted into SWRL format. To realize the automatic safety rules checking, automatic checking mechanism of integrating BIM, and safety rules was proposed. Meanwhile, an intelligent results presentation based on Natural Language Processing was designed. Finally, the construction process safety rule checking system was developed, and the effectiveness of the system was verified by a case study.

Two innovative methods were proposed in this paper: (1) ontology technology and BIM were put forward for safety rule checking for the dynamic construction process of the residential building; (2) the intelligent presentation mechanism based on the Natural Language Processing was presented. It not only provided ideas for the reuse and sharing of the safety knowledge but also proposed a solution for immediate and intelligent presentation of the safety problems on dynamic problems, which made a certain contribution to the improvement of efficiency of construction safety. The NLP-based presentation method for construction safety rule checking results will facilitate the access of construction safety managers to the relevant information, which will then improve the efficiency of construction safety management in practice. However, there are certain limitations of the paper that needs to be addressed in future work: (1) this paper only aimed at the construction process of civil residential buildings to establish the construction safety ontology, it is not comprehensive and extensive enough; (2) this paper selects only common safety regulations and clauses.

Author Contributions: S.W. and Y.D. were in charge of the conceptualization, methodology and formal analysis under the supervision of Q.S., H.D. and J.C.P.C. All authors have read and agreed to the published version of the manuscript.

Funding: This study was financed in part by the Science and Technology Program of Guangzhou, Grant No. 201804020069, and the support by Guangdong Science Foundation, Grant No. 2022A1515010174.

Institutional Review Board Statement: Not applicable.

Informed Consent Statement: Not applicable.

Data Availability Statement: Not applicable.

Acknowledgments: The authors would like to acknowledge by the Science and Technology Program of Guangzhou, Grant No. 201804020069, and the support by Guangdong Science Foundation, Grant No. 2022A1515010174.

Conflicts of Interest: The authors declare no conflict of interest.

References

1. Yan, W. *Construction Hazards Management Research Based on BIM*; Shanghai Institute of Technology: Shanghai, China, 2016.
2. Luo, C.; Xie, X. Comparison study of accident-causing theories. *J. Saf. Sci. Technol.* **2007**, *3*, 111–115. Available online: https://en.cnki.com.cn/Article_en/CJFDTotal-LDBK200705028.htm (accessed on 3 November 2021).
3. Zhi, P. *A Research in the Safety Control. of Reverse Construction*; Tianjin University: Tianjin, China, 2005.
4. Liu, G.; Lei, L. Analysis of heinrich accident-causing theory and the factors of safety ideology. *Saf. Environ. Eng.* **2013**, *20*, 138–142. Available online: https://en.cnki.com.cn/Article_en/CJFDTotal-KTAQ201301032.htm (accessed on 3 November 2021).
5. Bird, F. A handbook of personnel management practice: Michael Armstrong, 2nd edition, Kogan Page, London, 1984, £9.95 paperback. *Int. J. Hosp. Manag.* **1985**, *4*, 48. [CrossRef]
6. Li, J.; Su, X.; Qian, P. The methodology of developing domain ontology. *Comput. Agric.* **2003**, *12*, 7–10. Available online: http://en.cnki.com.cn/Article_en/CJFDTOTAL-JSJN200307002.htm (accessed on 3 November 2021).
7. Wang, H.; Boukamp, F. Ontology-based representation and reasoning framework for supporting job hazard analysis. *J. Comput. Civ. Eng.* **2011**, *25*, 442–456. [CrossRef]
8. Lu, Y.; Li, Q.; Zhou, Z.; Deng, Y. Ontology-based knowledge modeling for automated construction safety checking. *Saf. Sci.* **2015**, *79*, 11–18. [CrossRef]
9. Zhang, S.; Boukamp, F.; Teizer, J. Ontology-based semantic modeling of construction safety knowledge: Towards automated safety planning for job hazard analysis (JHA). *J. Comput. Civ. Eng.* **2015**, *52*, 29–41. [CrossRef]

10. Ding, L.; Zhong, B.T.; Luo, H. Construction risk knowledge management in BIM using ontology and semantic web technology. *Saf. Sci.* **2016**, *87*, 202–213. [CrossRef]
11. Zhang, J.; El-Gohary, N. Automated information transformation for automated regulatory compliance checking in construction. *J. Comput. Civ. Eng.* **2015**, *29*, B4015001. [CrossRef]
12. Zhang, J.; El-Gohary, N. Extending building information models semiautomatically using semantic natural language processing techniques. *J. Comput. Civ. Eng.* **2016**, *30*, C4016004. [CrossRef]
13. Zhang, J.; El-Gohary, N. Integrating semantic NLP and logic reasoning into a unified system for fully-automated code checking. *Autom. Constr.* **2017**, *73*, 45–57. [CrossRef]
14. Lin, J.; Hu, Z.; Zhang, J.; Yu, F. A natural-language-based approach to intelligent data retrieval and representation for cloud BIM. *Comput. Aided Civ. Infrastruct. Eng.* **2016**, *31*, 18–33. [CrossRef]
15. Jiao, H. *Modeling and Application of Ontology-Based Knowledge Base for Risks of Subway Construction*; Southeast University: Nanjing, China, 2015.
16. Zeng, W. *Pre-Control Study of Building Construction Safety Accident Based on DFS*; Southeast University: Nanjing, China, 2015.
17. Zhang, J. *Research on Ontology-Based Retrieval of Construction Domain under BIM Environment*; Dalian University of Technology: Dalian, China, 2013.
18. Ministry of Housing and Urban-Rural Development of the People's Republic of China. *Construction Safety Inspection Standards (JGJ 59-2011)*; China Building Industry Press: Beijing, China, 2012.
19. Ministry of Housing and Urban-Rural Development of the People's Republic of China. *Technical Specifications for Safe Operation at Height during Construction (JGJ 80-91)*; China Planning Press: Beijing, China, 1992.
20. Ministry of Housing and Urban-Rural Development of the People's Republic of China. *Technical Regulations for Safety in the Use of Construction Machinery (JGJ 33-2012)*; China Building Industry Press: Beijing, China, 2013.
21. Ministry of Housing and Urban-Rural Development of the People's Republic of China. *Technical Specifications for Temporary Electricity Safety at construction Sites (JGJ 46-2005)*; China Building Industry Press: Beijing, China, 2010.
22. Ministry of Housing and Urban-Rural Development of the People's Republic of China, General Administration of quality supervision, inspection and quarantine of the People's Republic of China. *Technical Specifications for Fire Safety on Construction Site of Construction Projects (gb50720-2011)*; China Planning Press: Beijing, China, 2011.
23. *GB8624-2012*; State General Administration of Quality Supervision, Inspection and Quarantine of the People's Republic of China, China National Standardization Administration Committee Classification of Combustion Performance of Building Materials and Products. China Standard Publishing House: Beijing, China, 2012.
24. Yu, S. *Research on SWRL-Based Semantic Relevant Discovery Based Ontology Mapping Approach*; Nanjing University of Aeronautics and Astronautics: Nanjing, China, 2009.
25. Horrocks, I. SWRL—A Semantic Web Rule Language Combining OWL and RuleML. 2003. Available online: http://www.daml.org/2003/11/swrl (accessed on 3 November 2021).
26. Leacock, C.; Chodorow, M. *Combining Local Context and Wordnet Similarity for Word Sense Identification*, 1st ed.; MIT Press: Cambridge, MA, USA, 1998.
27. Rosales-Huamaní, J.A.; Castillo-Sequera, J.L.; Montalvan-Figueroa, J.; Andrade-Choque, J. A Prototype of Speech Interface Based on the Google Cloud Platform to Access a Semantic Website. *Symmetry* **2018**, *10*, 268. [CrossRef]

buildings

MDPI

Article

Automated Selection and Localization of Mobile Cranes in Construction Planning

Hongling Guo [1], Ying Zhou [1], Zaiyi Pan [1], Zhitian Zhang [1], Yantao Yu [2,*] and Yan Li [1]

1 Department of Construction Management, Tsinghua University, Beijing 100086, China; hlguo@tsinghua.edu.cn (H.G.); zhouying17@mails.tsinghua.edu.cn (Y.Z.); pzy2456@126.com (Z.P.); zhangzt18@mails.tsinghua.edu.cn (Z.Z.); liyan1@cindata.cn (Y.L.)

2 Department of Civil and Environmental Engineering, The Hong Kong University of Science and Technology, Hong Kong 999077, China

* Correspondence: ceyantao@ust.hk

Abstract: Accurate selection and location of mobile cranes is a critical issue on construction sites, being able to contribute to the improvement of the safety and efficiency of lifting operations. Considering the complexities and dynamics of construction sites, this study aimed to develop a useful approach for automated selection and localization of mobile cranes based on the simulation of crane operations. First, the information required for crane selection and localization is analyzed and extracted from BIM (building information modeling). Then, mainly considering the crane capacity, the initial crane type is selected with candidate location points. Based on the simulation of lifting operation at the candidate points, feasible location points and crane types are determined through three constraint checks (i.e., environment constraint, operation constraint, and safety constraint). Besides, two kinds of efficiency optimization, namely lifting time minimization and crane movement minimization, are presented to figure out the best location points from the feasible points. Finally, the proposed approach is validated using a case study. This research contributes to not only crane operation planning but also automatic construction simulation, thus supporting the implementation of intelligent construction in the future.

Keywords: mobile crane; automated selection; automated localization; virtual construction

Citation: Guo, H.; Zhou, Y.; Pan, Z.; Zhang, Z.; Yu, Y.; Li, Y. Automated Selection and Localization of Mobile Cranes in Construction Planning. *Buildings* **2022**, *12*, 580. https://doi.org/10.3390/buildings12050580

Academic Editor: Lucio Soibelman

Received: 23 March 2022
Accepted: 27 April 2022
Published: 29 April 2022

Publisher's Note: MDPI stays neutral with regard to jurisdictional claims in published maps and institutional affiliations.

Copyright: © 2022 by the authors. Licensee MDPI, Basel, Switzerland. This article is an open access article distributed under the terms and conditions of the Creative Commons Attribution (CC BY) license (https://creativecommons.org/licenses/by/4.0/).

1. Introduction

As the most commonly used and shared site resources with large moving and heavy loads, mobile cranes are involved in all kinds of lifting works [1,2]. Considering the complication of hoisting, especially heavy weight lifting, a detailed hoisting scheme should be formulated before operation [3]. According to relevant statistics, the design of the hoisting scheme is a very important part of lifting work on construction sites, accounting for 60–80% of the whole lifting time [4]. It mainly includes crane selection, crane localization, and operation planning. However, the traditional hoisting scheme design mostly relies on the experience of site engineers, requiring a lot of two-dimensional (2D) drawings with limitations, such as a high error rate, low design efficiency, low safety level, and poor data synchronization. This calls for an urgent in-depth study to figure out the above critical problems.

In recent years, the rapid development of virtual construction and simulation technologies has provided a new solution. The application of virtual construction technology to the hoisting scheme design helps promote high-accuracy three-dimensional (3D) visualization, and advanced simulation and optimization. It is used extensively for the path planning of mobile cranes, training operators, or the detection of spatial conflicts [5–10]. Moreover, the simulated results further support the following crane selection or localization, respectively, especially the effects of different factors (i.e., wind effect [11], ground bearing pressure [12,13], rope breakage [14], and so on) on the operational feasibility. However, crane selection and localization are essentially two mutually prerequisite steps that cannot be separated independently while previous research has treated them less as a whole process [15]. Due to the difference in the performance parameters between various types of cranes, not only the load but also the spatial relationship between a lifting object and a crane on a site should be considered to ensure that the selected crane meets the lifting requirements regarding the height and distance. Lifting collision is another indispensable problem in the selection of cranes. Moreover, mobile cranes can move to different locations on construction sites to meet the specific hoisting requirements mentioned above. Therefore, the hoisting scheme design needs to consider different location point combinations of lifting objects, minimizing the movement of the crane, and ensuring the safety of lifting processes.

Aimed at the above problems, this research aimed to develop a method to automatically select and locate mobile cranes in virtual construction to aid in hoisting scheme design. The rest of this paper is structured as follows. First, a literature review is outlined to analyze the limitations of current research in Section 2. Then, the research methodology is presented in Section 3. After that, the method for the automated selection and localization of mobile cranes is proposed in Section 4 and further tested based on a case study in Section 5. Finally, a conclusion is drawn in Section 6.

2. Literature Review

2.1. Crane Selection

Previous research on the selection of mobile cranes has mainly focused on the crane load and the distance between crane booms and buildings. Al-Hussein et al. [16] proposed a method to select a feasible crane by calculating whether the load of the crane and the gap between the crane and building components meet basic requirements based on independent 2D drawings involving a lot of manual input. Moselhi et al. [17] developed a selection and positioning system of a mobile crane to establish 3D models based on the coordinates and size of hoisting objects and obstacles and then simulated the position of the mobile crane and potential collisions in a virtual environment to identify the optimal crane. Based on the above research, a systematic method for the selection of mobile cranes was proposed [18]. It selected the optimal crane based on different constraints from feasible cranes that meet the minimum clearance with buildings, crane working radius, and crane load. For instance, the selection of a truss boom crane was optimized with the lifting radius while that of a hydraulic telescopic boom crane was achieved using the main boom length and working radius. Besides the crane load and safety distance, other factors, such as pressure and cost, have been considered in the selection of mobile cranes. For example, a study investigated the effect of the ground bearing pressure on the selection of a crane type by comparing the pressure of the crane on the ground with the ground bearing pressure [19]. Hasan et al. [20] calculated the pressure of each leg of a crane at different horizontal swing angles and ensured safety by checking the leg pressure during crane operation. Han et al. [21,22] considered the lifting capacity, working range, lifting height, and the first lifting weight to select a crane with the lowest cost in a feasible list of cranes. Furusaka and Gray [23] proposed an algorithm to realize the most economical combination of different cranes, with a new definition of the minimum total cost of a mobile crane, including leasing, assembly, and disassembly. Based on the feasible crane locations, Han et al. [24] proposed a 3D-based crane evaluation system to select the most suitable crane type under two circumstances (i.e., the fixed and unfixed crane), requiring module information inputs. Artificial intelligence (AI), on the other hand, has been used to select cranes in recent years. For example, the genetic algorithm (GA), as a heuristic random search technique, was used to determine the optimum location of a crane by considering safety, clearance, site conditions, etc. [25]. A discrete event simulation model was developed to realize the automatic planning of crane operation and the selection of the optimal crane type [26]. A system called PRECISE was developed to select the optimal crane type, which minimized the number of mobile crane operations [27].

The existing methods enable rapid calculation and analysis of the geometric and mechanical parameters of mobile cranes and construction sites, and realize the selection of the optimum crane for a specific construction project by considering several specific factors. Based on experience, this is more efficient and accurate than traditional methods. However, the selection methods in previous research mainly considered a 2D static environment. Although some methods have been combined with 3D models, they still need significant manual input or modeling [24]. In some studies, 3D modeling was even used as a visualization tool for animation demonstration after selection, which does not provide enough assistance in hoisting scheme design. Moreover, there is no comprehensive and systematic consideration of various factors, such as the site environment, operation, safety, and economy, in crane selection, with less consideration of the following progress (e.g., crane localization and path planning) [15].

2.2. Crane Localization

In terms of crane localization, three constraints (i.e., environmental constraint, operation constraint, and safety constraint) have mainly been considered in previous research. To deal with these constraints, different methods have been applied, which can be classified as the 2D-based method and the 3D simulation method. The 2D-based method uses mathematical formulas to calculate the feasible location areas or location points for a mobile crane. Although some methods consider spatial factors and establish 3D models, they are still based on the mathematical analysis of 2D drawings, thus being classified as a 2D-based method [28]. Lei [29] and Ding [30] obtained the appropriate location areas of a mobile crane by defining the outer boundary constraints of existing buildings or obstacles but without consideration of 3D spatial factors. Al-Hussein et al. [16] judged whether a crane satisfied the environmental constraints by comparing the building coordinates around the proposed location points and calculating the distance between the crane boom and the building. Considering that a crane should be located in the center of all buildings, Olearczyk et al. [3,4] proposed a method for calculating the geometric center coordinates of all buildings and analyzed the maximum working radius of a crane and the distance between its boom and the buildings to ensure the operability and safety of lifting work. Moreover, Hermann et al. [31] proposed another method to determine the location of mobile cranes in prefabricated building construction. First, a project manager defines a line and generates several random points on the line. Then, the distance between the center of each building and each random point is calculated to select the maximum distance for the random points, and finally, the above two steps are repeated to select the minimum distance from these maximum distances. The point with the minimum distance is the desired location point of the crane. Based on the 2D-based method, the feasible location area of the mobile crane is firstly obtained by considering its working radius or minimum operating distance. Then, other factors, including time and cost, are further considered to determine the optimal location point. However, this requires manual input of many parameters [32,33], which takes a significant amount of time, especially for complex spatial calculations. In addition, the accuracy of the obtained location points may be affected because some location points with collisions are not deleted while some feasible points are eliminated by mistake.

Considering the limitations of the 2D-based method, some research presents discrete simulation-based 3D methods [28,34–36]. Firstly, feasible location points are determined based on the construction environment and the reachability of cranes. Through simulation of the lifting operation, location points with spatial collisions or other unsatisfied constraints are deleted to obtain the optimal location points. Tantisevi et al. [28,34] selected feasible location points initially by judging the accessibility of a boom; screened out feasible location points using the bounding box, ray tracing algorithm, and conflict analysis method; and then obtained the optimal location points by traversing all operations of the crane with the minimization of crane relocalization. The first two steps of the method proposed by Wang et al. [35] are similar to those proposed by Tantisevi [28,34]. However, the relocalization of the mobile crane is not considered while the total weight of a crane and lifting components is minimized from a safety perspective. In addition to the localization of an individual crane, some research has considered the operation of double cranes or multiple cranes. For example, Zi et al. [36] used a parallel robot method to solve the localization problem of multiple cranes based on the multi-point localization method and 3D grid method. Recently, integer programming [37,38] has been introduced to solve the location optimization problem of multiple cranes, with some assumptions disagreeing with reality, as limited by the computation resource.

Compared with the 2D-based method, the 3D simulation method not only considers spatial factors but also applies various simulation methods in spatial conflict analysis, making collision detection more dynamic and efficient. However, the changes in the lifting capacity during actual operations are not considered. The maximum load capacity of a mobile crane changes with the lifting radius. In addition, existing 3D simulation methods still require many manual inputs or modeling, neglecting the effect of other stages.

2.3. Research Gaps

As crane selection and localization are two interdependent processes [15], recently, some research [39–41] has attempted to consider the two processes simultaneously. Referring to the research gap that neglecting the impact of multiple construction stages leads to excessive costs, Yeoh and Chua [39] reframed crane selection and localization as a four-dimensional set cover problem (4D-SCP) to minimize the investment, with the limitation of static cranes. Focusing on multiple working cranes, Lin et al. [40] compared the working efficiency of different crane combinations to determine the optimal one with a value engineering model. However, collision detection, which is unavoidable between adjacent cranes, was considered less. Boo et al. [41] proposed a multi-objective optimization model for the selection of the tower crane number, types, and locations, which minimized cost and conflict, using a multi-objective optimization model. As a matter of fact, the calculation of the gap between a crane boom and an obstacle needs to consider the position of the boom in the elevation, and the existing selection algorithm inputs the crane location as a known item. Meanwhile, the determination of the crane location requires the information of the selected crane, which traps the calculation into a circular loop. Existing research does not take this iterative problem into account as a whole process. In addition, most of the relevant research only considers the hoist of a single component; however, it is a multi-component lifting process. Therefore, a method for automated selection and localization of mobile cranes is proposed in this research by combining the specification of the lifting operation and characteristics of mobile cranes with BIM (building information modeling) and virtual construction to ensure the safety, operability, and efficiency of lifting processes. Besides, the proposed method contributes to the automated simulation of construction processes, thus increasing the safety performance of lifting operations and reducing construction costs.

3. Methodology

Automated selection and localization of mobile cranes are two interdependent processes. As the first step of a hoisting scheme design, crane selection is based on the crane location and calculation to determine whether crane parameters meet the requirements. Therefore, this research regards the selection and localization of a mobile crane as a comprehensive planning process, taking both into consideration when analyzing the constraints and information requirements.

Figure 1 shows the process of automatically selecting and locating a mobile crane. Due to the complex environment of construction sites, the location of the mobile crane should meet the requirements of the site layout, that is, the environmental constraints. According to the construction plan, the plane range of an obstacle must be eliminated from the alternative location area. Then, the lifting requirements of the mobile crane, namely the operation constraints, including the transport distance, transport height, and the weight of lifting objects, should be considered. Thirdly, safety constraints should be considered in the operation of the mobile crane. The selection and localization of the mobile crane can be simulated in a virtual environment to identify and remove collisions between booms and buildings. Once these constraints are analyzed, suitable crane types and feasible location points can be achieved. Traversing the generated alternative location points, the optimal combination of the crane type and location point can be determined. Moreover, time and cost are two important aspects of project management, and they should be analyzed and reduced through further optimization. On the one hand, a crane with a lower rental price is more advantageous when it meets the above constraints. On the other hand, relocation and lifting are two time-consuming activities in crane operation, and minimizing the relocation and lifting time can effectively reduce the total lifting time and improve the lifting efficiency. Aiming to meet the requirements of the constraints and optimization, the constraint conditions should be digitized first. The BIM model can be converted into relevant parameters, which describe the conditions, to support the automated selection and localization of the mobile crane through appropriate algorithms. The spatial structure of each component of a crane can be further simulated as well. Based on the process of automated selection and localization of the mobile crane, the key algorithms for the three constraints and efficiency optimization are developed and tested using a virtual experiment in the following sections.

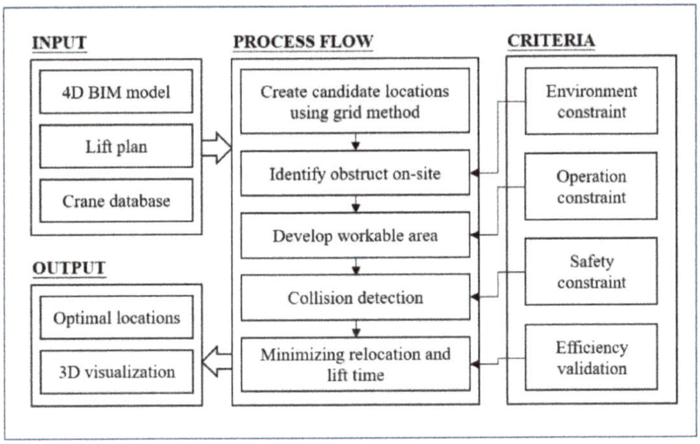

Figure 1. Logic framework of automated selection and localization of a mobile crane [42].

4. Method for Automated Selection and Localization of a Crane

4.1. Information Requirement and Extraction

The analysis of the above constraints and optimization requires a large amount of project information, most of which is contained in the 4D (four-dimensional) BIM model of a project. The 4D BIM model contains the specific location and external contour of the buildings and facilities on a construction site. This research takes prefabricated building construction as an example to explore the information requirements and extraction of crane selection and localization. For a prefabricated building, the lifting components usually involve walls, slabs, beams, etc. Relevant information can be obtained directly or indirectly from its IFC (Industry Foundation Classes) format BIM model. Table 1 displays the information regarding the building components that is extracted from the BIM model, which is required for crane selection and localization. Regarding the three constraints mentioned above, the information for each constraint is summarized in Table 2. Note that some information can be obtained directly while other information can be obtained through further calculation based on the parsed data from the BIM model, referring to our previous research [43,44]. In addition, the selection and localization calculation of a crane required its performance parameters, which are stored in the crane database and are applicable to different construction projects. According to the above constraints and optimization, the required performance parameters are summarized in Table 3.

Table 1. Parameter information of building components for crane selection and localization.

Type	Position (x, y, z), $\vec{R}$	Size	Others
Wall	Vertical: starting position Horizontal: wall center line, core layer center line, surface layer center line, surface layer internal/external line, core layer internal/external line	Length L, Depth D, Height H	ID, Floor number F, Height of floor HF, Volume V, Time T
Slab	Endpoint position	Depth D	ID, Floor number F, Height of floor HF, Volume V, Time T
Column	Column center point	Section size b*h, Height H	ID, Floor number F, Height of floor HF, Volume V, Time T
Beam	Vertical: starting position Horizontal: section center point	Section size b*h, Height H	ID, Floor number F, Height of floor HF, Volume V, Time T
Stairs	Endpoint position	-	ID, Floor number F, Height of floor HF, Volume V, Time T
Roof	Endpoint position	-	ID, Floor number F, Height of floor HF, Volume V, Time T

Table 2. Information required for the calculation of constraints.

Constraint	Required Information	Whether Directly Obtainable from the Parsed BIM Model	Data Required from the BIM Model
Environmental constraint	Outside boundary of a building's first floor	Needs conversion calculation	Coordinates and size of the first floor wall
	Boundary of main road and temporary facilities	Obtained directly	-
Operation constraint	Original and target positions of a lifting component	Needs conversion calculation	Relative position, size, direction, and height of the component
	Weight of the component	Needs conversion calculation	Material and volume of the component
Safety constraint	Height of the building, the outside boundary, and location of a crane jib	Needs conversion calculation	Time, coordinates of slab, wall and column, and location of crane

4.2. Initial Type Selection and Candidate Location Generation

As the selection and localization of mobile cranes are two indispensable steps that cannot be separated independently, an initial crane type should be selected with constraint checks. It means that appropriate cranes need to be re-selected if the initially selected crane type does not meet the constraints. The selection of the crane type mainly considers the maximum lifting weight and the cost of the crane. The weight of a lifting component can be calculated based on its volume and density (see Table 2), which are obtained directly from the IFC format BIM model. After all components C_k required for lift are traversed, the weights of different components CG_k are compared to determine the maximum weight CG_{max}. Then, CG_{max} and the maximum lifting capacity of the crane G_{max}, which is stored in the crane database, are compared to construct Equation (1). According to the results, all cranes that conform with Equation (1) are added to the alternative crane type database. As for the price, the crane type with the lowest price is selected as the initial crane type to carry out the subsequent calculation of crane localization by assuming G' = hook weight + rigging weight + weight of the other accessories:

$$G_{max} \geq CG_{max} + G' \quad (1)$$

Table 3. Parameter information of a crane required for crane selection and localization.

Constraints/ Optimization	Required Information
Environmental constraint	Outside boundary of a carne after the full extension of legs Rotation radius of the turntable
Operation constraint	Rated lifting weight corresponding to the length and radius of different boom
Minimum lifting time	Boom variation time Maximum elevation of boom Maximum speed of rotation
Lowest price of crane	Market price of crane

The location is determined according to the above three constraints and the optimization conditions. As the safety constraint and efficiency validation are analyzed by traversing each candidate location and simulating the lifting operation, the candidate locations should be represented as a set of discrete points rather than consecutive points in some areas. In this research, the candidate locations of a crane refer to a set of points resulting from the grid method. It divides a construction site into different square grids of the same size, which depends on the size of the crane [21]. The apex of each grid represents the location point of the mobile crane, where the rotation center of the crane is projected to the ground. The position of the point is represented by a three-dimensional Cartesian coordinate system (x, y, z). The size of the crane can be used as a reference for the width of the grid [26]. If the grid width is less than that of the crane, the adjacent points of the grid point will be covered when the crane is located at a certain point, and the difference between the adjacent points is small. Otherwise, if the width of the grid is larger than that of the crane, it is too sparse for the location of the crane. In addition, the candidate location points also consider the soil condition the crane sits on, where location points with a total load exceeding the land bearing capacity are deleted, which may lead to mobile cranes tipping over.

4.3. Environment Constraint Check

After generating the candidate points, a series of constraints are calculated to gradually remove the points that do not meet the requirements of the candidates; thus, a set of feasible points is obtained. The environmental constraint is tested first to define the boundaries of the buildings, temporary facilities, and main roads in the construction site, and then to eliminate the candidate points where the crane cannot be located to identify its feasible point set FL_1. For the buildings on the construction site, their flat coverage areas can be determined by the location of the ground walls. In IFC format BIM models, the location point of a wall component (x, y, z) is the starting position of the wall in the longitudinal direction and in the transverse direction. According to the modeling choices in BIM, they may be located at the center line of the wall, that of the core layer, that of the surface layer, the interior/exterior of the surface layer, or that of the core layer. An example is shown in Figure 2. The location point is located exterior to the surface layer, and points A_1 to A_6 are the vertices of the walls W1-W6 of a certain building. For wall W1, the starting point coordinates $A_1(x_1, y_1, z_1)$, the direction $\vec{R}(a, b, 0)$, and the length L can be extracted from the IFC format BIM model, and the line of the wall W1 is expressed by Equation (2). Similarly, the pattern enclosed by the walls W1-W6 is represented by Equation (3):

$$y - y_1 = \frac{b}{a}(x - x_1) \tag{2}$$

$$\begin{cases} y - y_1 = \frac{b}{a}(x - x_1), & x \in (x_1, x_2),\ y \in (y_2, y_1) \\ & \cdots\cdots \\ y - y_6 = \frac{b}{a}(x - x_6), & x \in (x_1, x_6),\ y \in (y_1, y_6) \end{cases} \tag{3}$$

where x_i and y_i are the coordinates of the wall Wi.

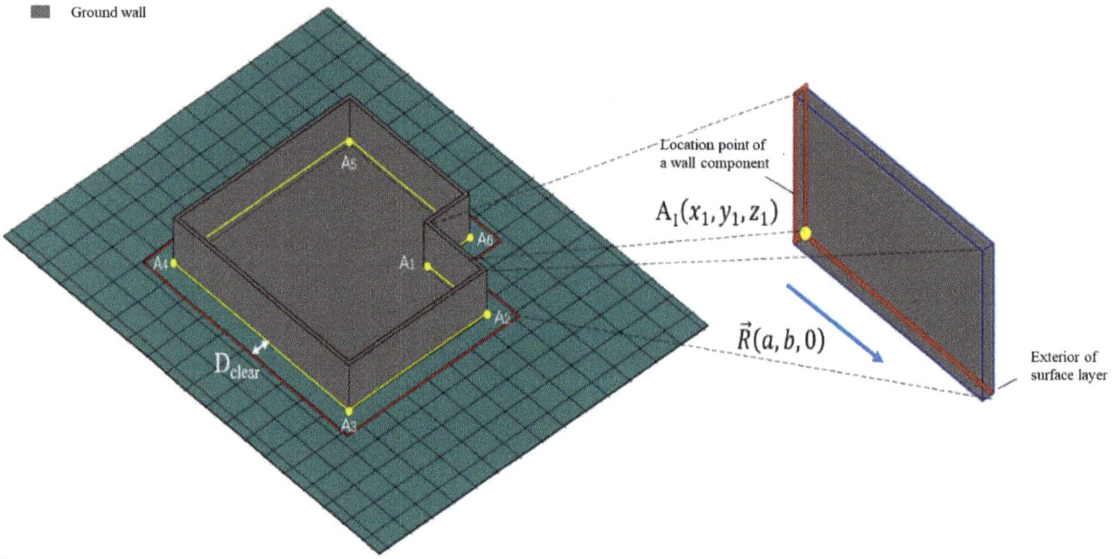

Figure 2. Building coverage area.

Moreover, an initial alternative point represents the rotation center of the crane. Thus, a certain distance from unreachable areas must exist, such as buildings, temporary facilities, and traffic routes, on the construction site. The rotation radius of the turntable is added, with a certain safety distance as the minimum distance D_{clear} between the location point and unreachable areas. The rotation radius of the turntable is determined based on the crane parameters. The safety distance is not specified in the national standard for crane operations. Thus, it can be set as a user's input parameter. The distance from a feasible location point to the nearest unreachable area is calculated. If the distance is less than D_{clear}, the feasible point will be eliminated from the point set.

4.4. Operation Constraint Check

The operation constraint test is carried out to identify whether the selected crane and the feasible alternative points meet the operation requirements. First, the maximum working radius and the maximum boom length of the crane are determined according to the weight of the building components. As shown in Table 4, the rated lifting capacity of the same crane is different under different boom lengths and working radiuses. Due to the limitation of torque, the larger the boom length or the working radius is, the weaker the lifting capacity is. Therefore, according to the original and target positions of the component and the maximum working radius, the location range of the crane can be determined.

Table 4. Lift capacity of a QY100H-3 mobile crane [42,45] (T).

Working Radius (m)	Boom Length (m)								
	13.0	**17.8**	**22.5**	**27.2**	**31.9**	**36.6**	**41.3**	**46.0**	**50.4**
3.0	100	80.0							
3.5	93.0	77.0	62.0						
4.0	88.0	72.0	62.0						
4.5	79.0	67.0	61.0	42.0					
5.0	72.0	62.0	60.0	42.0	40.0				
5.5	65.0	58.0	56.0	42.0	39.0				
6.0	59.0	55.0	52.0	42.0	37.5	31.5			
6.5	54.0	52.0	48.2	40.5	35.8	31.0			
7.0	50.0	49.0	45.0	39.0	34.5	29.5			
7.5	46.0	45.0	42.5	37.0	33.0	28.7			
8.0	42.0	41.0	40.5	35.5	31.8	27.6	23.5		
9.0	36.5	35.5	35.0	32.5	29.5	25.7	22.0	18.5	
10.0	32.0	31.0	30.5	30.0	27.5	24.0	20.8	17.5	
11.0		27.5	26.5	27.5	25.7	22.6	19.5	16.5	14.0
12.0		23.5	23.3	24.5	24.0	21.2	18.9	15.9	13.2
14.0		17.5	17.0	18.5	19.5	18.8	16.9	14.5	12.2
16.0			13.0	14.2	15.0	16.0	15.2	13.2	11.2
18.0			10.0	11.2	12.0	12.6	13.2	12.0	10.2
20.0				9.0	9.7	10.3	10.9	11.0	9.3
22.0				7.2	7.9	8.5	9.0	9.4	8.7
24.0					6.2	7.0	7.6	7.9	8.0
26.0					5.0	5.8	6.3	6.5	6.9
28.0						4.9	5.2	5.6	5.8
30.0						3.9	4.3	4.8	4.9
32.0						3.0	3.6	3.9	4.2
34.0							2.8	3.2	3.6
36.0							2.2	2.7	2.9
38.0								2.2	2.4
40.0								1.8	1.9
42.0									1.6

The maximum working radius R_{max} and the maximum boom length L_{max} of the crane can be automatically determined by the following procedure. From small to large, the rated lifting capacity $G_{m,n}$ corresponding to a different working radius R_n and boom length L_m is invoked and compared with the maximum weight CG_{max} of the lifting components to check whether Equation (4) is satisfied. When $G_{m,n}$ under a certain boom length does not satisfy Equation (4), the comparison between the rated lifting weight of the next level of the boom length and the maximum weight CG_{max} starts, until the comparison of each level of the boom length is conducted. Based on this, the maximum working radius R_{max} and the maximum boom length L_{max} of the crane is determined:

$$G_{m,n} \geq CG_{max} + G', \tag{4}$$

where $G_{m,n}$ is the rated lifting capacity of a crane under a working radius R_n and boom length L_m; and m and n are the level of the boom length and working radius, respectively.

Once the maximum working radius is determined, the plane working range of the crane is also determined. The components whose horizontal distance from the crane is within the maximum working radius can be lifted. Therefore, the distance between the location point of the crane and the original or target position of components should be less than the maximum working radius R_{max}. This research focuses on the prefabricated lifting components; thus, the target position of a lifting component corresponds to its design position in the building model, and its original location is determined by the hoisting scheme, which is presented in the 4D BIM model. Considering that the positions of different components are represented differently in the BIM model, and their target positions for a crane are represented by their centers of gravity, different coordinate conversions are needed.

1. For a column component, its position coordinates in the BIM model are its bottom center coordinates. Thus, only the z-axis coordinate needs to be re-calculated to determine its target position coordinates. Taking column C_1 as an example, based on its position coordinates (x_1, y_1, z_1) in the BIM model and its height H and floor height HF obtained from the BIM model, the coordinates of its gravity center are determined as $(x_1, y_1, HF + H/2)$.
2. For a regular rectangular slab, the coordinates of its gravity center can be calculated from the coordinates of each endpoint of the slab in the model. Taking slab P as an example, based on two diagonal endpoint coordinates (x_1, y_1, z_1) and (x_2, y_2, z_2), the height H, the floor height HF, and the slab depth D from the BIM model, the coordinates of its gravity center are $(x_1 + x_2/2, (y_1 + y_2)/2, HF + H + D/2)$.
3. For wall and beam components in the form of tension, since the position coordinates in the BIM model are the coordinates of the tension starting point, it is necessary to determine the changed plane coordinates of its gravity center. Taking the wall W_1 as an example, based on the starting point coordinates (x_1, y_1, z_1), direction $\vec{R}(a, b, 0)$, length L, height H, and floor height HF from the BIM model, the coordinates of its gravity center (x_1', y_1', z_1') can be calculated using Equation (5):

$$\begin{cases} x_1' = x_1 + \frac{L}{2} \times \frac{a}{\sqrt{a^2+b^2}} \\ y_1' = y_1 + \frac{L}{2} \times \frac{b}{\sqrt{a^2+b^2}} \\ z_1' = HF + \frac{H}{2} \end{cases} \quad (5)$$

After the original and target positions of the building components are determined, for each component C_k to be lifted, its original and target positions are taken, respectively, as the centers of two circles and the maximum working radius R_{max} of the crane is the radius of the circles. As a result, the overlapping area of the two circles is obtained, i.e., the feasible location area of the crane. This is because the distance between any point in the overlapping area and the two centers is less than the maximum working radius of the crane. As an example, Figure 3 shows an intersection for lifting a wall component, involving feasible location points. By following the original feasible point set FL_1, the feasible point set $FL_{2,k}$ for component C_k can be obtained. By traversing all components, the original feasible points that do not belong to any feasible point set $FL_{2,k}$ are eliminated to obtain a feasible point set FL_2. On the other hand, if the two circles do not intersect, meaning that the working radius of the crane does not meet the lifting requirements, the crane needs to be eliminated. Another crane with an increased load capacity would be loaded for the above constraint tests to be performed again.

4.5. Safety Constraint Check

The above two constraint tests screen out the basic feasible location points of the crane based on a 2D plane, which meets the basic requirements, such as accessibility and operability. However, safety problems may exist in crane operations and need to be solved. By traversing all feasible points $FL_{2,k}$ of a component C_k, the collisions between the crane boom and other objects during operation can be identified to prevent safety accidents. The spatial position of the crane boom at the beginning and the end of lifting operations should firstly be determined. For each feasible point of a component C_k, since there are various combinations of boom lengths and working radiuses, to simplify the calculation, the maximum boom length L_{max} obtained in the previous section is selected as its working boom length for calculation. If the maximum boom length does not meet the requirements of the safety constraints, collisions will occur during operations and the relevant location point should be eliminated. If all location points fail to meet the requirements, it means that the crane does not meet the safety requirements. As an example, Figure 4 shows the position of a lifting boom at the end of lifting work in the plan and space. The plane position can be determined using the plane coordinates of an alternative feasible point P_{ij} (x_i,y_j) and the final position of components (x',y') using Equation (6). In terms of the spatial position, the angle of the boom θ is calculated using the distance between the target position of the component and the feasible location point of the crane and the length of the boom L_{max} based on Equation (7). Moreover, the BIM model is used to obtain construction environment information, including the plane positions of building components when component C_k is lifted at time t_k. Meanwhile, all components', such as beams, columns, walls, and slabs, which might intersect with the crane boom in the plane, maximum heights are selected:

$$ax + by + c = 0, \tag{6}$$

where a, b, and c are constants.

$$\theta = \cos^{-1}\frac{r}{L_{max}}, \tag{7}$$

where $r = \sqrt{(x_i - x')^2 + (y_j - y')^2} + d_{OO'}$; $d_{OO'}$ is the horizontal distance between the rotation center line of the crane boom and the center point of the turntable.

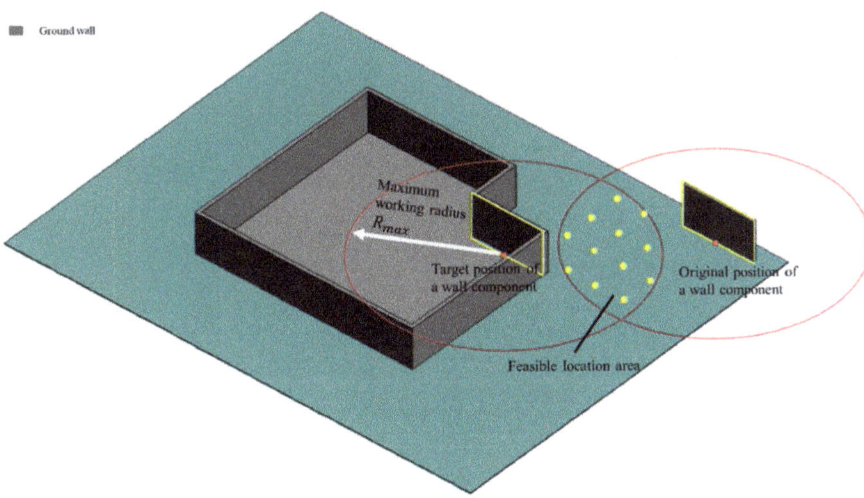

Figure 3. Operation constraint check.

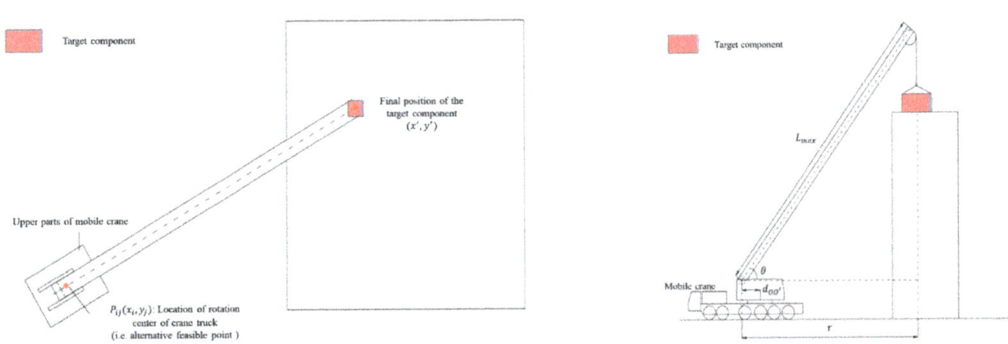

(a) (b)

Figure 4. Crane position in the plan and space. (a) Lifting boom in the plan, (b) Lifting boom in space.

Figure 5 shows an example of the hoist of component C_k. The boom intersects with the exterior wall of the building in the vertical plane at time t. Relevant parameters include the height of the intersection wall H_w, the floor height HF, the coordinates of the starting point of the wall $A_1(x_1, y_1, z_1)$, the direction $\vec{R}(a_1, b_1, 0)$, the wall depth D, and the component height H_c. Assuming that the center line of the wall is the positioning line, the center line of the wall can be expressed by Equation (8). Equations (6) and (8) can be combined to obtain the plane coordinates of the intersection point (x_w, y_w). Then, the plane distance between the location point P_{ij} of the crane and the surface of the wall can be calculated using Equation (9). Finally, Equation (10) is used to identify whether the lifting boom collides with the wall or the component itself. If Equation (10) is satisfied, collisions can be avoided effectively; otherwise, a collision problem will occur. By traversing all feasible points of each component C_k, the above-mentioned safety constraints are calculated to eliminate the points with collisions; thus, the feasible location point set $FL_{3,k}$ is obtained:

$$y - y_1 = \frac{b_1}{a_1}(x - x_1), \tag{8}$$

$$d = \sqrt{(x_i - x_w)^2 + (y_j - y_w)^2} + d_{OO'} - \frac{D}{2\sin\beta}, \tag{9}$$

where $\beta = \tan^{-1}\frac{b}{a} - \tan^{-1}\frac{b_1}{a_1}$.

$$\begin{cases} (d \times \tan\theta + H_{base} - (HF + H_w))\cos\theta \geq d_{clear} \\ (r \times \tan\theta + H_{base} - (HF + H_w + H_c))\cos\theta \geq d_{clear} \end{cases}, \tag{10}$$

where d_{clear} is the minimum distance between the crane boom and the obstacle; H_{base} is the distance between the rotation axis of the crane boom and the ground; and θ is the elevation angle of the boom.

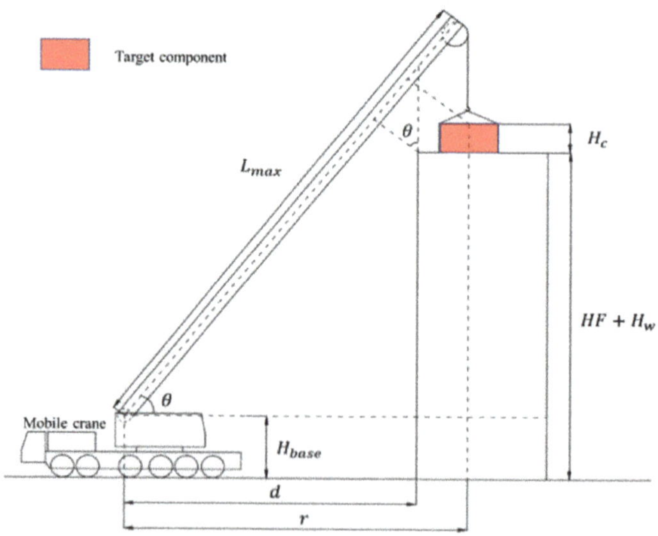

Figure 5. Safety constraint check.

4.6. Lifting Efficiency Optimization

Lifting efficiency is affected by two factors (i.e., the lifting time of a single component and relocation times of a mobile crane). The former consists of the boom variable amplitude time and boom extension time while the latter mainly depends on the lifting schedule stored in 4D BIM.

(1) Minimization of single lifting time

The lifting time of a crane at each location point depends on its work parameters, including the boom variable amplitude time, the boom extension time, the maximum rotation speed, the maximum lifting speed, etc. In a construction site, the boom is preferably stretched without loads, and the lifting height of the hook is independent of the location point and only related to the target installation position of the lifting components. Therefore, the calculation of a single lifting time mainly considers the boom amplitude time and the turntable rotation time. For the point in the feasible point set $FL_{3,k}$ of each component C_k, the boom pitch rotation angle difference $d\theta$ and the turntable rotation angle ω of the crane are firstly calculated. As mentioned above, the initial position coordinates (x_1, y_1, z_1) and final position coordinates (x_2, y_2, z_2) of the component; the boom elevation angles θ_1 and θ_2 of the crane at the beginning and ending operation, respectively; and the position of the lifting boom on the plane $a_1x + b_1y + c_1 = 0$ and $a_2x + b_2y + c_2 = 0$ are obtained. Therefore, the values of $d\theta$ and ω can be calculated using Equations (11) and (12). Then, the lifting time at the point is calculated using Equation (13). The lifting efficiency is calculated successively for several feasible points of the same component and stored in the database:

$$d\theta = |\theta_1 - \theta_2|, \tag{11}$$

$$\omega = \left|\tan^{-1}\frac{b_1}{a_1} - \tan^{-1}\frac{b_2}{a_2}\right| \tag{12}$$

$$T_L = \frac{T_{raising}}{\varphi} \times d\theta + v_{swing} \times \omega, \tag{13}$$

where $T_{raising}$ is the lifting time of the boom; φ is the maximum elevation angle of the lifting boom; and v_{swing} is the rotation speed of the turntable.

(2) Minimization of relocation times

The principle of minimizing relocation times is to find the coincident location points in the feasible points for different lifting components, so that more lifting operations can be performed with the same crane location. It should be noted that due to the complexity of lifting operation, relocation is unavoidable in practical construction even with the minimizing principle. Therefore, according to the lifting schedule, which is stored in 4D BIM, the method used in this research is to iteratively traverse the feasible point set of all components FL_3 and check its attributes. Since the attribute k is added to the feasible point of each component C_k in the operation constraint test, the feasible points with the greatest attribute value is acceptable. The specific procedure is presented as follows.

1. Check the attribute value k of each feasible point P_{ij} in turn, and add points with the same attribute to the same group.
2. Find the group $Group_m$ that contains the most contiguous attributes. Since the attribute value k is the number of components and also represents the lifting order of components, the lifting order should be considered to ensure that the attribute value is as continuous as possible.
3. Compare the minimum lifting time when the crane is located at the points of multiple $Group_m$, and select the group with the shortest lifting time as the final $Group_m$. If no multiple $Group_m$ exist, go straight to the next step.
4. Take point $P_{ij,k}$ in $\mathrm{Group_m}$ as the location point of the crane for the hoist of the component C_k, and release the attribute value k contained in other points at the same time.
5. Identify the remaining feasible point with the greatest attribute k and repeat the above steps.
6. Identify the crane location point combinations with the minimum number of movements after completing the above steps for all feasible points, calculate the lifting time of each point in the same group of location points, and take the point combination with the shortest lifting time.

Through the above steps, the optimal location point of the mobile crane with minimum relocations and the shortest lifting time can be identified.

5. Case Study

To test the feasibility and validity of the proposed method, a crane selection and localization system were developed based on Unity 3D, including the input module, selection module, positioning module, and optimization module, and a case of a 3-story prefabricated building project was adopted, involving the integrated selection and localization of a mobile crane under a lifting scenario of multi-components. In this case, the BIM model of the building was constructed in advance, with various types of precast components (see Figure 6a). In addition, site roads, temporary facilities, and relevant coordinates were directly extracted from the parsed IFC format BIM model, marked in the construction site layout (see Figure 6b). Table 5 lists some candidate mobile cranes and their basic performances stored in the crane database.

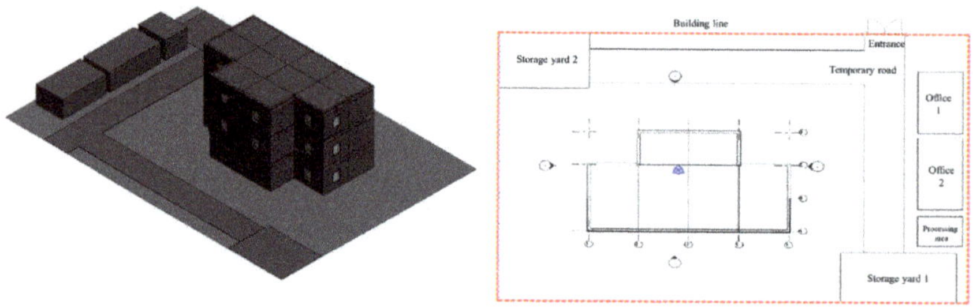

(a) (b)

Figure 6. BIM model and site layout (**a**) BIM model of the building with precast components. (**b**) Roads and temporary facilities in the construction site layout.

Table 5. Basic performances of the candidate cranes in the database.

Crane Model	Maximum Lifting Capacity (T)	Maximum Boom Length (m)	Maximum Working Radius (m)	Cost (Yuan/per Machine)
QY16D	16	30.5	22	1100
QY20G	20	32.27	28	1360
QY25K-I	25	33	30	1800
QY50K-II	50	42.7	32	2960
QY65K	65	42	30	3500
QY70K	70	44.5	36	4780
QY80K	80	45	36	5000
QY90K	90	55	50	5800
QY100K-I	100	51	42	6833
QY130K	130	58	56	6900
QY160K	160	62	52	7000

5.1. Data Input

The IFC format BIM model was firstly imported into the system using its input module to automatically obtain the attributes of all precast components, and the coordinates of the obstacles and stacking yards in the construction site. Referring to the range of building, facilities, and site, an initial candidate location point set for the mobile crane was generated automatically to support the subsequent selection and localization of the crane.

5.2. Initial Crane Type Selection

According to the maximum weight of the precast components, 4 types of mobile cranes were selected based on Equation (1), including QY90K, QY100K-I, QY130K, and

QY160K, and added to the alternatives. QY90K was automatically selected as the initial crane type by the selection module because of its lowest price. Then, the performance parameters of the selected mobile crane were extracted from the crane database to support the subsequent calculation.

5.3. Constraint Checks

The environmental constraint check, operation constraint check, and safety constraint check were performed orderly using the positioning module to check the feasibility of the selected crane type. The points satisfying the three constraints were set as the feasible location points, which meant these components could be hoisted safely. However, it was found that in terms of the initial crane type QY90K, all constraints with the lifting components could not be satisfied. Thus, the crane type had to be eliminated from the alternatives. Figure 7 shows the scenarios in which crane QY90K was inconsistent with the three constraints. Specifically, Figure 7a shows the crane was located within the boundaries of the existing buildings, which is defined by the minimum clearance, Figure 7b shows the weight of the lifting component that exceeded the rated lifting capacity, and Figure 7c shows there was a collision between the crane boom and the building during lifting operation. Another alternative crane type QY100K-I with the minimum cost was loaded as a new initial type for repeated constraint testing. As a result, this crane type conformed with all of the constraints, and the feasible location points were obtained (see Figure 8).

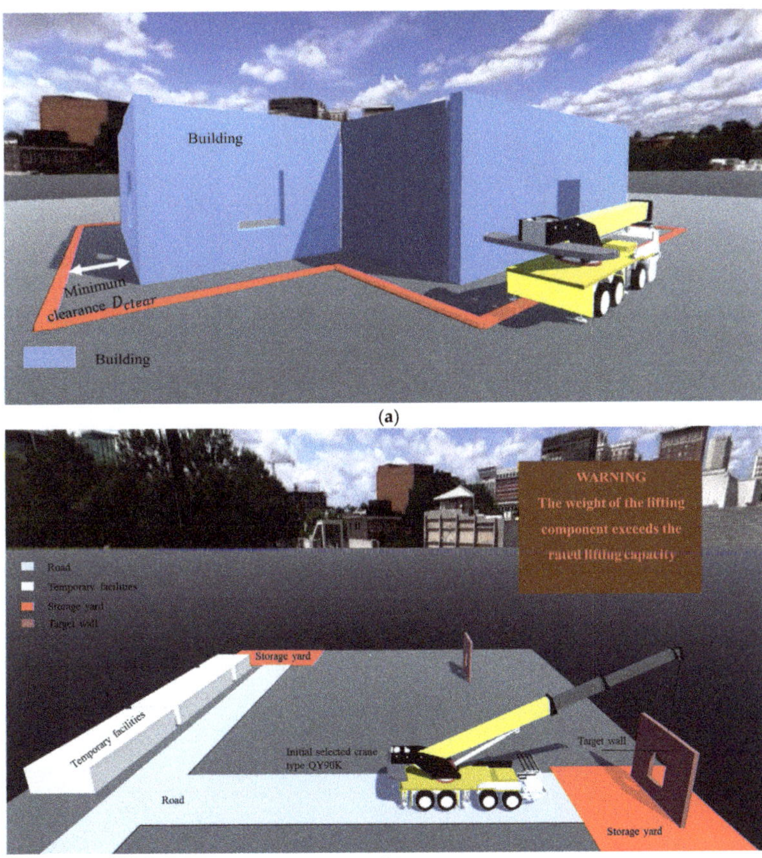

(a) (b)

Figure 7. *Cont.*

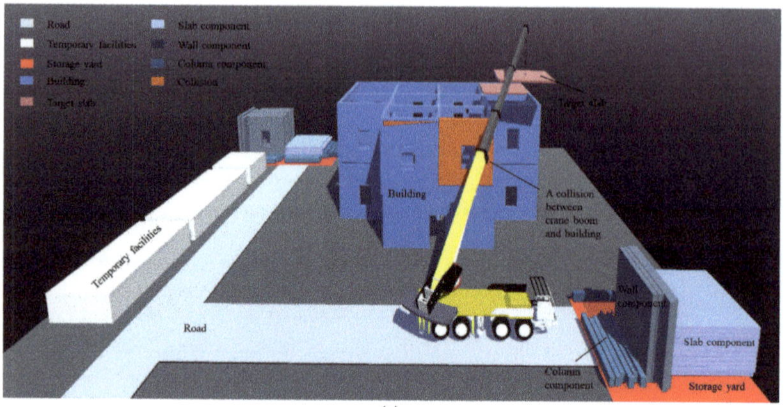

(c)

Figure 7. The initially selected type QY90K, which does not satisfy the constraints. (**a**) Unsatisfied environmental constraint. (**b**) Unsatisfied operation constraint. (**c**) Unsatisfied safety constraint.

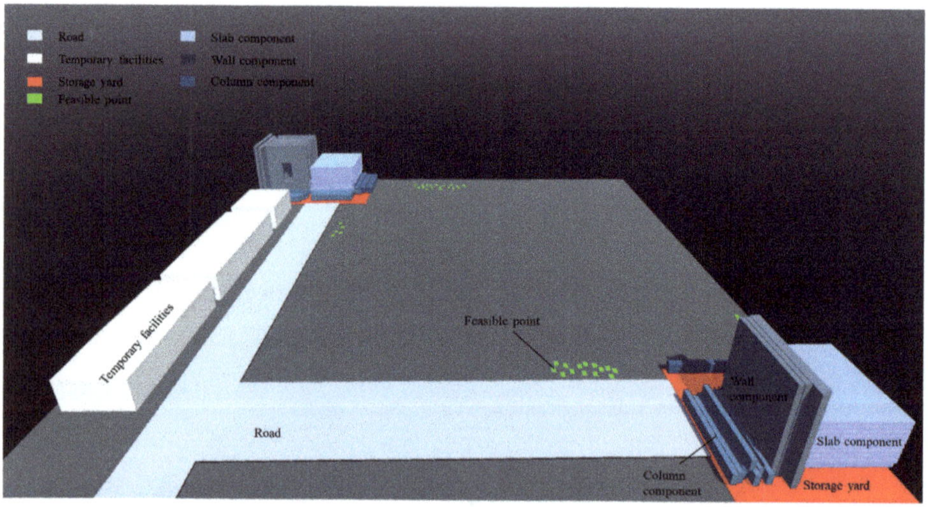

Figure 8. Feasible location points for a certain component (crane type QY100K-I).

5.4. Optimal Combination of Feasible Location Points

Based on the obtained feasible points for all lifting components, the combination with the minimum relocations of 2 (i.e., the optimal location point 1 and 2 in Figure 9) and the shortest lifting time was determined using the optimization module. The optimal locations were highlighted in the construction site layout. As shown in Figure 9, the highlighted combination of feasible points represents the optimal locations for the selected mobile crane QY100K-I.

It is shown from this case study that the proposed method is workable. Moreover, the selection and localization processes of the mobile crane can be automatically conducted by the developed system. Thus, it makes the selection and localization of the mobile crane more efficient than before, not only saving cost and time but also improving the safety performance. Furthermore, this proposed method has the potential to support the automatic simulation of construction processes, which further supports the implementation of intelligent construction.

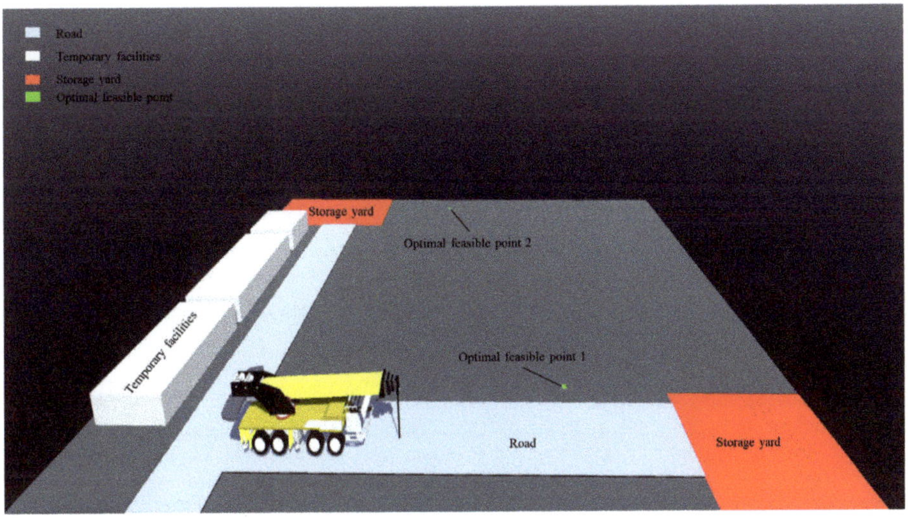

Figure 9. Optimal combination of feasible location points for crane QY100K-I.

6. Conclusions

The promotion of virtual construction technologies has the potential to improve the efficiency and safety of crane operation by producing a more precise, comprehensive, and dynamic operation plan. However, in terms of the selection and localization of the mobile crane, it seldom considers the combination of different feasible location points, the minimization of crane movement, and the spatial safety requirements during lifting processes. Consequently, this study developed a method to automatically select and locate the mobile crane, mainly involving three constraints, i.e., the environmental constraint, operation constraint, and safety constraint, and two optimization methods covering the least cost, minimum relocations, and shortest lifting time. Relevant parameter information can be directly obtained or indirectly transformed from the 4D BIM model. A case study was also presented to test the feasibility and validity of the proposed method. The result shows that the method was feasible and valid in the virtual simulation environment. This benefits not only on-site crane operations but also automatic construction simulation. Moreover, compared with other construction machinery, mobile cranes have more constraints and more complicated operation. Thus, it is easier to extend the method to other construction machinery.

In addition, there are still some limitations. On the one hand, the proposed method was only tested in a virtual construction scenario and the optimized result has not yet been used in real construction. On the other hand, the method only considered the construction scenario with a single mobile crane. Thus, future research will focus on the use of optimized results from the method in real construction scenarios to test its performance, with more criteria (i.e., wind effect, ground bearing pressure, and rope breakage) considered. Moreover, multi-crane construction scenarios will also be taken into consideration in the future.

Author Contributions: Conceptualization, H.G. and Y.Y.; methodology, Y.Z. and Z.P.; validation, Y.L. and Z.Z.; formal analysis, Z.Z.; investigation, Y.Z.; writing—original draft preparation, H.G. and Y.Z.; writing—review and editing, Y.Y. and Y.Z.; visualization, Y.Z. and Y.Y.; supervision, H.G. and Y.Y.; project administration, H.G.; funding acquisition, H.G. All authors have read and agreed to the published version of the manuscript.

Funding: This research was funded by the National Natural Science Foundation of China, grant number 51578318 and 51208282.

Institutional Review Board Statement: Not applicable.

Informed Consent Statement: Not applicable.

Data Availability Statement: Some or all data, models, or code that support the findings of this study are available from the corresponding author upon reasonable request.

Acknowledgments: We would like to thank Tsinghua-Glodon BIM Research Center for supporting this research. This paper is an extension of a conference paper *"Automated Method for Optimizing Feasible Locations of Mobile Cranes Based on 3D Visualization"* [42] in Creative Construction Conference (CCC) 2017.

Conflicts of Interest: The authors declare no conflict of interest.

References

1. Kang, S.; Miranda, E. Planning and visualization for automated robotic crane erection processes in construction. *Autom. Constr.* **2006**, *15*, 398–414. [CrossRef]
2. Lacey, A.; Chen, W.; Hao, H.; Bi, K. Structural response of modular buildings–An overview. *J. Build. Eng.* **2018**, *16*, 45–56. [CrossRef]
3. Olearczyk, J.; Al-Hussein, M.; Bouferguene, A.; Telyas, A. 3D-Modeling for Crane Selection and Logistics for Modular Construction On-Site Assembly. In *Computing in Civil Engineering*; ASCE: Reston, VA, USA, 2012. [CrossRef]
4. Olearczyk, J.; Al-Hussein, M.; Bouferguène, A. Evolution of the crane selection and on-site utilization process for modular construction multilifts. *Autom. Constr.* **2014**, *43*, 59–72. [CrossRef]
5. Lei, Z.; Han, S.; Bouferguene, A.; Taghaddos, H.; Hermann, U.; Al-Hussein, M. Algorithm for Mobile Crane Walking Path Planning in Congested Industrial Plants. *J. Constr. Eng. Manag.* **2015**, *141*, 05014016. [CrossRef]
6. Lin, Y.; Yu, H.; Sun, G.; Shi, P. Lift Path Planning without Prior Picking/Placing Configurations: Using Crane Location Regions. *J. Comput. Civ. Eng.* **2016**, *30*, 04014109. [CrossRef]
7. Kang, S.; Miranda, E. Computational methods for coordinating multipole construction cranes. *J. Comput. Civ. Eng.* **2008**, *22*, 252–263. [CrossRef]
8. Fang, Y.; Teizer, J. A Multi-User Virtual 3D Training Environment to Advance Collaboration Among Crane Operator and Ground Personnel in Blind Lifts. In *Computing in Civil and Building Engineering*; ASCE: Reston, VA, USA, 2014. [CrossRef]
9. Chen, Y.-C.; Chi, H.-L.; Kang, S.-C.; Hsieh, S.-H. Attention-Based User Interface Design for a Tele-Operated Crane. *J. Comput. Civ. Eng.* **2016**, *30*, 04015030. [CrossRef]
10. Al-Hussein, M.; Niaz, M.A.; Yu, H.; Kim, H. Integrating 3D visualization and simulation for tower crane operations on construction sites. *Autom. Constr.* **2006**, *15*, 554–562. [CrossRef]
11. Mara, T.G. Effects of a Construction Tower Crane on the Wind Loading of a High-Rise Building. *J. Struct. Eng.* **2010**, *136*, 1453–1460. [CrossRef]
12. Ali, G.M.; Kosa, J.; Bouferguene, A.; Al-Hussein, M. Competitive Assessment of Ice and Frozen Silt Mat for Crane Ground Support Using Finite-Element Analysis. *J. Constr. Eng. Manag.* **2021**, *147*, 04021038. [CrossRef]
13. Ali, G.M.; Olearczyk, J.; Bouferguene, A.; Al-Hussein, M. Implementation of Combined Loading to Calculate Ground Bearing Pressure under Crawler Crane Tracks. *J. Constr. Eng. Manag.* **2021**, *147*, 04021051. [CrossRef]
14. Yu, G.Y.H. Forensic investigation on crane accidents. *Int. J. Forensic Eng.* **2017**, *3*, 319–341. [CrossRef]
15. Wang, R.D.; Zayed, T.; Pan, W.; Zheng, S.; Tariq, S. A system boundary-based critical review on crane selection in building construction. *Autom. Constr.* **2020**, *123*, 103520. [CrossRef]
16. Al-Hussein, M.; Alkass, S.; Moselhi, O. An algorithm for mobile crane selection and location on construction sites. *Constr. Innov.* **2001**, *1*, 91–105. [CrossRef]
17. Moselhi, O.; Alkass, S.; Al-Hussein, M. Innovative 3D-modelling for selecting and locating mobile cranes. *Eng. Constr. Arch. Manag.* **2004**, *11*, 373–380. [CrossRef]
18. Al-Hussein, M.; Alkass, S.; Moselhi, O. Optimization Algorithm for Selection and on Site Location of Mobile Cranes. *J. Constr. Eng. Manag.* **2005**, *131*, 579–590. [CrossRef]
19. Wu, D.; Lin, Y.; Wang, X.; Wang, X.; Gao, S. Algorithm of Crane Selection for Heavy Lifts. *J. Comput. Civ. Eng.* **2011**, *25*, 57–65. [CrossRef]
20. Hasan, S.; Al-Hussein, M.; Hermann, U.H.; Safouhi, H. Interactive and Dynamic Integrated Module for Mobile Cranes Supporting System Design. *J. Constr. Eng. Manag.* **2010**, *136*, 179–186. [CrossRef]
21. Han, S.H.; Hasan, S.; Bouferguène, A.; Al-Hussein, M.; Kosa, J. Utilization of 3D Visualization of Mobile Crane Operations for Modular Construction On-Site Assembly. *J. Manag. Eng.* **2015**, *31*, 04014080. [CrossRef]
22. Han, S.; Al-Hussein, M.; Hasan, S.; Gökçe, K.U.; Bouferguene, A. Simulation of mobile crane operations in 3D space. In Proceedings of the 2012 Winter Simulation Conference, Berlin, Germany, 9–12 December 2012.
23. Furusaka, S.; Gray, C. A model for the selection of the optimum crane for construction sites. *Constr. Manag. Econ.* **1984**, *2*, 157–176. [CrossRef]
24. Han, S.; Bouferguene, A.; Al-Hussein, M.; Hermann, U. 3D-based crane evaluation system for mobile crane operation selection on mod-ular-based heavy construction sites. *J. Constr. Eng. Manag.* **2017**, *143*, 04017060. [CrossRef]

25. Zaki, T.M.; Hosny, O.; Nassar, K. An automated model for selecting the optimum mobile crane model and on-site position using genetic algorithms. In Proceedings of the Canadian Society for Civil Engineering's 5th International/11th Construction Specialty Conference, Vancouver, BC, Canada, 8–10 June 2015; p. 128.
26. Taghaddos, H.; AbouRizk, S.; Mohamed, Y.; Hermann, U. Simulation-Based Multiple Heavy Lift Planning in Industrial Construction. In *Construction Research Congress*; ASCE: Reston, VA, USA, 2010; pp. 349–358. [CrossRef]
27. Raynar, K.A.; Smith, G.R. Intelligent Positioning of Mobile Cranes for Steel Erection. *Comput. Civ. Infrastruct. Eng.* **1993**, *8*, 67–74. [CrossRef]
28. Tantisevi, K.; Akinci, B. Simulation-Based Identification of Possible Locations for Mobile Cranes on Construction Sites. *J. Comput. Civ. Eng.* **2008**, *22*, 21–30. [CrossRef]
29. Lei, Z.; Taghaddos, H.; Olearczyk, J.; Al-Hussein, M.; Hermann, U. Automated Method for Checking Crane Paths for Heavy Lifts in Industrial Projects. *J. Constr. Eng. Manag.* **2013**, *139*, 04013011. [CrossRef]
30. Ding, L.; Zhou, Y.; Akinci, B. Building Information Modeling (BIM) application framework: The process of expanding from 3D to computable nD. *Autom. Constr.* **2014**, *46*, 82–93. [CrossRef]
31. Hermann, U.; Hendi, A.; Olearczyk, J.; Al-Hussein, M. An Integrated System to Select, Position, and Simulate Mobile Cranes for Complex Industrial Projects. In Proceedings of the Construction Research Congress 2010, Banff, AB, Canada, 8–10 May 2010; pp. 267–276.
32. Han, S.; Lei, Z.; Bouferguene, A.; Al-Hussein, M.; Hermann, U. 3D Visualization-Based Motion Planning of Mobile Crane Operations in Heavy Industrial Projects. *J. Comput. Civ. Eng.* **2016**, *30*, 04014127. [CrossRef]
33. Kang, S.; Miranda, E. Numerical methods to simulate and visualize detailed crane ctivities. *Comput. Aided Civ. Infrastruct. Eng.* **2009**, *24*, 169–185. [CrossRef]
34. Tantisevi, K.; Akinci, B. Automated generation of workspace requirements of mobile crane operations to support conflict de-tection. *Autom. Constr.* **2007**, *16*, 262–276. [CrossRef]
35. Wang, X.; Liu, J.; Liu, F.; Gao, S. Collision-free locating of mobile cranes in 3D lifting system. In Proceedings of the ASME 2010 International Design Engineering Technical Conferences and Computers and Information in Engineering Conference, Montreal, QC, Canada, 15–18 August 2010; pp. 479–486.
36. Zi, B.; Lin, J.; Qian, S. Localization, obstacle avoidance planning and control of a cooperative cable parallel robot for multiple mobile cranes. *Robot. Comput. Manuf.* **2015**, *34*, 105–123. [CrossRef]
37. Ji, Y.; Leite, F. Optimized planning approach for multiple tower cranes and material supply points using mixed-integer programming. *J. Constr. Eng. Manag.* **2020**, *146*, 04020007. [CrossRef]
38. Briskorn, D.; Dienstknecht, M. Mixed-integer programming models for tower crane selection and positioning with respect to mutual interference. *Eur. J. Oper. Res.* **2018**, *273*, 160–174. [CrossRef]
39. Yeoh, J.K.W.; Chua, D.K.H. Optimizing Crane Selection and Location for Multistage Construction Using a Four-Dimensional Set Cover Approach. *J. Constr. Eng. Manag.* **2017**, *143*, 04017029. [CrossRef]
40. Lin, J.; Fu, Y.; Li, R.; Lai, W. An Algorithm for Optimizing the Location and Type Selection of Attached Tower Cranes Based on Value Engineering. In *ICCREM 2020: Intelligent Construction and Sustainable Buildings*; ASCE: Reston, VA, USA, 2020; pp. 106–117. [CrossRef]
41. Yoon, S.; Park, M.; Jung, M.; Hyun, H.; Ahn, S. Multi-objective optimization model for tower crane layout planning in modular con-struction. *Korean J. Constr. Eng. Manag.* **2021**, *22*, 36–46.
42. Pan, Z.; Guo, H.; Li, Y. Automated Method for Optimizing Feasible Locations of Mobile Cranes Based on 3D Visualization. *Procedia Eng.* **2017**, *196*, 36–44. [CrossRef]
43. Zhou, Y.; Guo, H.; Ma, L.; Zhang, Z.; Skitmore, M. Image-based onsite object recognition for automatic crane lifting tasks. *Autom. Constr.* **2020**, *123*, 103527. [CrossRef]
44. Hongling, G.; Ying, Z.; Xiaotian, Y.; Zhubang, L.; Fan, X. Automated mapping from an IFC data model to a relational database model. *J. Tsinghua Univ.* **2021**, *61*, 152–160.
45. Available online: https://cn.auto-che.com/b/puyuan/truck-crane.html (accessed on 22 March 2022).

buildings

MDPI

Article

Visual Relationship-Based Identification of Key Construction Scenes on Highway Bridges

Chen Wang, Jingguo Lv *, Yu Geng and Yiting Liu

School of Geomatics and Urban Spatial Informatics, Beijing University of Civil Engineering and Architecture, Beijing 102612, China; 2108160220002@stu.bucea.edu.cn (C.W.); 201904020125@stu.bucea.edu.cn (Y.G.); 202003020130@stu.bucea.edu.cn (Y.L.)
* Correspondence: lvjingguo@bucea.edu.cn

Abstract: Highway bridges play an important role in traffic construction; however, accidents caused by bridge construction occur frequently, resulting in significant loss of life and property. The identification of bridge construction scenes not only keeps track of the construction progress, but also enables real-time monitoring of the construction process and the timely detection of safety hazards. This paper proposes a deep learning method in artificial intelligence (AI) for identifying key construction scenes of highway bridges based on visual relationships. First, based on the analysis of bridge construction characteristics and construction process, five key construction scenes are selected. Then, by studying the underlying features of the five scenes, a construction scene identification feature information table is built, and construction scene identification rules are formulated. Afterward, a bridge key construction scene identification model (CSIN) is built; this model comprises target detection, visual relationship extraction, semantic conversion, scene information fusion, and identification results output. Finally, the effectiveness of the proposed method is verified experimentally. The results show that the proposed method can effectively identify key construction scenes for highway bridges with an accuracy rate of 94%, and enable the remote intelligent monitoring of highway bridge construction processes to ensure that projects are carried out safely.

Keywords: construction scene identification; visual relationship detection; scene rules; deep learning; neural networks; highway bridges

Citation: Wang, C.; Lv, J.; Geng, Y.; Liu, Y. Visual Relationship-Based Identification of Key Construction Scenes on Highway Bridges. *Buildings* **2022**, *12*, 827. https://doi.org/10.3390/buildings12060827

Academic Editor: Heap-Yih Chong

Received: 29 April 2022
Accepted: 9 June 2022
Published: 14 June 2022

Publisher's Note: MDPI stays neutral with regard to jurisdictional claims in published maps and institutional affiliations.

Copyright: © 2022 by the authors. Licensee MDPI, Basel, Switzerland. This article is an open access article distributed under the terms and conditions of the Creative Commons Attribution (CC BY) license (https://creativecommons.org/licenses/by/4.0/).

1. Introduction

Highway bridge construction is an important element of road transport, and plays an increasingly significant role in the development of the transportation sector. In the actual bridge construction process, the complex operating on-site environments, large numbers of construction personnel, and irregular operation of equipment often lead to major safety accidents [1], resulting in significant life and economic losses to societies and families [2]. Therefore, the identification of workers, equipment, and the behavioral relationship between workers and equipment at bridge construction sites, and thus, the inference of the current construction scene, has important application value for construction safety prevention.

The earliest methods used for construction safety monitoring relied primarily on manual monitoring during construction and safety assessment after completion [3]. However, owing to factors such as a wide working area, the large number of people on the construction site, and the complexity of the equipment used, a reliance only on manual point-to-point monitoring is often time-consuming and labor-intensive, and the monitoring results are prone to error.

Most current researchers use deep learning methods in artificial intelligence (AI) for safety monitoring during construction processes [4], with a focus on target detection of construction workers wearing helmets and holding equipment [5]. However, this method ignores the interrelationship between workers and construction objects, leading to a lack of early warning capability for safety monitoring when workers perform non-compliant construction operations.

In recent years, more researchers have focused on visual relationship detection in deep learning, which aims to determine the topological relationship between targets in a scene [6,7] and generate the triplet form of subject–predicate–object. This approach can more accurately represent and describe construction scene information and contextual relationships. R-CNN [8] was used by VDR [9] to obtain the target candidate frame, and the relationship likelihood score of the triplet was obtained by a visual model and a semantic model for relationship prediction. Different from VRD, VTransE [10] was an end-to-end model that maps the visual features of targets into a low-dimensional relational space, using transfer vectors to represent the relationships between targets. The textual representation of subject/object was used by CAI [11] as contextual information to establish a visual relationship detection model. Features are the basis of target identification, so more features are incorporated into the DR-Net model to count the occurrence probability of subjects, predicates, and objects by visual features, spatial structure features, and relational features [12]. In order to better understand the relationship between targets, ViP-CNN [13] was used to establish the association between subjects, predicates, and objects on visual features by passing information between different models at the same layer. Zoom-net [14] was used for deep information transfer between local target features and global predicate relation features to the achieve deep integration of subjects and predicates. At present, visual relationship detection has been applied to a variety of image understanding tasks, such as image understanding in construction scenes. Wu et al. [5] performed relationship detection between workers and equipment by obtaining the head pose and body orientation of the worker. Kim et al. [15] reconstructed individual behaviors using object types of interactions between workers and equipment to improve construction scene identification. Xiong et al. [16] applied visual relationship detection in construction to a video surveillance system, enabling further improvement with respect to the immediate effectiveness of construction safety warnings. The above methods are able to identify specific targets and interrelationships between targets in construction scenes, but fail to further realize scene identification and understanding on this basis, and thus cannot achieve automation and intelligence in safety monitoring during construction. In addition, owing to the relatively high complexity of construction scenes, it is easy to encounter the problem of missing and incorrect detection of targets.

Visual relationship detection fully presents all information in an image and solves the problem of object relationship fragmentation caused by using target detection algorithms alone. However, there are only a few applications of visual relationship detection in highway bridge construction. In order to achieve intelligent safety monitoring of the bridge construction process and to complete construction scene identification and understanding, this paper proposes a visual relationship-based method for construction scene identification on highway bridges. The method combined the construction characteristics of highway bridges, and is based on the idea of deep learning. In this method, scene identification rules are formulated according to the target features and interrelationships in the construction scenes, and a scene identification model is then built based on the rules to complete the textual output of key scene information. The main work of this paper is as follows:

(1) Selection of key construction scenes on bridges. There are numerous bridge construction processes. Therefore, in this study, five key construction scenes of a bridge were selected based on an analysis of its construction characteristics and construction process.

(2) Formulation of identification rules for key construction scenes on bridges. A feature is the basis of scene identification. This study examines the underlying features that can distinguish the categories of key construction scenes, and establishes a feature information table and a tree diagram for the identification of key construction scenes on highway bridges. On this basis, the identification rules under different construction scenes are formulated.

(3) Building an identification model for key construction scenarios on bridges. In the target detection module, a feature pyramid network (FPN) and color moments are introduced to perform the multiscale detection of targets and obtain construction personnel identity information, while reducing the rate of missing and incorrect detection of targets. In the visual relationship extraction module, feature vectors are introduced to connect subjects, objects, and predicates in construction scenes in order to determine the interaction relationship between targets. In the semantic conversion module, frequency baselines are introduced to count the number of predicates in the construction scene, and the probability distribution of construction personnel actions is then obtained. In the scene information fusion module, an image–text encoder is introduced to combine the image results with the detection results to obtain the correspondence between the images and text. In the scene identification results output module, a rule consistency matching strategy is introduced to match the detected feature results with the formulated rules, and the category information of key construction scenes of highway bridges is then obtained.

(4) Validation of scene identification method. Experimental validation was performed using a homemade key construction scene identification dataset on a highway bridge. In addition, the accuracy, precision, recall, and other evaluation indexes were used to evaluate the accuracy of the proposed scene identification method. Moreover, we performed a comparative analysis with other visual relationship-detection methods to prove the effectiveness of the proposed method.

2. Proposed Method

2.1. Selection of Key Construction Scenes on Bridges

2.1.1. Analysis of Bridge Construction Characteristics

In highway bridge engineering, there is a degree of difference between its production and general industrial production, which includes the following three perspectives.

(1) Large span of engineering structures. Highway bridge projects are often used to connect two distant areas; therefore, the bridge body has a long span. Furthermore, gantry cranes are essential types of equipment for the transport and installation of bridge bodies, but are more dangerous.

(2) More open-air and high-altitude operations. The fixed nature of highway bridge locations makes construction workers often face open-air work and to work from heights. As the distance of construction workers from the ground increases, the risk factor also increases layer-by-layer.

(3) High periodicity and repetitiveness. Bridge projects involve the use of similar types of structures, the same part of the sub-section construction, as well as other factors during the construction process. Therefore, they need to be carried out in a step-by-step manner, such as embedding steel casing, fixed formwork installation, concrete pouring, etc., which gives the bridge construction a certain periodicity and repetitiveness.

Owing to the aforementioned characteristics of highway bridge construction, there are a number of difficulties and safety hazards. To reduce the occurrence of accidents, it is necessary to monitor the bridge construction scene in real time. However, the bridge construction process is complex and varied; therefore, five key scenes were selected for this study.

2.1.2. Key Construction Scenes on Bridges

The construction process of a highway bridge consists mainly of in situ construction and assembly construction. That is, the formwork and stand are set up at the location of the entire bridge, followed by the welding of the reinforcement and concrete pouring. After the concrete reaches its target strength, the formwork and stand are removed. Finally, prefabrication of the beams and bridge deck construction is carried out near the bridge site. The flow of the construction process is illustrated in Figure 1.

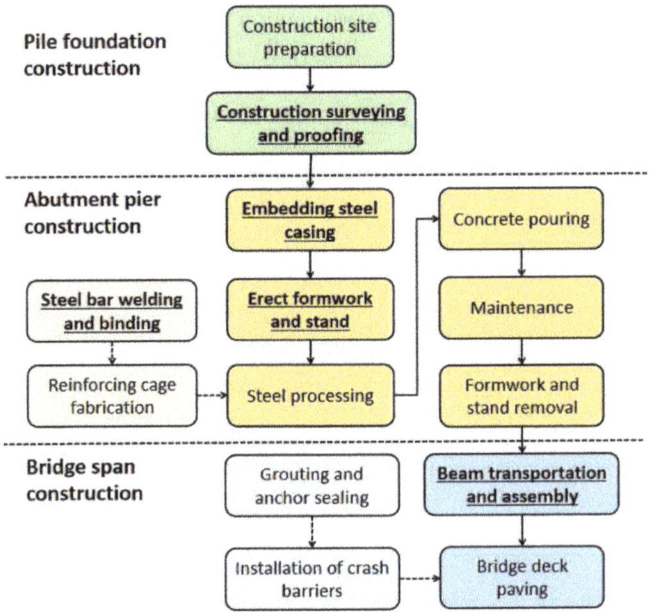

Figure 1. Schematic diagram of the bridge construction process.

Bridge construction is divided into three main stages: pile foundation construction, abutment pier construction, and bridge span structure construction, each of which is a complex and tedious construction process. In this study, five key construction scenes (indicated by bold underlining in Figure 1) were selected for analysis and identification.

(1) Construction surveying and proofing. Before the construction of the bridge project, the technicist should first carry out measurement lofting and data calculation on-site to provide the construction direction for the entire project. This is the premise and foundation for ensuring the quality of bridge projects.

(2) Embedding of steel casing. During bridge construction, to achieve the required load-bearing capacity, the steel casing needs to be embedded to ensure the verticality of the bridge and prevent collapse caused by the falling of debris around it. This directly affects the stability of the bridge pile foundations.

(3) Erection of formwork and stand. When pouring the superstructure of the bridge on-site, the first step is to erect a stand at the location of the bridge hole to support the formwork and poured reinforced concrete. This is an important construction step in bridge engineering.

(4) Steel bar welding and binding. Steel processing is an extremely important step in bridge construction, and the welding and binding of steel bars are basic links in steel processing to ensure the stability of steel installation. This, in turn, affects the structural safety of the entire bridge.

(5) Beam transportation and assembly. The weight and volume of the equipment involved in the beam transportation and assembly stages are large, such as gantry cranes and bridge erectors, which are prone to accidents if not operated carefully. This is a major source of danger during the bridge construction process.

To ensure the stability and safety of bridge structures, it is necessary to strengthen the management of the construction process, particularly during the key construction scenes. The first step in management is to accurately identify the current scene information and monitor hazards according to the interrelationship between workers and equipment. Based on this idea, this study proposes identification rules for key construction scenes on bridges.

2.2. Formulation of Identification Rules for Key Construction Scenes on Bridges

Scene identification can be achieved by extracting the underlying features of different instances in an image and the spatial location relationship between them, and by inferring the relationship to output, the current scene information of that image. Based on this idea, this study designed identification rules for key construction scenes on highway bridges. Table 1 presents the information of bridge key construction scenes identification features: Table 1 A presents the key-scene construction equipment information, and Table 1 B presents the key-scene construction personnel and construction material information. In both tables, the underlying features required to identify the five key scenes are marked as "√". Where, ①–⑤ denote five key construction scenes on the bridge. ① denotes construction surveying and proofing, ② denotes embedding steel casing, ③ denotes erect formwork and stand, ④ denotes steel bar welding and binding, and ⑤ denotes beam transportation and assembly.

Figure 2 shows the rules of bridge key construction scene identification. It describes the logical relationship of the underlying features of construction personnel (blue), construction equipment (green), and construction materials (orange). The left shows the five key construction scenes and the corresponding construction equipment for each scene.

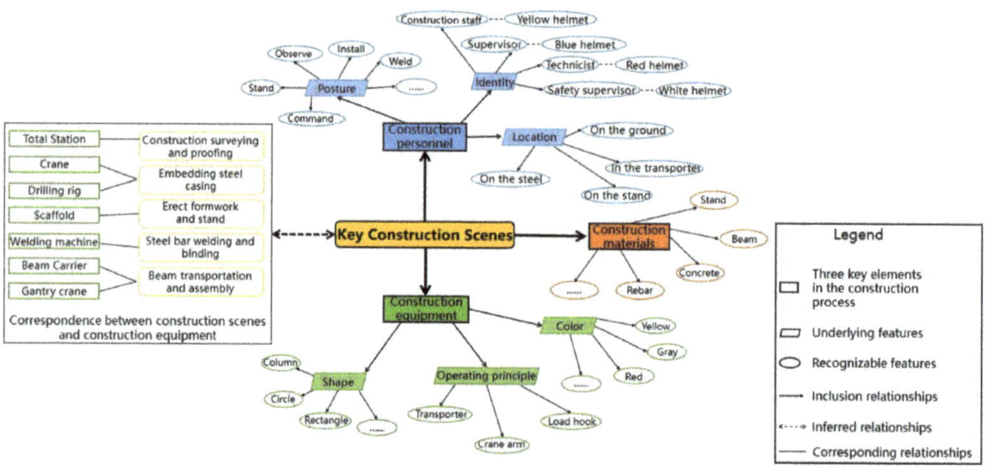

Figure 2. Identification rules tree diagram of bridge key construction scenes.

(1) Construction personnel include three underlying features: posture, identity, and location. ① Posture features include seven kinds: observe, command, stand, etc.; ② Identity features include construction stuff (yellow helmet), supervisor (blue helmet), etc.; and ③ Location features include four types of location information such as on the ground, on the stand, etc.

(2) Construction equipment includes three underlying characteristics of shape, color, and working principle. ① Shape characteristics include cylindrical, round, etc.; ② Color characteristics include red, yellow, etc.; ③ Working principles include arm, hook, etc.

(3) Construction materials include rebar, concrete, etc.

Table 1. (**A**) Construction equipment in five scenes; (**B**) Construction personnel and construction materials in five scenes.

(A)

Scenes	Construction Equipment															
	Name							Shape			Color			Principle		
	Total Station	Crane	Drilling Rig	Scaffold	Welding Machine	Beam Carrier	Gantry Crane	Column	Circle	Rectangle	Red	Yellow	Gray	Transporter	Crane Arm	Load Hook
①	√									√						
②		√	√					√								
③				√												√
④					√			√					√			
⑤						√	√							√	√	

(B)

Scenes	Construction Personnel														Construction Materials			
	Posture						Identity				Location							
	Observe	Command	Stand	Install	Weld	Transport	Yellow Helmet	Blue Helmet	Red Helmet	White Helmet	On the Ground	In the Transporter	On the Stand	On the Steel	Rebar	Stand	Beam	Concrete
①	√							√	√		√							
②		√					√				√							
③			√	√			√						√			√		
④					√		√							√	√			
⑤						√				√		√					√	

For example, the worker wearing a blue helmet stands on the ground observing the total station and instructing the worker wearing a yellow helmet, it can be inferred that the current scene is "construction surveying and proofing"; the worker wearing a yellow helmet holds a cutting machine to weld long objects, and it can be inferred that the current scene is "steel bar welding and binding".

2.3. *Building of Identification Model for Key Construction Scenes on Bridges*

The authors in [17] proposed a relationship detection model named RelDN, and this study draws on the idea of constructing a CSIN network model with CNN (Convolutional Neural Network) [18] and DCR (Deep Convolutional Relationship) [19] as the basic framework for bridge construction scene identification. The structure of CSIN is shown in Figure 3. There are four parts. The first part is image input and feature extraction part, which is composed of the convolutional neural network CNN to extract the underlying features of construction images. The second part is the feature processing part, which mainly includes the target detection module, visual relationship extraction module, and semantic conversion module to obtain different feature score charts. The third part is the feature fusion part, which is composed of the scene information fusion module to fuse image features and text features. The fourth part is the result output part, which is composed of the scene identification result output module to obtain the current construction scene information.

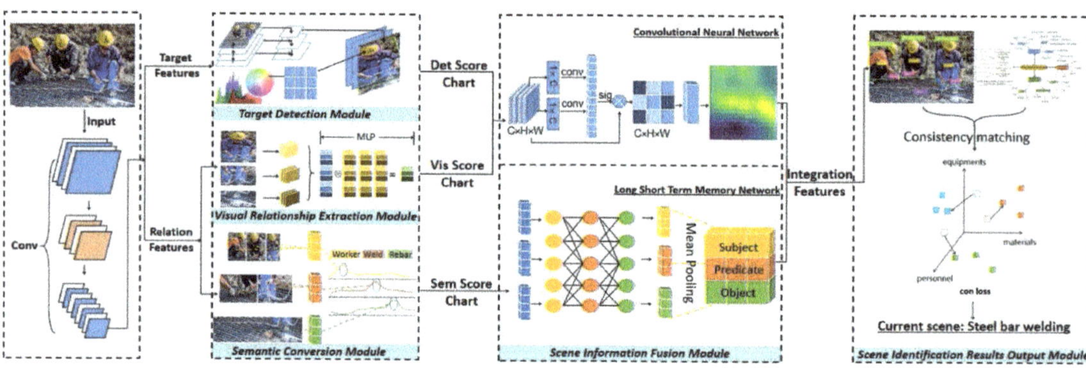

Figure 3. Structure schematic diagram of bridge construction scene identification model. (The source of the identifiable image in Figure 3 is shown in the Supplementary Materials).

2.3.1. FPN-Based Target Detection Module

Detecting and locating various types of targets in construction images are the basis for achieving construction scene identification; therefore, this study first needs to extract and capture feature information, such as the location and category of construction personnel, using the target detection module. To address the problem of target size difference and target miss detection in construction scene images owing to the imaging angle, a detection method that can cope with such multi-scale variation is needed. The feature pyramid network (FPN) [20] can feature extraction for each scale of the image, increasing the perceptual field of the bottom layer of the feature map. So, the FPN is able to obtain more contextual information when performing small target detection at the bottom layer, reducing the rate of missing and incorrect detection. Therefore, this study adds an FPN in the target detection module, which makes shallow networks focus more on detailed information, and high-level networks focus more on semantic information.

In addition, standardized coordinates were used to encode the bounding box between targets to obtain position information and to complete the prediction of the position relationship.

$$\Delta b_1, b_2 = (\frac{x_1 - x_2}{W_2}, \frac{y_1 - y_2}{H_2}, \log\frac{W_1}{W_2}, \log\frac{H_1}{H_2}) \quad (1)$$

$$c(b) = (\frac{x}{W_{img}}, \frac{y}{H_{img}}, \frac{x+W}{W_{img}}, \frac{y+H}{H_{img}}, \frac{WH}{W_{img}H_{img}}) \quad (2)$$

where b_1, b_2 are the bounding boxes between the two targets, $\Delta b_1, b_2$ denote the increments between the bounding box coordinates, and (x, y, W, H) is the coordinate information of the bounding box. In addition, $c(b)$ denotes the normalized coordinate feature of the bounding box, and W_{img}, H_{img} are the width and height of the input image, respectively.

To address the problem of mismatch between helmet type and construction personnel identity, in this paper, the color characteristics of the safety helmet are extracted using the color moment method [21]. The first-order moments describe the average color of the safety helmet, the second-order moments describe the color variance, and the third-order moments describe the offset of the color. Thus, the color moments can present comprehensive color characteristics of the safety helmet to achieve the purpose of corresponding with the identity of workers. The correspondence between helmet color and worker identity is shown in Table 2. The formulae for calculating the first-order, second-order, and third-order moments are as follows:

$$M_1 = \frac{1}{N}\sum_{j=1}^{N} P_{ij} \quad (3)$$

$$M_2 = (\frac{1}{N}\sum_{j=1}^{N}(P_{ij} - M_1)^2)^{\frac{1}{2}} \quad (4)$$

$$M_3 = (\frac{1}{N}\sum_{j=1}^{N}(P_{ij} - M_1)^3)^{\frac{1}{3}} \quad (5)$$

Table 2. Matching relationship between safety helmet color and construction personnel identity.

Color Classification	Red Helmet	Yellow Helmet	Blue Helmet	White Helmet
Worker status	Technicist	Construction staff	Supervisor	Safety supervisor

2.3.2. Visual Relationship Extraction Module Based on Feature Vectors

The visual relationship detection branch is used to capture deeper visual features in construction images, including the construction scene content, interrelationships, and logical relationships between different objects. The visual relationship extraction module focuses on obtaining the interaction probability values between the construction action sender (subject), construction action (predicate), and construction object (object), as shown in Figure 2. This module generates a set of class vector logits conditioned on region-of-interest (ROI) feature maps and passes the fused feature map information so that the network can fully learn and perceive the visual and semantic intersection information in the construction scene. A multilayer perceptron (MLP) is used to connect the feature vectors of the subject, predicate, and object to obtain the probability values of the interaction relationships between different entity targets in the construction scene. The formula is as follows:

$$f(x) = G(b^{(2)} + W^{(2)}(s(b^{(1)} + W^{(1)}x))) \quad (6)$$

where W is the connection weight, b is the bias, G is the softmax function, and s is a sigmoid function.

To improve the processing efficiency of the network for visual information and reduce the computational cost of the network, two cross-layer connections [22] are constructed in the visual relation extraction module. Then the subject/object ROI features extracted by the detection module are mapped to the predicate class vector logits to facilitate the transfer and flow of information in the network.

2.3.3. Semantic Conversion Module Based on Frequency Baseline

A construction scene graph contains not only intuitive visual information, but also deep semantic information. Scene graphs are one of the methods used to construct the visual relations of images [23]. The main idea is to divide the visual relations between all objects in an image into a triadic subject–predicate–object form, which is used as a whole learning task [24,25]. The semantic conversion module focuses on outputting the relationship information between the subject and object. This module draws on the idea of a scene graph to generate a set of binary relational feature maps of the ROI and passes the semantic information extracted by the relationship detection branch to a higher level of cyberspace. The interrelationships and attribute information between different objects are then captured by calculating the frequency of predicates between subjects and objects. This predicate is generally limited and regular; for example, the relationship between construction workers and scaffolding is generally workers "install" scaffolding or workers "stand" on scaffolding, but not other predicates such as "wear". Therefore, to improve the processing and learning efficiency of the semantic conversion model, a frequency baseline was set based on the number of occurrences of the predicate [26]. For any pair of training images, the prediction probability distribution was obtained by counting the number of occurrences of subject s and object o in the real box with the set frequency baseline.

$$\omega(s,o) = 1 - p(pred = \varnothing|s,o) \tag{7}$$

where $p(pred|s,o)$ denotes the probability of predicate distribution between subject s and object o, and $p(pred = \varnothing|s,o)$ denotes that there is no interrelationship between subject s and object o.

To prevent the network from incorrectly inferring two targets that are close but not interrelated, a loss function L is designed when subject s and object o are interrelated to maximize the bounding box distance between the two targets determined by the predicate.

$$\begin{aligned} L = & \frac{1}{N}\sum_{i=1}^{N}\frac{1}{|P(O_i^+)|}\sum_{p\in P(O_i^+)}\max(0,\alpha - m^s(i,p)) \\ & +\frac{1}{N}\sum_{j=1}^{N}\frac{1}{|P(O_j^+)|}\sum_{p\in P(O_j^+)}\max(0,\alpha - m^o(j,p)) \end{aligned} \tag{8}$$

where P() is the specific set of predicates associated with the input, p represents the predicate class, and O_i^+, O_j^+ denote the set of targets whose relationship is p. In addition, α is the threshold value, m^s, m^o denotes the confidence of the subject and object, and i, j denotes the index of the subject and object.

2.3.4. Scene Information Fusion Module Based on Image-Text Encoder

After the target detection module and visual relationship extraction module, the image information of the construction personnel and the image information of the subject and object in the scene were obtained. Moreover, the text information of the predicate in the scene was obtained after the semantic conversion module. The key step in realizing scene identification is to combine image information with text information. In this study, the scene information fusion module was formulated by referring to the method of correlation description between images and text in the literature [27].

First, the detection, visual, and semantic scores obtained by the three modules are softmax normalized to obtain the target relationship probability P^{pre}.

$$P^{pre} = \text{softmax}(f_{Det} + f_{Vis} + f_{Sem}) \tag{9}$$

where f_{Det}, f_{Vis}, and f_{Sem} denote the output relationship probabilities of the target detection module, visual relationship extraction module, and semantic conversion module, respectively.

After obtaining the target relationship probabilities, the output image results and detection results are encoded to the same dimension by the image encoder ϕ through convolutional neural networks (CNNs) [28], and the text encoder φ through the long short-term memory network (LSTM) [29]. Then, the cosine similarity between the paired image results and the detection results was calculated to construct the ranking loss function. The ranking loss function of encoder L_{rank} is as shown in Equation (10).

$$\begin{aligned} L_{rank} = {}^{\min}_{\theta} \sum_{x}\sum_{k} \max\{0, \alpha - s(\phi(x), \varphi(t)) + s(\phi(x), \varphi(t_k))\} \\ + \sum_{t}\sum_{k} \max\{0, \alpha - s(\phi(x), \varphi(t)) + s(\phi(x_k), \varphi(t))\} \end{aligned} \tag{10}$$

where θ denotes all parameters in the image encoder and text encoder, α is the boundary value, and s is used to calculate the cosine similarity between the image embedding vector $\phi(x)$ and the detection result embedding vector $\varphi(t)$; x_k, t_k denote the mismatched images and texts, respectively.

2.3.5. Scene Identification Results Output Module Based on Rule Consistency Matching

The four modules above are all intermediate results, which can be expressed as "features," while the final goal of this study is to output a textual expression that is consistent with the scene image to be detected. The textual output of scene identification is obtained by matching the integration features acquired from the scene information fusion module with scene identification rules (Figure 2). In this paper, the method of reference [30] is referred to, and the loss function L_{con} is used to calculate the consistency between the integration features and the rules. L_{con} is calculated as shown in Equation (11).

$$L_{con} = \left(\frac{1}{a}\sum_{i}^{a}\frac{u_i^T x_i^t}{||u_i||\cdot||x_i^t||} + \frac{1}{b}\sum_{j}^{b}\frac{v_i^T y_i^t}{||v_i||\cdot||y_i^t||} + \frac{1}{c}\sum_{j}^{c}\frac{w_i^T z_i^t}{||w_i||\cdot||z_i^t||}\right)^2 \tag{11}$$

where (u_i,x_i^t), (v_i,y_i^t), and (w_i,z_i^t) appear in pairs and denote subject-construction personnel matching, object-construction equipment matching, and predicate-posture matching, respectively; and a, b, and c represent the number of instances of the three construction elements (mentioned in Figure 2).

In the training process, given a dataset $D = \left\{(I_k, S_k)_{k=1}^{N}\right\}$ containing N image-text, a batch of images is sampled from the dataset for training, and the final loss function is a weighted sum L of the ranking loss and the consistency loss.

$$L = \sum_{k}^{N_b} L_{rank}(I_k, S_k) + \lambda_{con} \sum_{k}^{N_b} L_{con}(I_k, S_k) \tag{12}$$

where I denotes the image, S denotes the text, and λ_{con} is a hyperparameter with an adjustable balance.

2.3.6. Method Flow-Chart

The overall method flow-chart is shown in Figure 4. Firstly, the bridge construction scene images are input into the convolutional neural network, then the geometric features and color features in the shallow layer of the image are extracted by the operations of convolution, pooling, and full connection to form the feature maps. Then the extracted feature maps are fed into the target detection module, the visual relationship extraction module, and the semantic conversion module through the target detection branch and the relation detection branch, respectively, to obtain information on parameters such as location, category, probability values, and attributes of image targets to form a detection score chart, visual score chart, and semantic score chart. Afterward, the three score charts are fed into the convolutional neural network and the long short-term memory network, respectively, for image coding and text coding to obtain the integration features. Finally, the integration features are matched with the scene identification rules for consistency, and the current scene information is obtained and output in the text form.

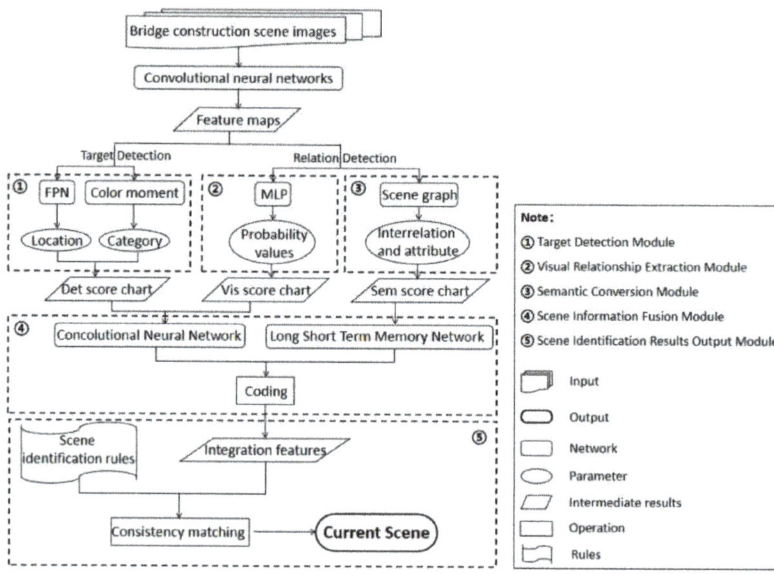

Figure 4. Method flow-chart.

3. Experiment

3.1. Experimental Configuration

3.1.1. Dataset

To fully learn the target features, semantic features, and visual features in different construction scenes, and to identify key construction scenes on bridges, a large amount of data is required for training. Because there is no dataset for construction scene identification that satisfies the needs of this study, a scene-based construction identification dataset for highway bridges is built in this study. For the five key scenes mentioned in Section 2.1.2, the construction scene images are intercepted by online bridge construction monitoring videos considering various factors such as the target size variation, location distribution, and similar color interference. In addition, LabelImg is used to label visual information such as the location and category of targets, as well as the semantic information of the interrelationship between targets in the images. So, the model can fully learn and understand the logical relationships embedded in the images. The specific information of the bridge key construction scene identification dataset constructed in this study is presented in Table 3, containing a total of 465 images. This dataset was constructed from three aspects: subject, object, and predicate. Furthermore, 60% of the images were selected as the training set and the remaining 40% were selected as the test set, including 37 images for each of the five key construction scenes.

Table 3. Number and interrelationship of images in each construction scene.

Bridge Construction Scene	Number of Images	Visual Relationship		
		Subject	Predicate	Object
Construction surveying and proofing	95	Worker	Observe	Total station
Embedding steel casing	90	Crane	Conduct	Steel case
Erect formwork and stand	90	Worker	Install	Scaffold
Steel bar welding and binding	100	Worker	Weld	Rebar
Beam transportation and assembly	90	Beam carrier	Transport	Beam

3.1.2. Evaluation Indicators

To verify the accuracy of the proposed bridge key construction scene identification method, indicators such as accuracy (Acc), precision (P), and recall @K (R@K) were used to evaluate the results of the experiments. The main formulae are shown in Equations (13)–(15).

$$\text{Accuracy} = \frac{\text{TP} + \text{TN}}{\text{TP} + \text{TN} + \text{FP} + \text{FN}} \tag{13}$$

$$\text{Precision} = \frac{\text{TP}}{\text{TP} + \text{FP}} \tag{14}$$

$$\text{Recall@K} = \frac{\text{TP@K}}{(\text{TP@K}) + (\text{FN@K})} \tag{15}$$

Among them, true positive (TP) and true negative (TN) are correct detection results, false positive (FP) is wrong detection, and false negative (FN) is missed detection.

3.1.3. Implementation Details

The configuration of this experimental platform is the Windows 10 operating system and CUDA 10.1 computing platform; the algorithm framework is TensorFlow-GPU1.12.0 and Keras2.13; the programming language is Python 3.6.13. To obtain a better training effect, the size of the image input network was set to 800 pixels in the training phase. Then, the batch was set to 1, the number of iterations was 10,000, and the initial learning rate was 0.001.

3.2. Identification Results and Accuracy Analysis for Key Construction Scenes on Bridges

To verify the effectiveness of the bridge key construction scene identification method proposed in this study, two parameters, namely the identification effect and identification accuracy, were evaluated and analyzed.

3.2.1. Scene Identification Results and Analysis

To verify the scene identification effect of the proposed method, experiments were conducted on the test set. Figure 5 shows some of the data in the test set, including five key construction scenes: (a) shows three technicists wearing red helmets to operate the total station and recording; (b) shows that the steel case is controlled by the crane arm, and the crane is operated by two construction workers wearing yellow helmets; (c) shows three construction personnel in yellow helmets welding steel bars with electric welders; (d) shows two construction workers in yellow helmets standing on the support to install the scaffold; and (e) shows a construction worker in a yellow helmet directing the beam transporter to transport the beam.

Figure 5. Partial data in the test set (five key construction scenes). (The sources of the identifiable images in Figure 4 are shown in the Supplementary Materials.)

The scene identification method proposed in this paper was applied to the test set for the experiment. Figure 6 and Table 4 show the scene identification results.

By analyzing the identification results in Figure 6 and the information in Table 4, it was found that the proposed scene identification method can correctly output the final scene category information (black box in the upper left corner in Figure 6). The information is derived from the intermediate results by reasoning through the formulated scene identification rules. The intermediate results consisted of two parts: the target detection result (green) and the visual relationship detection result (yellow for the subject, purple for the object, and pink for the predicate).

From the target detection results, we can see that the proposed method can distinguish different identity types according to the color of the helmet worn by workers, such as the detection result for workers wearing helmets in Figure 6; (a) is a "technicist", while the workers wearing yellow helmets in (b–e) are detected as "construction staff". In particular, the method proposed in this paper can still accurately detect the type and location information of the relatively small-sized workers appearing on the left side of (e) (this result will be analyzed in Comparison Results and Analysis of Target Detection Module). The visual relationship detection results show that the proposed method can correctly identify the subject, predicate, and object, and can connect the above three through the red line segment to reflect the correlation between them. Finally, the final scene identification results were obtained from the above two intermediate results using inference rules.

Figure 6. Partial results of the bridge key construction scene identification method on the test set. (The sources of the identifiable images in Figure 5 are shown in the Supplementary Materials.)

Table 4. Information table showing identification results for key construction scenes on bridges.

Figure	Intermediate Results				Final Results
	Target Detection Results	Visual Relationship Detection Results			Construction Scene Category Information
		Subject	Predicate	Object	
A	Technicist	Worker	Observe	Total station	Construction surveying and proofing
B	Construction staff	Crane	Conduct	Steel case	Embedding steel casing
C	Construction staff	Worker	Weld	Rebar	Erect formwork and stand
D	Construction staff	Worker	Install	Stand	Steel bar welding and binding
E	Construction staff	Beam carrier	Transport	Beam	Beam transportation and assembly

3.2.2. Scene Identification Accuracy and Analysis

To verify the scene identification accuracy of the method in this study, it was evaluated using a confusion matrix, as shown in Table 5. Each scene category contains 37 test images.

Table 5. Table of identification accuracy results for key construction scenes on bridges, where: ① denotes construction surveying and proofing, ② denotes embedding steel casing, ③ denotes erection of formwork and stand, ④ denotes steel bar welding and binding, and ⑤ denotes beam transportation and assembly.

		True Value					Precision (%)
		①	②	③	④	⑤	
Predicted Value	①	37					100
	②		35				100
	③			31	3		91.2
	④			6	34		85.0
	⑤		2			37	94.9
Recall (%)		100	94.6	83.8	91.9	100	
			Accuracy (%) = 94				

Comparing the data in the table, it can be seen that the identification accuracy and recall in the "construction surveying and proofing" are the highest, and 100% recognition can be achieved. The identification accuracy and recall in the "erection of formwork and stand" and "steel bar welding and binding" were lower, with accuracies of 91.2% and 85.0%, and recall values of 83.8% and 91.9%, respectively. From an analysis of the reasons, we found that the processes of "erect formwork and stand" and "steel bar welding and binding" have high similarity. It is obvious from (c,d) of Figure 5 that the above two scenes have confusing targets, so the identification accuracy is slightly lower than that of the other scenes. The identification accuracy of "beam transportation and assembly" is higher because the size of the beam transporter is larger than the targets in other scenes, which is easy to identify. The identification accuracy of the "construction surveying and proofing" is the highest because the target in this scene is clear and the background is simple, which is not easily disturbed by other information. However, in general, the scene identification accuracy of this study reached 94%, which can complete the identification of key construction scenes on bridges.

Based on the experimental results obtained, it can be concluded that the key construction scene identification method proposed in this paper has a good scene understanding ability. This method can fully learn the semantic and visual information in the graph, perform target localization and relationship detection, and accurately output the category information of the scene.

3.3. Experimental Results and Analysis of Identification Model CSIN for Key Construction Scenes on Bridges

The CSIN model proposed in this study plays an important role in the scene identification process. To verify its effectiveness, two aspects of the model, namely the overall performance and internal modules, were evaluated and analyzed.

3.3.1. Experimental Results and Analysis of the Performance for the Scene Identification Model

The PR curves were plotted using precision and recall, which can visually describe the model performance. Figure 7 shows the PR curves generated when the IoU threshold is set to 0.5, 0.6, and 0.7, where the horizontal and vertical coordinates represent recall and precision, respectively. From an analysis of the three curves in the figure, it can be seen that when the IoU threshold is set to 0.5, the PR curve is closer to the upper right; that is, the precision and recall are both higher. The area formed by the PR curve and coordinate axis gradually decreased as the IoU threshold increased. When the IoU is 0.6 and 0.5, the two PR curves start to decrease significantly at recall >0.7, and when the IoU is 0.7, the PR curves start to decrease around recall =0.5. This indicates that the proposed CSIN model had the best detection effect when the IoU threshold was 0.5.

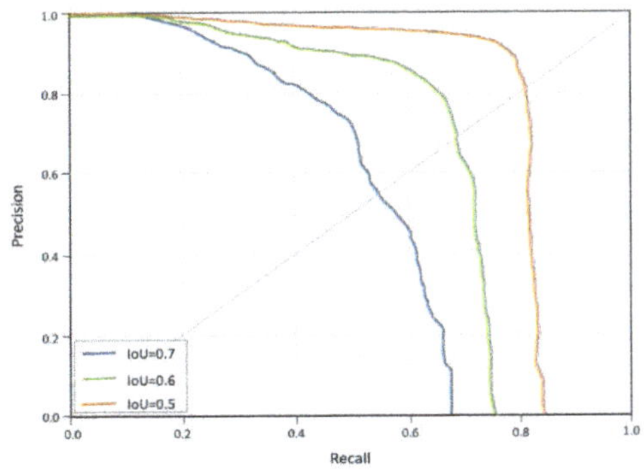

Figure 7. PR curves under different IoU thresholds.

3.3.2. Comparative Experimental Results and Analysis of Specific Modules

The CSIN model proposed in this study includes two important modules: target detection and visual relationship extraction. To verify the effectiveness of the proposed algorithms in these two modules, comparative experiments were conducted separately.

Comparison Results and Analysis of Target Detection Module

The basis of achieving scene identification is to correctly detect the classification and location information of targets in construction scenes; therefore, the effect of adding FPN in the detection module was tested in this study.

As shown in Figure 8, the second column is the true value, which is the feature map obtained from the original image after the grayscale comparison operation, and it is used to highlight specific regions of the foreground targets of the image. Further, the green box is the target with a smaller size. The third column is the convolutional heat map, which is used to delineate the target regions. The last two columns are the feature visualizations obtained by the two methods after channel-dimension averaging. Based on the results, the images (e,j) obtained by our method are generally clearer than the images (d,i) without FPN. In addition, both sets of feature maps contain targets of smaller size (red and yellow boxes), where the red boxes are marked by the multi-scale target detection without an FPN, and their response value is low when compared with the real value of the green boxes; that is, there is a missed detection. It is evident from the yellow boxes that the response value of the features is higher after the FPN is applied, and the human shape can be roughly detected. It can be concluded that the CSIN model proposed in this study, which applies an FPN for multi-scale target detection, is effective.

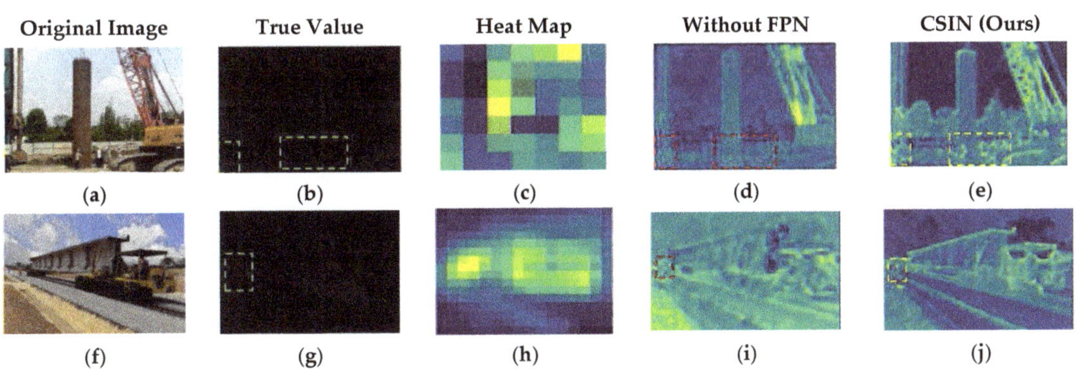

Figure 8. Visualization of the convolutional feature map obtained by channel-dimension averaging. (The sources of the identifiable images in Figure 5 are shown in the Supplementary Materials.)

Comparison Results and Analysis of the Visual Relationship Detection Module

Visual relationship detection is a prerequisite for achieving deep perception and understanding of construction scenes, and the degree of detection accuracy determines the merit of the CSIN. Therefore, in this paper, four mainstream algorithms are selected for visual relationship detection in the visual relationship extraction module for comparison experiments. Among them, VRD [9] is one of the earliest algorithms used for visual relationship detection and often appears as a comparison model; both Large-Scale [31] and the proposed method adopt the semantic module and visual module for feature extraction in model design; Motifs [26], Graph R-CNN [32], and the proposed method are all based on the basic model of scene graph for visual relationship detection. So, these four algorithms are selected for comprehensive comparison experiments. Their effectiveness can be assessed based on two aspects: subject/object localization accuracy and predicate detection accuracy, where subject/object localization focuses more on the target detection ability of the model, whereas predicate detection focuses more on the relationships.

Table 6 lists the visual relation detection results of the different algorithms. In general, the CSIN model in this study has better performance for the target detection of subjects and objects; for predicate detection, the CSIN model is not significantly different from other algorithms. Almost all of the models had the highest detection results for Recall@100.

Table 6. Comparison results with other visual relationship detection algorithms.

	Subject Detection			**Object Detection**			**Predicate Detection**		
Recall at (%)	20	50	100	20	50	100	20	50	100
Large-Scale	20.7	27.9	32.5	36.0	36.7	36.7	66.8	68.4	68.4
Motifs	21.4	27.2	30.3	32.9	35.8	36.5	58.5	65.2	67.1
VRD	-	0.3	0.5	-	11.8	14.1	-	27.9	35.0
Graph R-CNN	-	28.5	35.9	-	29.6	31.6	-	54.2	59.1
CSIN (ours)	35.9	37.8	42.4	36.1	36.7	37.0	65.3	67.9	69.3

Specifically, the subject detection accuracy in Recall@100 reached 42.4%, which is at least 6.5% or so higher than that of other algorithms. The object detection accuracy in Recall@100 reached 37.0%, which was also slightly higher than those of the other detection algorithms. However, the relationship detection was slightly lower than those of the other algorithms. Specifically, the predicate detection accuracies of Recall@20 and Recall@50 are 65.3% and 67.9%, respectively, which are lower than the predicate detection accuracy of the large-scale algorithm. This is because the large-scale algorithm is for the location and relationship detection of larger size targets. What is more, the predicate detection in this study does not show obvious superiority; this may be because bridge constructions are characterized by complex scenes and ambiguous relationships between workers and equipment, and it is relatively difficult to distinguish relationships. Subsequent experiments could be further improved to address this problem.

In summary, the localization accuracy results for the subject and object show that the use of an FPN can improve the detection accuracy of the target. The CSIN model proposed in this study works well for relationship detection and can effectively infer scene information during the construction process.

4. Discussion

In this part, we discuss three main points: robustness of scene identification rules, stability of the CSIN model detection frame, and generalization capabilities of the CSIN model.

4.1. Robustness of Bridge Construction Scene Identification Rules

The bridge construction scene identification rules developed in this paper adopt the idea of consistency matching. The logical relationship between construction personnel, construction equipment, and construction materials in the construction scenes is considered, which satisfies the needs of this paper to a certain extent. However, the lack of some reasoning strategies, such as inductive reasoning [33] and deductive reasoning [34], makes the constraint relationships among construction activities unable to be further refined into rules. BIM technology is used to obtain construction information [35,36] and obtain the constraint relationship between construction activities, so as to deduce the logical sequence between construction activities, which can improve the robustness of the scene identification rules to a certain extent.

4.2. Stability of the CSIN Model Detection Frame

In this paper, the CSIN model applies FPN for the target detection of construction personnel with high detection accuracy. The identity and location information of construction personnel can be obtained accurately in many cases. However, when shadows appear in the construction image, the accuracy of the CSIN model detection frame is affected to some extent. As shown in Figure 6d, the worker identification frame in the lower-left corner only detected the worker's head and hands, which may be due to the fact that the worker's legs blended into the shadow. Since FPN cannot distinguish shadows and targets better, it will affect the stability of the detection frame to a certain extent when shadows appear in the construction image. To solve this problem, generative adversarial networks (GAN) [37] or texture features of shadows in HSV space [38] for shadow suppression can be considered to eliminate the interference of shadows on image targets.

4.3. Generalization Capabilities of the CSIN Model

The CSIN model proposed in this paper was experimented on with self-made datasets. It has been verified that the model can complete target detection, visual relationship detection, and output construction scene information as text, realizing the automation and intelligence of identification in the key construction scenes on bridges. In the CSIN model, CNN and DCR are used as the base networks for target detection and relationship detection, respectively, which have been proved to have certain generalization abilities in related literature [39,40]. In addition, the underlying features of scene identification rules in this paper, such as color features, geometric features, and posture features, will not change greatly with different scenes, so they are portable. Therefore, the CSIN model can be applied to other types of construction and infrastructure projects, such as housing construction, road construction, etc. However, in port and tunnel construction, its generalization ability needs to be further verified due to the influence of datasets.

5. Conclusions

The construction process of highway bridges is tedious, and site environments are complex; thus, the realization of bridge construction scene identification helps relevant departments to carry out safety control. Therefore, based on the idea of visual relationships, this paper proposes the identification method of key construction scenes on highway bridges. This method can provide automated intelligent monitoring during the construction process and provide more applications for visual relationship detection in bridge construction. Firstly, the characteristics of bridge construction are analyzed and five key construction scenes are selected as research objects. Then, the scene identification rules are formulated from the three aspects of construction personnel, construction equipment, and construction materials. Following this, the CSIN model is built: FPN and color moments are first introduced to obtain the image features of construction workers, and solve the problem of missing and incorrect detection of target; then, through the division of subject–predicate–object triplet and image-text coding, the semantic features and visual features of construction scene can be obtained; finally, the integration features are matched with the scene identification rules for consistency, and the category information of the construction scene is further obtained. Finally, the method in this paper is verified; the experimental results show that compared with other algorithms, the CSIN model obtained better results, especially on Recall@100.

Although the method in this paper has addressed the above problems, there are still two limitations. One is that the method is only experimentally validated in five key construction scenes, and research on other bridge construction scenes has not been carried out. The other is that the method involves fewer large equipment and construction materials, such as the lack of detection of large cranes, pile-driving machines, concrete, long bars, and other targets. Therefore, for the construction monitoring of different bridge types, such as girder bridges, arch bridges, rigid bridges, suspension bridges, cable-stayed bridges, and combined system bridges, it is necessary to further increase the identifiable elements in the construction scenes to enrich the bridge construction scene categories.

In our study, we found that the production of the dataset was time-consuming and laborious. In future work, we will combine efficient methods such as crowdsourcing labeling technology to produce targeted visual relationship detection datasets, so as to improve work efficiency. In addition, we will further optimize the CSIN model, combined with the relevant construction safety standards to realize the safety monitoring and safety assessment of bridge construction based on the existing methods. Thus, we will form a complete set of methods for intelligent monitoring and safety assessment of bridge construction, and extend it to other construction scenes.

Supplementary Materials: The following supporting information can be downloaded at: https://www.mdpi.com/article/10.3390/buildings12060827/s1.

Author Contributions: Conceptualization, C.W. and J.L.; Writing—Original Draft Preparation, C.W.; Writing—Review and Editing, C.W. and J.L.; data collection, Y.G.; data analysis, Y.L. All authors have read and agreed to the published version of the manuscript.

Funding: This research received no external funding.

Institutional Review Board Statement: Not applicable.

Informed Consent Statement: Not applicable.

Data Availability Statement: Some or all data are available from the corresponding author by request.

Acknowledgments: The authors are very thankful to those who made suggestions and comments to helpe improve the manuscript quality.

Conflicts of Interest: The authors declare that the research was conducted in the absence of any commercial or financial relationships that could be construed as a potential conflict of interest.

References

1. Ramos-Hurtado, J.; Rivera, M.-L.; Mora-Serrano, J.; Deraemaeker, A.; Valero, I. Proposal for the Deployment of an Augmented Reality Tool for Construction Safety Inspection. *Buildings* **2022**, *12*, 500. [CrossRef]
2. Vasavi, S.; Sravanthi, G.L.; Ram, B.S.; Gokhale, A.A. Predictive analytics of bridge safety for intelligent transportation system using ensemble model. *Mater. Today Proc.* **2021**, *45*, 5608–5616. [CrossRef]
3. Zhou, J.; Li, X.; Xia, R.; Yang, J.; Zhang, H. Health monitoring and evaluation of long-span bridges based on sensing and data analysis: A survey. *Sensors* **2017**, *17*, 603. [CrossRef] [PubMed]
4. Munawar, H.S.; Ullah, F.; Shahzad, D.; Heravi, A.; Qayyum, S.; Akram, J. Civil Infrastructure Damage and Corrosion Detection: An Application of Machine Learning. *Buildings* **2022**, *12*, 156. [CrossRef]
5. Wu, J.; Cai, N.; Chen, W.; Wang, H.; Wang, G. Automatic detection of hardhats worn by construction personnel: A deep learning approach and benchmark dataset. *Autom. Constr.* **2019**, *106*, 102894. [CrossRef]
6. Lee, S.; Kim, H.; Lieu, Q.X.; Lee, J. CNN-based image recognition for topology optimization. *Knowl. Based Syst.* **2020**, *198*, 105887. [CrossRef]
7. Zhang, L.; Wang, Y.; Chen, H.; Li, J.; Zhang, Z. Visual relationship detection with region topology structure. *Inf. Sci.* **2021**, *564*, 384–395. [CrossRef]

8. Girshick, R.; Donahue, J.; Darrell, T.; Malik, J. Rich feature hierarchies for accurate object detection and semantic segmentation. In Proceedings of the IEEE Conference on Computer Vision and Pattern Recognition, Columbus, OH, USA, 23–28 June 2014; pp. 580–587.
9. Lu, C.; Krishna, R.; Bernstein, M.; Fei-Fei, L. Visual relationship detection with language priors. In Proceedings of the European Conference on Computer Vision, Amsterdam, The Netherlands, 11–14 October 2016; Springer: Berlin/Heidelberg, Germany, 2016; pp. 852–869.
10. Zhang, H.; Kyaw, Z.; Chang, S.-F.; Chua, T.-S. Visual translation embedding network for visual relation detection. In Proceedings of the IEEE Conference on Computer Vision and Pattern Recognition, Honolulu, HI, USA, 21–26 July 2017; pp. 5532–5540.
11. Zhuang, B.; Liu, L.; Shen, C.; Reid, I. Towards context-aware interaction recognition for visual relationship detection. In Proceedings of the IEEE International Conference on Computer Vision, Venice, Italy, 22–29 October 2017; pp. 589–598.
12. Dai, B.; Zhang, Y.; Lin, D. Detecting visual relationships with deep relational networks. In Proceedings of the IEEE Conference on Computer Vision and Pattern Recognition, Honolulu, HI, USA, 21–26 July 2017; pp. 3076–3086.
13. Li, Y.; Ouyang, W.; Wang, X.; Tang, X. Vip-cnn: Visual phrase guided convolutional neural network. In Proceedings of the IEEE Conference on Computer Vision and Pattern Recognition, Honolulu, HI, USA, 21–26 July 2017; pp. 1347–1356.
14. Yin, G.; Sheng, L.; Liu, B.; Yu, N.; Wang, X.; Shao, J.; Loy, C.C. Zoom-net: Mining deep feature interactions for visual relationship recognition. In Proceedings of the European Conference on Computer Vision (ECCV), Munich, Germany, 8–14 September 2018; pp. 322–338.
15. Kim, J.; Chi, S.; Seo, J. Interaction analysis for vision-based activity identification of earthmoving excavators and dump trucks. *Autom. Constr.* **2018**, *87*, 297–308. [CrossRef]
16. Xiong, R.; Song, Y.; Li, H.; Wang, Y. Onsite video mining for construction hazards identification with visual relationships. *Adv. Eng. Inform.* **2019**, *42*, 100966. [CrossRef]
17. Zhang, J.; Shih, K.J.; Elgammal, A.; Tao, A.; Catanzaro, B. Graphical contrastive losses for scene graph parsing. In Proceedings of the IEEE/CVF Conference on Computer Vision and Pattern Recognition, Long Beach, CA, USA, 15–20 June 2019; pp. 11535–11543.
18. Albawi, S.; Mohammed, T.A.; Al-Zawi, S. Understanding of a convolutional neural network. In Proceedings of the 2017 International Conference on Engineering and Technology (ICET), Antalya, Turkey, 21–23 August 2017; pp. 1–6.
19. Peng, Y.; Chen, D.Z.; Lin, L. Visual Relationship Detection with A Deep Convolutional Relationship Network. In Proceedings of the 2020 IEEE International Conference on Image Processing (ICIP), Abu Dhabi, United Arab Emirates, 25–28 October 2020; pp. 1461–1465.
20. Lin, T.-Y.; Dollár, P.; Girshick, R.; He, K.; Hariharan, B.; Belongie, S. Feature pyramid networks for object detection. In Proceedings of the IEEE Conference on Computer Vision and Pattern Recognition, Honolulu, HI, USA, 21–26 July 2017; pp. 2117–2125.
21. Keen, N. Color moments. *Sch. Inform. Univ. Edinb.* **2005**, 3–6. Available online: https://homepages.inf.ed.ac.uk/rbf/CVonline/LOCAL_COPIES/AV0405/KEEN/av_as2_nkeen.pdf (accessed on 28 April 2022).
22. He, K.; Zhang, X.; Ren, S.; Sun, J. Deep residual learning for image recognition. In Proceedings of the IEEE Conference on Computer Vision and Pattern Recognition, Las Vegas, NV, USA, 27–30 June 2016; pp. 770–778.
23. Zuo, G.; Tong, J.; Liu, H.; Chen, W.; Li, J. Graph-Based Visual Manipulation Relationship Reasoning Network for Robotic Grasping. *Front. Neurorobotics* **2021**, *15*, 719731. [CrossRef] [PubMed]
24. Kuznetsova, P.; Ordonez, V.; Berg, A.; Berg, T.; Choi, Y. Collective generation of natural image descriptions. In Proceedings of the 50th Annual Meeting of the Association for Computational Linguistics: Long Papers, Jeju Island, Korea, 8–14 July 2012; Volume 1, pp. 359–368.
25. Kuznetsova, P.; Ordonez, V.; Berg, T.L.; Choi, Y. Treetalk: Composition and compression of trees for image descriptions. *Trans. Assoc. Comput. Linguist.* **2014**, *2*, 351–362. [CrossRef]
26. Zellers, R.; Yatskar, M.; Thomson, S.; Choi, Y. Neural motifs: Scene graph parsing with global context. In Proceedings of the IEEE Conference on Computer Vision and Pattern Recognition, Salt Lake City, UT, USA, 18–23 June 2018; pp. 5831–5840.
27. Li, W.; Duan, L.; Xu, D.; Tsang, I.W.-H. Text-based image retrieval using progressive multi-instance learning. In Proceedings of the 2011 International Conference on Computer Vision, Barcelona, Spain, 6–13 November 2011; pp. 2049–2055.
28. Gupta, D. Architecture of Convolutional Neural Networks (cnns) Demystified. *Anal. Vidhya* **2017**. Available online: https://www.analyticsvidhya.com/blog/2017/06/architecture-ofconvolutional-neural-networks-simplified-demystified/ (accessed on 28 April 2022).
29. Vinyals, O.; Toshev, A.; Bengio, S.; Erhan, D. Show and tell: Lessons learned from the 2015 mscoco image captioning challenge. *IEEE Trans. Pattern Anal. Mach. Intell.* **2016**, *39*, 652–663. [CrossRef] [PubMed]
30. Chen, H.; Ding, G.; Lin, Z.; Zhao, S.; Han, J. Cross-modal image-text retrieval with semantic consistency. In Proceedings of the 27th ACM International Conference on Multimedia, New York, NY, USA, 21–25 October 2019; pp. 1749–1757.
31. Zhang, J.; Kalantidis, Y.; Rohrbach, M.; Paluri, M.; Elgammal, A.; Elhoseiny, M. Large-scale visual relationship understanding. *Proc. AAAI Conf. Artif. Intell.* **2019**, *33*, 9185–9194. [CrossRef]
32. Yang, J.; Lu, J.; Lee, S.; Batra, D.; Parikh, D. Graph r-cnn for scene graph generation. In Proceedings of the European Conference on Computer Vision (ECCV), Munich, Germany, 8–14 September 2018; pp. 670–685.
33. Thagard, P. Naturalizing logic: How knowledge of mechanisms enhances inductive inference. *Philosophies* **2021**, *6*, 52. [CrossRef]
34. Brisson, J.; Markovits, H. Reasoning strategies and semantic memory effects in deductive reasoning. *Mem. Cogn.* **2020**, *48*, 920–930. [CrossRef] [PubMed]

35. Weldu, Y.W.; Knapp, G.M. Automated generation of 4D building information models through spatial reasoning. In Proceedings of the Construction Research Congress 2012: Construction Challenges in a Flat World, West Lafayette, Indiana, 21–23 May 2012; pp. 612–621.
36. de Soto, B.G.; Rosarius, A.; Rieger, J.; Chen, Q.; Adey, B.T. Using a Tabu-search algorithm and 4D models to improve construction project schedules. *Procedia Eng.* **2017**, *196*, 698–705. [CrossRef]
37. Zhang, L.; Long, C.; Zhang, X.; Xiao, C. Ris-gan: Explore residual and illumination with generative adversarial networks for shadow removal. *Proc. AAAI Conf. Artif. Intell.* **2020**, *34*, 12829–12836. [CrossRef]
38. Wu, M.; Chen, R.; Tong, Y. Shadow elimination algorithm using color and texture features. *Comput. Intell. Neurosci.* **2020**, *2020*, 2075781. [CrossRef] [PubMed]
39. Sandhya, N.; Marathe, A.; Ahmed, J.D.; Kumar, A.; Harshith, R. Convolutional Neural Network Based Approach to Detect Pedestrians in Real-Time videos. *Int. J. Innov. Technol. Explor. Eng.* **2020**, *10*, 303–308.
40. Yan, H.; Song, C. Multi-scale deep relational reasoning for facial kinship verification. *Pattern Recognit.* **2021**, *110*, 107541. [CrossRef]

buildings

MDPI

Article

Feature-Based Deep Learning Classification for Pipeline Component Extraction from 3D Point Clouds

Zhao Xu [1,*], Rui Kang [1] and Heng Li [2]

1 Department of Civil Engineering, Southeast University, Nanjing 210096, China; kr97102@seu.edu.cn
2 Department of Building and Real Estate, Hong Kong Polytechnic University, Kowloon, Hong Kong 999077, China; heng.li@polyu.edu.hk
* Correspondence: xuzhao@seu.edu.cn

Abstract: This paper proposes a novel method for construction component classification by designing a feature-based deep learning network to tackle the automation problem in construction digitization. Although scholars have proposed a variety of ways to achieve the use of deep learning to classify point clouds, there are few practical engineering applications in the construction industry. However, in the process of building digitization, the level of manual participation has significantly reduced the efficiency of digitization and increased the application restrictions. To address this problem, we propose a robust classification method using deep learning networks, which is combined with traditional shape features for the point cloud of construction components. The proposed method starts with local and global feature extraction, where global features processed by the neural network and the traditional shape features are processed separately. Then, we generate a feature map and perform deep convolution to achieve feature fusion. Finally, experiments are designed to prove the efficiency of the proposed method based on the construction dataset we establish. This paper fills in the lack of deep learning applications of point clouds in construction component classification. Additionally, this paper provides a feasible solution to improve the construction digitization efficiency and provides an available dataset for future work.

Keywords: deep learning; pipeline component extraction; point clouds; feature; CNN (convolutional neural network)

Citation: Xu, Z.; Kang, R.; Li, H. Feature-Based Deep Learning Classification for Pipeline Component Extraction from 3D Point Clouds. *Buildings* **2022**, *12*, 968. https://doi.org/10.3390/buildings12070968

Academic Editor: Ahmed Senouci

Received: 14 June 2022
Accepted: 5 July 2022
Published: 7 July 2022

Publisher's Note: MDPI stays neutral with regard to jurisdictional claims in published maps and institutional affiliations.

Copyright: © 2022 by the authors. Licensee MDPI, Basel, Switzerland. This article is an open access article distributed under the terms and conditions of the Creative Commons Attribution (CC BY) license (https://creativecommons.org/licenses/by/4.0/).

1. Introduction

The 3D reconstruction models are gradually replacing 2D drawings for information transmission and more in-depth processing to meet the demands of civil engineering digitization, according to the research by Ma and Liu [1]. Among a variety of 3D information data formats, the point cloud model is the research focus of many scholars. The 3D point cloud data are widely used in the construction industry for model reconstruction, geometry inspections and other applications, but there is still a research gap regarding the practical applications [2]. In the early stage of point cloud data processing, scholars mainly utilize traditional algorithms to deal with complex and irregular object reconstruction [3], complicated scenes with repetitive objects [4] and the updating of as-designed BIM to as-built BIM [5]. On the basis of shape extraction, some scholars further enriched the semantics of point cloud data and deepened the relationship between BIM [6,7] and point clouds via IFC (Industry Foundation Classes) extension [8]. This has had positive significance for data management in the field of civil engineering, and has greatly promoted the informatization process in the construction industry. However, such studies mainly focus on indoor scenes and generally require manual participation, which leads to a low level of automation and compatibility [9]. Thus, some scholars have turned to deep learning for resolution. With the maturity of deep learning algorithms, more network structure designs for the deep learning of point clouds have emerged, the feasibility of applying deep networks to point cloud learning has been verified and the design theory has been continuously improved [10]. The application of deep learning in civil engineering information management has developed from image damage detection [11] to 3D reconstruction segmentation [12]. In order to use deep learning to enhance the efficiency of engineering information acquisition, some scholars directly apply deep learning networks on IFC models [13] or use deep learning to assist in removing unwanted information [14]. Overall, most scholars focus their attention on deep learning to help in the acquisition and processing of 3D point clouds, so as to reduce the manual participation and achieve automation. This approach would be for the great promotion of efficient digital management, especially for complicated building scenarios such as MEP (Mechanical, Electrical and Plumbing) systems.

With the development of 3D reconstruction technology, the accuracy of the obtained data is gradually improved. At this time, the requirements for the accuracy and flexibility of the model become more stringent. In the realization of as-built BIM from point clouds, the extraction of building components plays a vital role, among which pipeline component extraction is an important task. The digitization of existing buildings can be completed by extracting and identifying point clouds, and then the management of important building components, such as MEP systems, can be completed. In the construction industry, the application of point cloud data in information sharing platforms is synchronized with the development of point cloud data processing methods [15]. To efficiently use point clouds, scholars have paid attention to semantic recognition, which is an essential step to identify the part that a point cloud belongs to. In reverse engineering, many algorithms are used in the mesh and point clouds for labeling, such as randomized cuts for the mesh [16], mesh labeling via CRF [17] and the octree-based method for point cloud segmentation [18]. These algorithms are designed using the geometry features and can work for certain datasets, but it is difficult for them to maintain high accuracy in various datasets. To address this problem, the concepts and structures of deep learning are introduced to design a self-learning algorithm for mesh labeling [19], and we further turn to point cloud labeling [20] due to the flexibility and integrity of the point cloud approach.

However, the currently proposed algorithms experiment with repetitive and limited datasets, which leads to difficulties for actual applications in engineering practice, with even fewer applications in MEP systems [9]. Among the datasets used by these algorithms, one part of the datasets is composed of small-volume point clouds generated by CAD models which have little noise interference, such as ModelNet40 [21], while the other part of the datasets is designed for scene segmentation and focuses on identifying ceilings, tables and chairs in indoor rooms, such as the Stanford 3D semantic parsing dataset [22]; or trees, roads and buildings in outdoor spaces, such as Semantic3D.net [23]. In addition to the problems with the datasets used for training, the practical engineering application of these algorithms is also restricted by the environment and other conditions. The point clouds obtained on site are incomplete because of interference. At the same time, they are affected by the speed of the data collection, meaning it is difficult to obtain complete attribute data, such as RGB (Red, Green, Blue) data.

To overcome the above problems, it is necessary to adjust the structure design of the deep learning algorithms to adapt to the needs of engineering applications, especially for MEP systems. Thus, based on previous studies, this paper proposes a new deep network structure and builds a dataset that emphasizes engineering scenarios for learning and training. The key to our approach is the surrounding description of a single point. Through the usage of the SHOT (signature of histograms of orientations) and spin image approaches, the attributes of a single point are expanded to fill the gaps in data collection.

The key contributions of our work are as follows: (1) we design a novel neural network architecture suitable for an imperfect point cloud collected from the construction project; (2) we introduce the concepts of SHOT and spin image in point cloud deep learning and improve the performance of point cloud labeling by depicting the distribution of points; (3) we establish a point cloud deep learning dataset for engineering application scenarios and pipeline component extraction.

The rest of this paper is organized as follows. Section 2 reviews related work and Section 3 introduces the overall methodology used for the point cloud labeling. In Section 4, the architecture of the network is shown to express how it works. Then, we demonstrate the performance in the experiments in Section 5 with pipelines as examples. Finally, the article is concluded in Section 6.

2. Literature Review

2.1. Information Extraction and 3D Reconstruction of Pipeline Components

As a more efficient data acquisition method, non-destructive tests (NDTs) replace traditional manual measurement methods in many application scenarios [24]. Among the many NDT methods, oblique photography, ultrasonic detection and laser scanning are the most representative ones. As a large-scale data collection method, oblique photography mainly uses drones for outdoor data collection. The subsequent data processing tends to distinguish buildings from the environment [25] or extract building outlines [26]. Ultrasonic detection is mainly used for internal damage detection, such as crack detection [27] and weld detection [28], and no reconstruction model is generated during the detection process. Compared with the first two types of NDTs, laser scanning is more suitable for construction component reconstruction in the field of civil engineering due to its accuracy and scanning range. Different types of laser scanning equipment can be applied for overall scene scanning or partial scanning to obtain stable and accurate point clouds. Using these point clouds, 3D models of buildings or components can be reconstructed to conduct measurement [29], recognition [30] and other data processing operations.

Processed point clouds are used in the field of civil engineering for quality and safety management during the construction phase and in building information management, which is known as BIM. Containing high-precision three-dimensional data, the point cloud can describe the settlement state of the construction site and the deformation state of the construction components, which is of great benefit for the collection of data for MEP systems from complicate pipelines. Some scholars have proposed algorithms, including the convex hull [31] and improved RANSAC algorithms [32], to extract primitive geometric information form a construction site or building for further as-built BIM applications.

As an important building information exchange technology, BIM requires more data to show a complete model with rich semantics to support the information required from design to operation and maintenance phases while gradually improving. With the expansion of the application range and the increase in importance, there are stricter requirements for the point cloud data used in the BIM platform [33]. In the process of using BIM for digital construction, pipeline management methods have also been developed, since the pipeline occupies an important position during maintenance. In buildings that use BIM technology for forward design, the management of pipelines can achieve efficient and diversified purposes. In addition to conventional pipeline information management and collision detection, some scholars have realized the prediction of pipeline corrosion on the basis of BIM. Tsai et al. [34] utilized the semantic information management function to store pipeline sensor information and to monitor and visualize the maintenance status. Under the premise of complete pipeline information in BIM, many technologies, e.g., IoT and RFID, can be used to assist in pipeline management [35,36]. Although BIM technology has been widely used in the construction industry, a large number of existing buildings need to be processed via as-built BIM technology to harness its superior functions for efficient management. For the realization of as-built BIM for pipelines, there has been continuous research for the purpose of reducing the manual participation proportion during processing. Patil et al. [37] paid attention to the Hough transform for the automatic detection of cylinder parameters in point clouds. Tran et al. [38] used traditional shape features for cylinder fitting to complete the pipeline extraction in a similar way.

Among these works for MEP systems and pipeline management, these proposed algorithms are usually used in ideal conditions, while the point clouds collected in actual engineering scenes, especially in complex MEP systems, are often noisy and incomplete.

The existing research has explored a feasible workflow to use the point clouds for building information acquisition and management. However, these traditional point cloud processing methods require a lot of manual participation and are limited to a few fixed models, leading to poor efficiency and compatibility when dealing with the pipelines.

2.2. Deep Learning in Point Cloud

As a data format that is widely used in 3D reconstruction, the point cloud has always been a research hotspot in graphics processing and computer vision. Recognizing the semantic information of the point cloud as an important part of its use has been a continuous concern for scholars [39]. Before deep learning gained widespread attention, graphical analyses of point clouds were already carried out. The point cloud data features, including spin image [40] and SHOT [41] features, are calculated to classify different parts to achieve segmentation. Traditional classification algorithms such as the principal component analysis [42] are also used to assist in point cloud processing. Although these characteristics can help distinguish the geometric features of a point cloud reconstruction model, they are too rigid to lose their self-learning ability and strong adaptability. The emergence of deep learning algorithms fills the gaps in this research field.

Deep learning for point clouds mainly focuses on the research fields of recognition and classification and is applied to road condition recognition for automatic driving [43], large-scale scene object classification [44] and indoor scene recognition [23]. According to the methods of deep learning data recognition, they can be divided into three types in general: Pointnet, Voxelnet and feature-based net.

Pointnet proposed by Qi et al. is an important algorithm in the field of point cloud deep learning [20]. Pointnet is highly efficient and robust and can deal with object classification, part segmentation and scene semantic parsing. Since Qi et al. provided both an authoritative theoretical analysis and experimental evaluation, many follow-up works aiming to improve the network structure have been based on Pointnet, such as Pointnet++ [10], Foldingnet [45] and dynamic graph CNN [46]. The novel deep net architecture proposed as Pointnet, which focuses on processing unordered 3D points, has proved its stability and efficiency via experiments. Compared with other algorithms, this algorithm is characterized by the extraction and processing of the main properties of a point cloud, which are the disorder, interactions among points and invariance under transformations. The core network structures of Pointnet are T-net, which is designed to strengthen the relevance of points in the early stage of data training, and the max pool layer, which is used for dealing with unordered points and keeping the invariance under transformations. Additionally, on the basis of a classification network, Qi et al. [20] further expanded the network structure to implement part and semantic segmentation by combining point features and global features. Intuitive explanations were developed for the robust and effective performance of Pointnet in their paper, and more studies are following the steps of Pointnet to pursue better performance and more flexible use in more fields.

However, because of the training samples used by Pointnet, which are relatively simple and small in size, and the lack of attention to local features, it is difficult to implement Pointnet in large-scale continuous scenes, especially construction engineering scenes.

Differing from Pointnet, which uses a small point cloud as the input, Voxelnet emphasize points connections. This algorithm is characterized in the early-stage data by voxel mark processing. Point clouds are allocated in voxels and form input data according to the distribution in the voxel [47] or voxel labels [48]. Voxelnet can achieve precise control because of the rasterization of point clouds and can handle variable point clouds in more flexible ways in different usage scenarios. Although forming a voxel grid in the most straight-forward way fully utilizes ConvNet, which was originally designed for 2D imaging by changing the input data, it also leads to many disadvantages. Since VoxelNet is mainly used in large-scale scenarios, it is often difficult to balance the workload and algorithm accuracy, and the stability is greatly affected by the voxels. Although Kd-net was proposed to further optimize the low efficiency of the voxels, it is still not a fundamental solution to the application of deep learning on 3D data based on the characteristics of the point cloud [49].

In addition to the above two types of deep learning networks, some scholars start from combining deep learning and traditional shape features by extracting features in advance and then using them as the inputs of networks [19,50]. These types of networks rely on the study foundation of traditional algorithms to provide reliable parameters for subsequent learning, balancing the robustness and flexibility. Nevertheless, there have been cases where traditional shape features are overused. For example, Guo et al. used more than 7 features in their paper, which led to excessive reliance on the traditional shape features and weakness of the learning network, together with increasing the workload at the preprocessing stage. Although these networks can provide stable performance, they have not brought about obvious improvements compared to traditional algorithms.

2.3. Deep Learning in Construction Industry

The introduction of digital technology such as deep learning has accelerated the construction digitization process and provided more efficient tools to process engineering data. A recent study [51] showed that deep learning technologies have been applied for prevalent construction challenges such as site management, budget and energy control and building quality monitoring. In terms of information management and prediction, Sun et al. [52] used the long–short-term memory (LSTM) neural network in deep learning for exchange rate forecasting and Ziari et al. [53] made full use of the deep learning network of natural language processing (NLP) to assist highway agencies in decision-making procedures, indicating that deep learning is gradually shifting from theory to practice and becoming a substitute for traditional algorithms in some empirical research fields. Additionally, during the process of image recognition, since ImageNet [54], which is used for large-scale image recognition, was proposed in 2012, deep-learning-related research has received more attention and there has been wide adoption of similar deep learning structures, e.g., convolutional neural networks. In the construction industry, the numerical prediction and image recognition functions of deep learning algorithms are mainly used to solve practical engineering problems. For instance, Deng et al. [55] and Nguyen et al. [56] separately proposed CNN model and DNN model to predict the strength of concrete for monitoring and design. Rahman et al. [57] designed an RNN (recurrent neural network) model for building energy prediction. Rafiei and Adeli [58] presented a novel machine learning model for estimating construction costs by combining the DBM-SoftMax (deep Boltzmann machine) layer and BPNN (back-propagation neural network). For image recognition functions, scholars tend to invest more in research on on-site worker behavior and construction quality and safety monitoring. For example, Fang et al. [59,60] proposed a CNN model to check workers' posture and helmet-wearing from videos. Kolar et al. [61] presented a CNN-based model to detect guardrails in 2D images for safety improvements. In researching applications related to quality issues, scholars have focused more on crack detection. Similar CNN-based approaches have been adopted for crack detection, where scholars have made efforts towards accuracy and robustness improvements [62,63].

Although scholars have done a lot of work by combining deep learning approaches for the digitalization of the construction industry, there has been very little work related to 3D scenes and BIM, despite the model recognition work carried out by Wang et al. in a BIM environment [64]. The adoption of BIM models has brought significant improvements to construction digitization over the years [65]; scholars are also striving to achieve BIM through the use of more advanced equipment and algorithms [66] but there is still a lot of work to be done to achieve efficiency improvements, especially for as-built BIM. At present, the achievement of as-built BIM requires a lot of human participation to manually label construction components, which can be automated by applying deep learning algorithms for point clouds. However, the datasets used in the related algorithms, such as the ShapeNet dataset [67], ModelNet40 dataset [21] and Stanford large-scale indoor space dataset [22], do not match the engineering application scenarios.

3. Methodology

In this paper, we present a point cloud classification method by using deep CNNs, as shown in Figure 1. First, the original data are processed by the preprocessing network to obtain the training features. The original input data are the global coordinate values of the point cloud. Inspired by PointNet, the network structure of this paper also adopts a combination of global features and local features for training, whereby the global features are formed by the global coordinates of a single point, which are entered into UnitNet, then according to the traditional algorithm the spin image, SHOT and normal of each point are calculated as local features and input into FeatureNet for training. Differing from PointNet's processing of global and local features, the different kinds of features in this paper are calculated separately in the early stage and trained by independent networks, which increases the irrelevance of the data. After the two kinds of features are trained on their own networks, 36-dimensional and 64-dimensional data are output, from which the new input data are merged as 100-dimensional data. Second, deep CNNs are built for feature extraction and final classification. Before being sent to FinalNet for training, a 2D feature matrix (n $\times$ 10 $\times$ 10) is transformed from processed features in the form of feature maps to facilitate subsequent feature extraction. Finally, the deep convolutional networks and the pooling layers are combined for training and learning, and two fully connected layers are applied in the end to obtain the classification results.

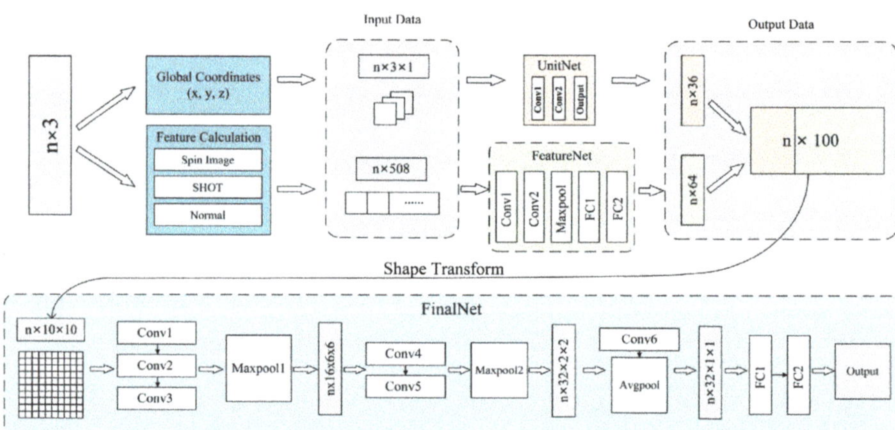

Figure 1. Network resolution.

The key parts of this paper are the preprocessing networks used for the extraction of local and global features by Unitnet and FeatureNet in the early stage, and the use of a deep convolutional network in FinalNet to weight the merged features for recognition in the later stage.

4. Network Architecture

4.1. Preprocessing Networks

In this paper, the network is used to classify the large-scene point clouds, especially the one that is collected at the construction site. Due to the different application scenarios, the data input to the neural network have the following characteristics: (1) Isolation of the input data. The point cloud is disordered, and in the recognition of large scenes, each point is input into the network separately, which highlights the weak connection and poor continuity between the data. (2) The particularity of the application scenario. Many segmentation and recognition algorithms for large scenes are based on indoor scenes, and many elements such as chairs can be obtained from existing model libraries and trained in advance. For a construction site, there are few large-scale point cloud datapoints available for use. (3) The lack of data features. Although the current laser scanning technology has reached a high level of accuracy and good RGB color rendering performance by taking photos for scanning, in order to ensure versatility and avoid interference during color collection under construction conditions, the original data obtained often only have coordinate values.

In order to solve the above problems, this paper sets up UnitNet and FeatureNet for data preprocessing for global features and local features.

4.1.1. Global Feature Extraction

UnitNet is designed for global coordinate transformation, where each point is input as a separate unit with normalized coordinate values as input channels. The network architecture is shown in Figure 2. The input unit is trained by a one-kernel convolutional network with 16 channels and 36 channels. Simultaneously, the BN (batch normalization) layer and ReLU layer are applied for parameter correction and activation during the training process.

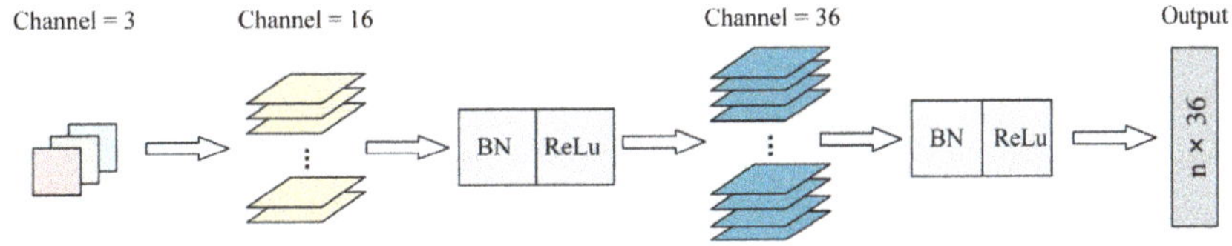

Figure 2. Architecture of UnitNet.

It can be seen from the network structure that the shape of the input unit is not changed, only the number of channels is gradually increased. In this way, the coordinate values of the input data can be further input into the subsequent network after transformation, while retaining the global characteristics of the original data. In this process, the coordinate function of the point is strengthened.

4.1.2. Local Feature Extraction

The raw data from the point cloud have strong independence, which makes it difficult to find the connections with the surrounding points when only using the coordinate value. This disadvantage can be overcome by traditional algorithms such as SHOT and spin image algorithms through the description of local points.

In this paper, SHOT (signature of histograms of orientations) and spin image are important features used as input data to train neural networks, which are originally designed for surface matching. Both features can be used to describe the distribution of surrounding points, whereby SHOT focuses on the locations of surrounding points and spin image focuses on the distribution density of the points.

There are two reasons for using SHOT and spin image as input data for this convolutional neural network: (1) Each point in the point cloud exists independently, and its own attributes need to be obtained through subsequent processing, except for the coordinates and RGB color that are obtained. SHOT and spin image can increase the correlation between points and attach local attributes to independent single points. (2) The CNN framework requires rich data for training and judgment, so algorithms that form a large amount of features are needed.

SHOT

SHOT (signature of histograms of orientations) was proposed as a local reference system for surface matching [41]. SHOT, which is also as a local 3D descriptor that has been used in many scenarios, balances the signature and histogram to maintain the descriptiveness as well as the robustness. After the concept of SHOT was put forward, some scholars further improved it with global structure frames [68] or textures [69]. In this paper, a local reference frame is first established to determine the locations of neighbor points, as shown in Figure 3. In the local reference frame, the spherical space is divided into 32 spatial bins, resulting from 8 latitude divisions (only 4 are shown in Figure 3), 2 longitude divisions and 2 radius divisions. According to the normal of the point in every spatial bin, n_{vi}, and the normal of the feature point, n_p, the cosine of the corresponding angle, θ_i, is calculated by $\cos\theta_i = n_p – n_{vi}$. Then, according to the calculated cosine value, 11-level histogram statistics are performed on the number of points falling into each spatial bin. Finally, after the data are normalized, a 352-dimensional ($8 \times 2 \times 2 \times 11$) feature is generated as an array to input to the neural network.

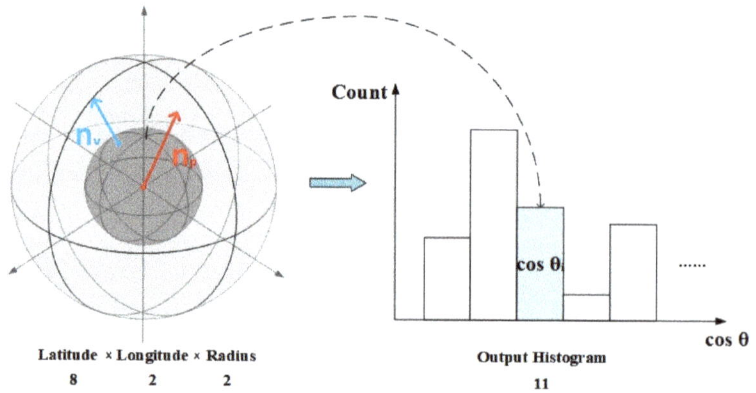

Figure 3. Generation principle of SHOT.

Spin Image

Spin image was proposed by Johnson et al. in 1999 for efficient object recognition and surface matching in 3D scenes. [40] It was first used in 3D meshes. Subsequently, the related algorithm further optimized the spin image approach [70,71]. With the rise of laser scanning technology, point clouds are gradually displayed as 3D models in more application scenarios. Thus, there are increasing studies exploring the application of the spin image approach in point clouds, such as for classification [72] and registration [73]. In addition to the traditional application technology route, a spin image also provides usable data features for deep learning in point cloud processing [19].

The spin image approach was proposed to perform surface matching by depicting the local distribution of other points around the feature point. As shown in Figure 4, first an oriented point with the surface normal is selected to define a coordinate system, where α is the radial coordinate and β is the elevation coordinate. Then, a 2D accumulator indexed by α and β is rotated, taking the normal as the axis. As the accumulator rotates, the number of points falling into the bin indexed by (α, β) gradually increases, forming the spin image, which can be described by Equation (1), where $S_O(x) : R^3 \to R^2$, p is the reference point, n is normal vector of p and x is the neighbor point.

$$S_O(x) \to (\alpha, \beta) = \left(\sqrt{\|x - p\|^2 - (n\cdot(x - p))^2}, n\cdot(x - p)\right) \tag{1}$$

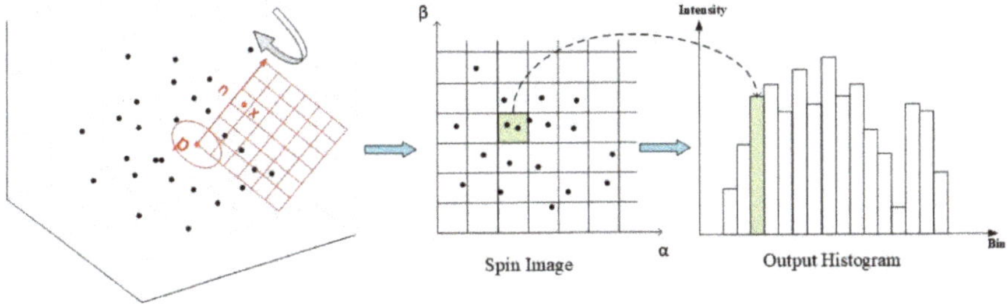

Figure 4. Generation principle of the spin image approach.

Finally, an intensity-related histogram is output as an array to compose the input data of the neural network.

FeatureNet

The output of the above algorithms forms a sequence of 508 elements, of which 352 are from SHOT, 153 are from the spin image and 3 are from the normal. Most of the elements are obtained by establishing a local spatial coordinate system and describing the surrounding points in blocks, which also leads to two problems: (1) Excessive data volume. The method of dividing the space into blocks and the output form of the histogram results in a large amount of local feature data, which is difficult to process. (2) The existence of invalid data. Since the points are not evenly distributed in all spaces, there are lots of zero-value data in the sequence, together with invalid data caused by noise.

The large data volume is often processed using a PCA algorithm for dimensionality reduction, as shown in other studies, but this issue cannot be solved in this paper due to the linearly independence of the feature data. Therefore, this paper proposes the use of FeatureNet for effective feature extraction to reduce the amount of data and eliminate invalid data. The structure of FeatureNet is shown in Figure 5.

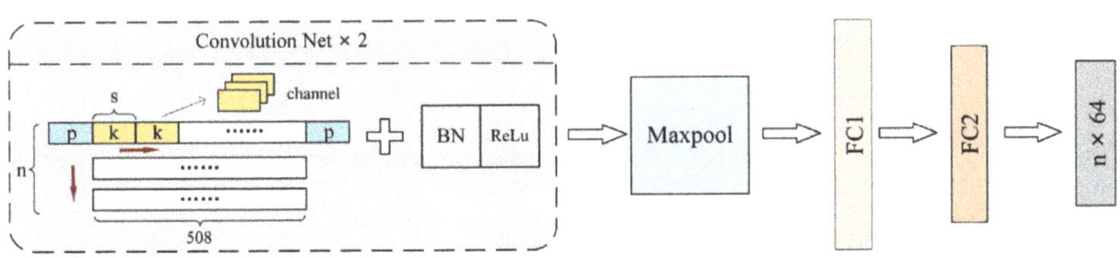

Figure 5. The FeatureNet structure proposed in this study.

In FeatureNet, two one-dimensional convolutional layers and a pooling layer are used to assign weights to each feature element to eliminate invalid data and zero-valued data, and then two fully connected layers are applied to perform dimensional compression to achieve the purpose of data cleaning. Through the processing of FeatureNet, 64-dimensional data are finally output, effectively compressing the dimensions of feature data.

4.2. FinalNet

FinalNet is the main structure that implements feature extraction in the deep convolutional networks. As shown in Figure 1, considering the deeper network structure leads to better performance [74] and improves the response speed of the network structure. FinalNet uses 6 convolutional layers to perform the feature extraction, 3 pooling layers to reduce the dimensions and 2 fully connected layers to make the final judgement.

FinalNet directly processes the feature map size formed by reshaping the sequence spliced from the local feature and global feature. Figure 6 shows the details for composing these two features. There is a weak connection between the two different features in the initial feature map formed via concatenation and deformation in the first stage. After processing the convolutional layer, connections between different feature elements are established through convolutional kernels.

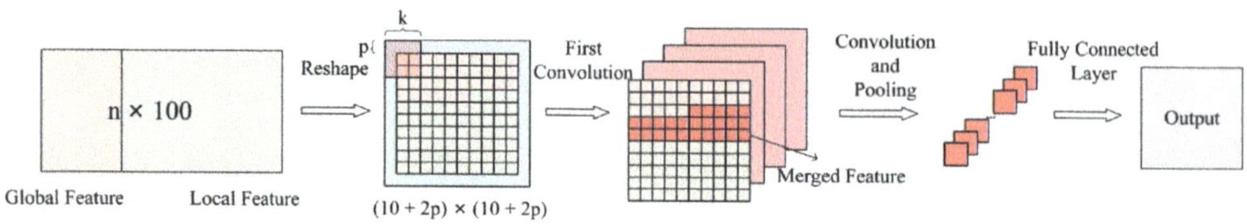

Figure 6. Feature combination process in FinalNet.

The merging result after the first convolution phase can be seen in Figure 6. Subsequently, with the deepening of the convolutional layers and the participation of the pooling layers, a fully integrated feature sequence is formed. Finally, the fully connected layers are applied to form non-linear combination and output the result.

5. Experiment

This paper focuses on the object classification of construction components, especially the components that require maintenance management in complex construction scenes. The existing datasets, such as ModelNet40 (Wu et al., 2015), which has 12,311 CAD models from 40 man-made object categories, and ShapeNet (Yi et al., 2016), which is normally used for indoor scenes in segmented learning, e.g., for tables and chairs, cannot meet the need for the classification of construction components in the field of civil engineering. Additionally, the point clouds generated by CAD models cannot reflect the actual state of the point clouds obtained via laser scanning due to the environmental interference and blind spots. Thus, this paper uses a self-built dataset via the laser scanning and labeling of the appropriate scenes, since there are few datasets that can be directly used for the related training and learning processes.

In this paper, the classification of pipelines, which are difficult to manage in the construction process and complicate the maintenance process, is shown as an example to prove the validity of the proposed network. The whole scene with the format of the point cloud and the pipelines that are manually extracted, labeled and used for training and learning can be seen in Figure 7.

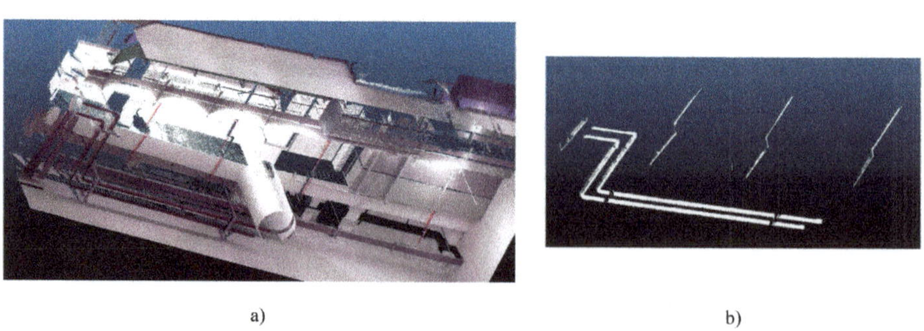

Figure 7. The point cloud dataset. (**a**) is the engineering scene dataset and (**b**) is showing extracted pipelines.

5.1. Architecture Design Validation

In this section, a control experiment is designed for the validation of the networks this paper has proposed. The networks that are trained and perform the pipeline classification in the control experiment are inspired by feature-based DNNs (Y. Fang et al., 2015; Guo et al., 2015), where traditional features are extracted and sent into deep networks directly. In some simple application scenarios (such as image recognition), a network that has enough neurons can achieve good results, even without an effective structural design. However, compared with ordered and feature-rich datasets, such as image datasets, the raw data from point clouds, which are unordered, lack features and are weakly connected with each other, meaning they have higher requirements. Thus, a control experiment is designed here for the validation of UnitNet and FeatureNet, whereby spin image and SHOT features are combined and sent to a deep convolution network directly after the PCA for dimensionality reduction, as shown in Figure 8. This control experiment used for comparison contains 11 convolutional layers, together with BN, ReLU and 2 pooling layers.

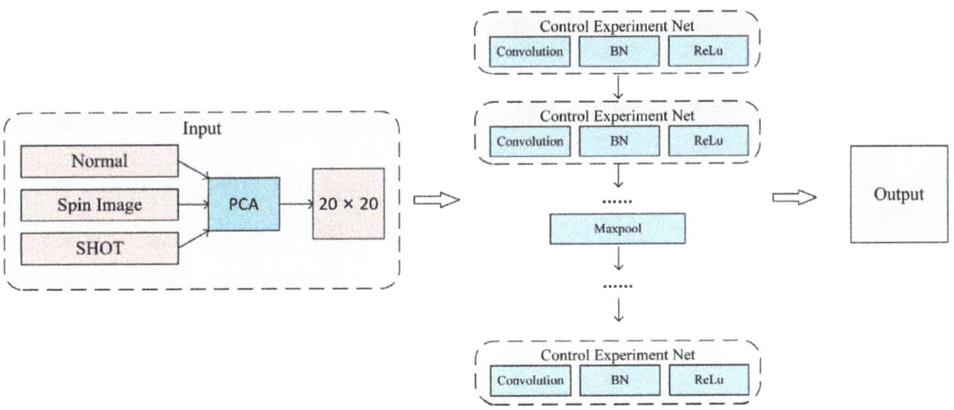

Figure 8. Network structure of the control experiment.

The results of the control experiment demonstrate the positive effects of the application of UnitNet and FeatureNet, as listed in Table 1. The train loss and train accuracy of the control experiment are better than the proposed method but lead to an overfitting result, as reflected by the test accuracy comparison.

Table 1. The results of the pipeline classification experiments.

	Train Loss	Train Accuracy	Test Accuracy
Control Experiment	0.0036	99.87%	82.37%
Proposed method	0.0054	99.75%	94.62%

5.2. Results of the Dataset and Comparison

Due to the dataset used and the specificity in the engineering of the classification object, there are few experimental comparison datasets available for other algorithms. The experimental results are mainly presented via a large-scale scene and a comparison with the control experiment. The results listed in Table 2 show that our method, which uses UnitNet and FeatureNet for structural improvement, significantly outperforms the deep network in the control experiment.

Table 2. Classification results for a large-scale dataset.

Dataset	Mean	Set 1	Set 2	Set 3	Set 4	Set 5	Set 6	Set 7	Set 8	Set 9
Points number		191,404	696,406	1,419,606	718,361	411,160	296,335	521,597	1,203,240	732,260
Control Experiment	84.06%	80.29%	88.07%	91.55%	92.73%	82.98%	86.10%	92.78%	78.59%	63.45%
Proposed method	98.03%	97.24%	97.50%	94.93%	98.91%	97.80%	98.17%	99.59%	98.54%	99.63%

The qualitative visible results are shown in Figure 9. Here, a comparison of the classification results of the two algorithms is shown for datasets 2 and 5. It can be seen more intuitively that our method can effectively avoid misjudgments by strengthening the connections between local features and global features to achieve better classification.

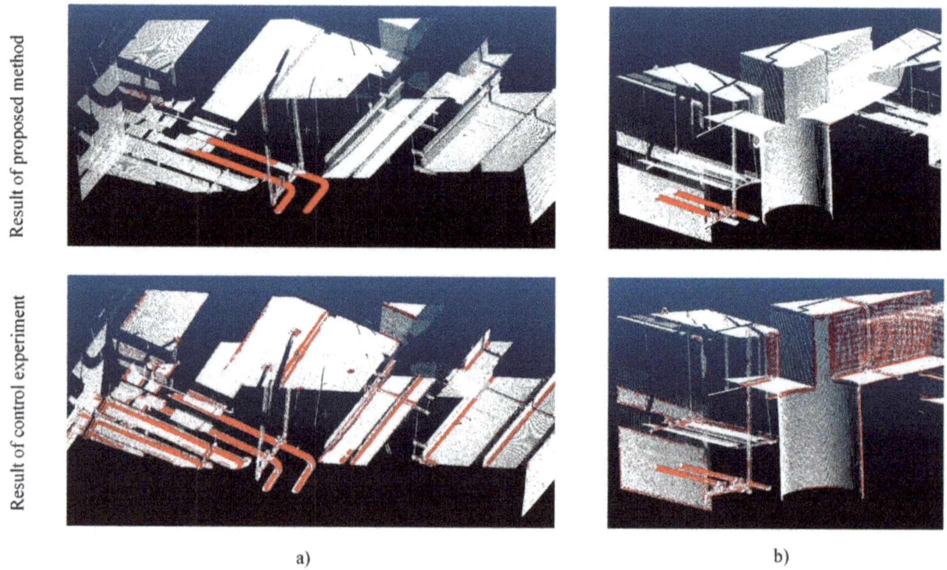

Figure 9. Visible result comparison for the intercepted datasets. (**a**) shows the result comparison on dataset 2 and (**b**) shows the result comparison on dataset 5.

According to Figure 9, the results show that the proposed method can significantly focus more on the pipeline components. This could be the result of applying UnitNet and FeatureNet to process features separately and the better combination of local features and global features carried out in FinalNet. The mean accuracy of the proposed method is over 98% and the visible results also indicate the effectiveness of the pipeline extraction process. All of these results of the experiments prove that the proposed method is workable and could be further studied and applied for MEP systems.

6. Conclusions and Discussion

In this work, we proposed a deep neural network that is designed especially for complex construction industry applications and demonstrated the effectiveness of our method. Firstly, we built a dataset based on the engineering situation of the construction industry. Then, we established a neural network structure in which local features and global features are processed separately in UnitNet and FeatureNet. We made full use of traditional shape features to enrich the features of simple point clouds and avoid excessive dependence on them through the structural design of the neural networks. Further, a feature map was proposed in FinalNet for feature fusion. Finally, by establishing a control experiment and comparing the results, the method proposed in this paper offers an effective and feasible solution for deep learning applications for classification in the construction industry.

This study makes three main contributions to the knowledge and engineering informatics. Firstly, we have optimized the combination of traditional shape features and deep learning networks, which improves the accuracy of the engineering information collected for 3D scenes and the feasibility of this method, as proven via experiments. Secondly, we have established a targeted dataset for the actual situation in construction and engineering, which provides a data basis for related research in the construction field in the future and expands the breadth of engineering information in three-dimensional space. Thirdly, we have provided efficient solutions to replace manual labor in the processes of 3D reconstruction and as-built BIM in the construction industry, since they are vital parts of the engineering information digitization process. Among these contributions, the most important is that this study provides a feasible and effective method of MEP system reconstruction and digitalization, and this method could be further studied for application in other construction fields.

Last but not least, although the proposed method achieved good performance in the experiment, there are still some problems to cover in future studies. Because the extraction of local features is based on the description of the neighboring points, the classification accuracy of the boundary part of the input data is relatively lower than other parts. Additionally, although this paper has initially established a dataset for construction engineering, there are comparatively few categories available. Regarding the experiments, the design of this work could also be improved for the simplification of input data, and the binary classification should be further extended to multiple classifications of other MEP components.

Author Contributions: Conceptualization, Z.X., R.K. and H.L.; methodology, Z.X., R.K. and H.L.; software, R.K.; validation, R.K; investigation, R.K.; writing—original draft preparation, Z.X., R.K. and H.L.; writing—review and editing, Z.X., R.K. and H.L.; visualization, Z.X. and R.K.; funding acquisition, Z.X. All authors have read and agreed to the published version of the manuscript.

Funding: This work was supported by the National Natural Science Foundation of China (NSFC-72071043), Natural Science Foundation of Jiangsu Province (BK20201280), MOE (Ministry of Education in China) and Project of Humanities and Social Sciences (20YJAZH114).

Institutional Review Board Statement: Not applicable.

Informed Consent Statement: Not applicable.

Data Availability Statement: The data are contained within the article. Additional supporting data presented in this study are available on request from the corresponding author.

Conflicts of Interest: The authors declare no conflict of interest.

References

1. Ma, Z.; Liu, S. A review of 3D reconstruction techniques in civil engineering and their applications. *Adv. Eng. Inform.* **2018**, *37*, 163–174. [CrossRef]
2. Wang, C.; Cho, Y.K.; Kim, C. Automatic BIM component extraction from point clouds of existing buildings for sustainability applications. *Autom. Constr.* **2015**, *56*, 1–13. [CrossRef]
3. Wang, L.; Zhao, Z.; Wu, X. A Deep Learning Approach to the Classification of 3D Models under BIM Environment. *Int. J. Control Autom.* **2016**, *9*, 179–188. [CrossRef]
4. Wang, Q.; Kim, M.-K. Applications of 3D point cloud data in the construction industry: A fifteen-year review from 2004 to 2018. *Adv. Eng. Inform.* **2019**, *39*, 306–319. [CrossRef]
5. Wang, Y.; Pan, G.; Wu, Z.; Han, S. Sphere-spin-image: A viewpoint-invariant surface representation for 3D face recognition. In Proceedings of the International Conference on Computational Science, Krakow, Poland, 6–9 June 2004.
6. Wang, Y.; Sun, Y.; Liu, Z.; Sarma, S.E.; Bronstein, M.M.; Solomon, J.M. Dynamic graph cnn for learning on point clouds. *ACM Trans. Graph.* **2019**, *38*, 1–12. [CrossRef]
7. Barazzetti, L. Parametric as-built model generation of complex shapes from point clouds. *Adv. Eng. Inform.* **2016**, *30*, 298–311. [CrossRef]

8. Xue, F.; Lu, W.; Chen, K.; Webster, C.J. BIM reconstruction from 3D point clouds: A semantic registration approach based on multimodal optimization and architectural design knowledge. *Adv. Eng. Inform.* **2019**, *42*, 100965. [CrossRef]
9. Gao, T.; Akinci, B.; Ergan, S.; Garrett, J. An approach to combine progressively captured point clouds for BIM update. *Adv. Eng. Inform.* **2015**, *29*, 1001–1012. [CrossRef]
10. Rodríguez-Gonzálvez, P.; Rodríguez-Martín, M.; Ramos, L.F.; González-Aguilera, D. 3D reconstruction methods and quality assessment for visual inspection of welds. *Autom. Constr.* **2017**, *79*, 49–58. [CrossRef]
11. Rodríguez-Moreno, C.; Reinoso-Gordo, J.F.; Rivas-López, E.; Gómez-Blanco, A.; Ariza-López, F.J.; Ariza-López, I. From point cloud to BIM: An integrated workflow for documentation, research and modelling of architectural heritage. *Surv. Rev.* **2018**, *50*, 212–231. [CrossRef]
12. Krijnen, T.; Beetz, J. An IFC schema extension and binary serialization format to efficiently integrate point cloud data into building models. *Adv. Eng. Inform.* **2017**, *33*, 473–490. [CrossRef]
13. Agapaki, E.; Brilakis, I. CLOI-NET: Class segmentation of industrial facilities' point cloud datasets. *Adv. Eng. Inform.* **2020**, *45*, 101121. [CrossRef]
14. Qi, C.R.; Su, H.; Mo, K.; Guibas, L.J. Pointnet: Deep learning on point sets for 3d classification and segmentation. In Proceedings of the IEEE Conference on Computer Vision and Pattern Recognition, Honolulu, HI, USA, 21–26 July 2017.
15. Qi, C.R.; Yi, L.; Su, H.; Guibas, L.J. Pointnet++: Deep hierarchical feature learning on point sets in a metric space. In Proceedings of the 31st International Conference on Neural Information Processing Systems, Long Beach, CA, USA, 4–9 December 2017.
16. Cao, M.-T.; Tran, Q.-V.; Nguyen, N.-M.; Chang, K.-T. Survey on performance of deep learning models for detecting road damages using multiple dashcam image resources. *Adv. Eng. Inform.* **2020**, *46*, 101182. [CrossRef]
17. Czerniawski, T.; Leite, F. Automated segmentation of RGB-D images into a comprehensive set of building components using deep learning. *Adv. Eng. Inform.* **2020**, *45*, 101131. [CrossRef]
18. Koo, B.; Jung, R.; Yu, Y. Automatic classification of wall and door BIM element subtypes using 3D geometric deep neural networks. *Adv. Eng. Inform.* **2021**, *47*, 101200. [CrossRef]
19. Saovana, N.; Yabuki, N.; Fukuda, T. Development of an unwanted-feature removal system for Structure from Motion of repetitive infrastructure piers using deep learning. *Adv. Eng. Inform.* **2020**, *46*, 101169. [CrossRef]
20. Hichri, N.; Stefani, C.; De Luca, L.; Veron, P.; Hamon, G. From point cloud to bim: A survey of existing approaches. *Int. Arch. Photogramm. Remote Sens. Spatial Inf. Sci* **2013**, *XL-5/W2*, 343–348. [CrossRef]
21. Golovinskiy, A.; Funkhouser, T. Randomized cuts for 3D mesh analysis. In Proceedings of the ACM SIGGRAPH Asia 2008, Singapore, 10–13 December 2008.
22. Kalogerakis, E.; Hertzmann, A.; Singh, K. Learning 3D mesh segmentation and labeling. In Proceedings of the ACM SIGGRAPH 2010, Los Angeles, CA, USA, 26–30 July 2010.
23. Woo, H.; Kang, E.; Wang, S.; Lee, K.H. A new segmentation method for point cloud data. *Int. J. Mach. Tools Manuf.* **2002**, *42*, 167–178. [CrossRef]
24. Guo, K.; Zou, D.; Chen, X. 3D Mesh Labeling via Deep Convolutional Neural Networks. *ACM Trans. Graph.* **2015**, *35*, 1–12. [CrossRef]
25. Guo, Y.; Wang, H.; Hu, Q.; Liu, H.; Liu, L.; Bennamoun, M. Deep learning for 3d point clouds: A survey. *IEEE Trans. Pattern Anal. Mach. Intell.* **2020**, *43*, 4338–4364. [CrossRef]
26. Wu, Z.; Song, S.; Khosla, A.; Yu, F.; Zhang, L.; Tang, X.; Xiao, J. 3d shapenets: A deep representation for volumetric shapes. In Proceedings of the IEEE Conference on Computer Vision and Pattern Recognition, Boston, MA, USA, 7–12 June 2015.
27. Armeni, I.; Sener, O.; Zamir, A.R.; Jiang, H.; Brilakis, I.; Fischer, M.; Savarese, S. 3d semantic parsing of large-scale indoor spaces. In Proceedings of the IEEE Conference on Computer Vision and Pattern Recognition, Las Vegas, NV, USA, 27–30 June 2016.
28. Tchapmi, L.; Choy, C.; Armeni, I.; Gwak, J.; Savarese, S. Segcloud: Semantic segmentation of 3d point clouds. In Proceedings of the 2017 International Conference on 3D Vision (3DV), Qingdao, China, 10–12 October 2017.
29. Lin, Y.; Jiang, M.; Yao, Y.; Zhang, L.; Lin, J. Use of UAV oblique imaging for the detection of individual trees in residential environments. *Urban For. Urban Green.* **2015**, *14*, 404–412. [CrossRef]
30. Liu, J.; Xu, G.; Ren, L.; Qian, Z.; Ren, L. Defect intelligent identification in resistance spot welding ultrasonic detection based on wavelet packet and neural network. *Int. J. Adv. Manuf. Technol.* **2017**, *90*, 2581–2588. [CrossRef]
31. Liu, P.; Li, Y.; Hu, W.; Ding, X.B. Segmentation and reconstruction of buildings with aerial oblique photography point clouds. *Int. Arch. Photogramm. Remote Sens. Spat. Inf. Sci.* **2015**, *40*, 109. [CrossRef]
32. Ibrahim, M.; Smith, R.; Wang, C.H. Ultrasonic detection and sizing of compressed cracks in glass-and carbon-fibre reinforced plastic composites. *Ndt E Int.* **2017**, *92*, 111–121. [CrossRef]
33. Kim, M.-K.; Wang, Q.; Park, J.-W.; Cheng, J.C.; Sohn, H.; Chang, C.-C. Automated dimensional quality assurance of full-scale precast concrete elements using laser scanning and BIM. *Autom. Constr.* **2016**, *72*, 102–114. [CrossRef]
34. Shi, S.; Wang, X.; Li, H. Pointrcnn: 3d object proposal generation and detection from point cloud. In Proceedings of the IEEE Conference on Computer Vision and Pattern Recognition, Long Beach, CA, USA, 15–20 June 2019.
35. Pu, S.; Vosselman, G. Knowledge based reconstruction of building models from terrestrial laser scanning data. *ISPRS J. Photogramm. Remote Sens.* **2009**, *64*, 575–584. [CrossRef]
36. Xu, Y.; Tuttas, S.; Hoegner, L.; Stilla, U. Geometric Primitive Extraction From Point Clouds of Construction Sites Using VGS. *IEEE Geosci. Remote Sens. Lett.* **2017**, *14*, 424–428. [CrossRef]

37. Rebolj, D.; Pučko, Z.; Babič, N.Č.; Bizjak, M.; Mongus, D. Point cloud quality requirements for Scan-vs-BIM based automated construction progress monitoring. *Autom. Constr.* **2017**, *84*, 323–334. [CrossRef]
38. Tsai, Y.-H.; Wang, J.; Chien, W.-T.; Wei, C.-Y.; Wang, X.; Hsieh, S.-H. A BIM-based approach for predicting corrosion under insulation. *Autom. Constr.* **2019**, *107*, 102923. [CrossRef]
39. Arulogun, O.; Falohun, A.; Akande, N. Radio frequency identification and internet of things: A fruitful synergy. *Br. J. Appl. Sci. Technol.* **2016**, *18*, 1–16. [CrossRef]
40. Domdouzis, K.; Kumar, B.; Anumba, C. Radio-Frequency Identification (RFID) applications: A brief introduction. *Adv. Eng. Inform.* **2007**, *21*, 350–355. [CrossRef]
41. Patil, A.K.; Holi, P.; Lee, S.K.; Chai, Y.H. An adaptive approach for the reconstruction and modeling of as-built 3D pipelines from point clouds. *Autom. Constr.* **2017**, *75*, 65–78. [CrossRef]
42. Tran, T.-T.; Cao, V.-T.; Laurendeau, D. Extraction of cylinders and estimation of their parameters from point clouds. *Comput. Graph.* **2015**, *46*, 345–357. [CrossRef]
43. Johnson, A.E.; Hebert, M. Using spin images for efficient object recognition in cluttered 3D scenes. *IEEE Trans. Pattern Anal. Mach. Intell.* **1999**, *21*, 433–449. [CrossRef]
44. Tombari, F.; Salti, S.; Di Stefano, L. Unique signatures of histograms for local surface description. In Proceedings of the European Conference on Computer Vision, Heraklion, Greece, 5–11 September 2010.
45. Wold, S.; Esbensen, K.; Geladi, P. Principal component analysis. *Chemom. Intell. Lab. Syst.* **1987**, *2*, 37–52. [CrossRef]
46. Li, B. 3d fully convolutional network for vehicle detection in point cloud. In Proceedings of the 2017 IEEE/RSJ International Conference on Intelligent Robots and Systems (IROS) 2017, Vancouver, BC, Canada, 24–28 September 2017.
47. Hackel, T.; Savinov, N.; Ladicky, L.; Wegner, J.D.; Schindler, K.; Pollefeys, M. Semantic3d. net: A new large-scale point cloud classification benchmark. *ISPRS Ann. Photogramm. Remote Sens. Spat. Inf. Sci.* **2017**, *IV-1/W1*, 91–98. [CrossRef]
48. Yang, Y.; Feng, C.; Shen, Y.; Tian, D. Foldingnet: Point cloud auto-encoder via deep grid deformation. In Proceedings of the IEEE Conference on Computer Vision and Pattern Recognition, Salt Lake City, UT, USA, 18–23 June 2018.
49. Zhou, Y.; Tuzel, O. Voxelnet: End-to-end learning for point cloud based 3d object detection. In Proceedings of the IEEE Conference on Computer Vision and Pattern Recognition, Salt Lake City, UT, USA, 18–23 June 2018.
50. Huang, J.; You, S. Point cloud labeling using 3d convolutional neural network. In Proceedings of the 2016 23rd International Conference on Pattern Recognition (ICPR), Cancun, Mexico, 4–8 December 2016.
51. Klokov, R.; Lempitsky, V. Escape from cells: Deep kd-networks for the recognition of 3d point cloud models. In Proceedings of the IEEE International Conference on Computer Vision, Venice, Italy, 22–29 October 2017.
52. Fang, Q.; Li, H.; Luo, X.; Ding, L.; Luo, H.; Rose, T.M.; An, W. Detecting non-hardhat-use by a deep learning method from far-field surveillance videos. *Autom. Constr.* **2018**, *85*, 1–9. [CrossRef]
53. Fang, Q.; Li, H.; Luo, X.; Ding, L.; Rose, T.M.; An, W.; Yu, Y. A deep learning-based method for detecting non-certified work on construction sites. *Adv. Eng. Inform.* **2018**, *35*, 56–68. [CrossRef]
54. Fang, Y.; Xie, J.; Dai, G.; Wang, M.; Zhu, F.; Xu, T.; Wong, E. 3d deep shape descriptor. In Proceedings of the IEEE Conference on Computer Vision and Pattern Recognition, Boston, MA, USA, 7–12 June 2015.
55. Akinosho, T.D.; Oyedele, L.O.; Bilal, M.; Ajayi, A.O.; Delgado, M.D.; Akinade, O.O.; Ahmed, A.A. Deep learning in the construction industry: A review of present status and future innovations. *J. Build. Eng.* **2020**, *32*, 101827. [CrossRef]
56. Sun, S.; Wang, S.; Wei, Y. A new ensemble deep learning approach for exchange rates forecasting and trading. *Adv. Eng. Inform.* **2020**, *46*, 101160. [CrossRef]
57. Ziari, H.; Sobhani, J.; Ayoubinejad, J.; Hartmann, T. Prediction of IRI in short and long terms for flexible pavements: ANN and GMDH methods. *Int. J. Pavement Eng.* **2016**, *17*, 776–788. [CrossRef]
58. Krizhevsky, A.; Sutskever, I.; Hinton, G.E. Imagenet classification with deep convolutional neural networks. *Adv. Neural Inf. Processing Syst.* **2012**, *25*, 1097–1105. [CrossRef]
59. Deng, F.; He, Y.; Zhou, S.; Yu, Y.; Cheng, H.; Wu, X. Compressive strength prediction of recycled concrete based on deep learning. *Constr. Build. Mater.* **2018**, *175*, 562–569. [CrossRef]
60. Nguyen, T.; Kashani, A.; Ngo, T.; Bordas, S. Deep neural network with high-order neuron for the prediction of foamed concrete strength. *Comput.-Aided Civ. Infrastruct. Eng.* **2019**, *34*, 316–332. [CrossRef]
61. Rahman, A.; Srikumar, V.; Smith, A.D. Predicting electricity consumption for commercial and residential buildings using deep recurrent neural networks. *Appl. Energy* **2018**, *212*, 372–385. [CrossRef]
62. Rafiei, M.H.; Adeli, H. Novel machine-learning model for estimating construction costs considering economic variables and indexes. *J. Constr. Eng. Manag.* **2018**, *144*, 04018106. [CrossRef]
63. Kolar, Z.; Chen, H.; Luo, X. Transfer learning and deep convolutional neural networks for safety guardrail detection in 2D images. *Autom. Constr.* **2018**, *89*, 58–70. [CrossRef]
64. Cha, Y.J.; Choi, W.; Büyüköztürk, O. Deep learning-based crack damage detection using convolutional neural networks. *Comput.-Aided Civ. Infrastruct. Eng.* **2017**, *32*, 361–378. [CrossRef]
65. Pathirage, C.S.N.; Li, J.; Li, L.; Hao, H.; Liu, W.; Ni, P. Structural damage identification based on autoencoder neural networks and deep learning. *Eng. Struct.* **2018**, *172*, 13–28. [CrossRef]
66. Eastman, C.M.; Eastman, C.; Teicholz, P.; Sacks, R.; Liston, K. *BIM Handbook: A Guide to Building Information Modeling for Owners, Managers, Designers, Engineers and Contractors*; John Wiley & Sons: Hoboken, NJ, USA, 2011.

67. Franz, S.; Irmler, R.; Rüppel, U. Real-time collaborative reconstruction of digital building models with mobile devices. *Adv. Eng. Inform.* **2018**, *38*, 569–580. [CrossRef]
68. Yi, L.; Kim, V.G.; Ceylan, D.; Shen, I.-C.; Yan, M.; Su, H.; Lu, C.; Huang, Q.; Sheffer, A.; Guibas, L. A scalable active framework for region annotation in 3d shape collections. *ACM Trans. Graph.* **2016**, *35*, 1–12. [CrossRef]
69. Shen, Z.; Ma, X.; Li, Y. A hybrid 3D descriptor with global structural frames and local signatures of histograms. *IEEE Access* **2018**, *6*, 39261–39272. [CrossRef]
70. Salti, S.; Tombari, F.; Di Stefano, L. SHOT: Unique signatures of histograms for surface and texture description. *Comput. Vis. Image Underst.* **2014**, *125*, 251–264. [CrossRef]
71. Assfalg, J.; Bertini, M.; Del Bimbo, A.; Pala, P. Content-based retrieval of 3-D objects using Spin Image Signatures. *IEEE Trans. Multimed.* **2007**, *9*, 589–599. [CrossRef]
72. Li, Z.; Zhang, L.; Tong, X.; Du, B.; Wang, Y.; Zhang, L.; Zhang, Z.; Liu, H.; Mei, J.; Xing, X.; et al. A three-step approach for TLS point cloud classification. *IEEE Trans. Geosci. Remote Sens.* **2016**, *54*, 5412–5424. [CrossRef]
73. He, Y.Q.; Mei, Y.G. An efficient registration algorithm based on spin image for LiDAR 3D point cloud models. *Neurocomputing* **2015**, *151*, 354–363. [CrossRef]
74. Szegedy, C.; Liu, W.; Jia, Y.; Sermanet, P.; Reed, S.; Anguelov, D.; Erhan, D.; Vanhoucke, V.; Rabinovich, A. Going deeper with convolutions. In Proceedings of the IEEE Conference on Computer Vision and Pattern Recognition, Boston, MA, USA, 7–12 June 2015.

buildings

MDPI

Review

Green Building Construction: A Systematic Review of BIM Utilization

Yu Cao , Syahrul Nizam Kamaruzzaman * and Nur Mardhiyah Aziz *

Faculty of Built Environment, University of Malaya, Kuala Lumpur 50603, Malaysia
* Correspondence: syahrulnizam@um.edu.my (S.N.K.); nurmardhiyah@um.edu.my (N.M.A.)

Abstract: As a multi-function method, Building Information Modeling (BIM) can assist construction organizations in improving their project's quality, optimize collaboration efficiency, and reduce construction periods and expenditure. Given the distinguished contributions of BIM utilization, there is a trend that BIM has significant potential to be utilized in the construction phase of green buildings. Compared with traditional buildings, green buildings have more stringent requirements, including environmental protection, saving energy, and residents' comfort. Although BIM is deemed an effective method to achieve the abovementioned requirements in the construction process of green buildings, there are few systematic reviews that explore the capabilities of BIM in the construction phase of green buildings. This has hindered the utilization of BIM in the construction of green buildings. To bridge this research gap and review the latest BIM capabilities, this study was developed to perform a systematic review of the BIM capabilities in the construction phase of green buildings. In this systematic review, the PRISMA protocol has been used as the primary procedure for article screening and review. The entire systematic review was performed from January 2022 to April 2022. In this process, 165 articles were included, reviewed, and discussed. Web of Science (WoS) and Scopus were adopted as the databases. Through this systematic review, it can be identified that BIM capabilities have significant advantages in project quality improvement, lifecycle data storage and management, collaboration optimization, planning, and schedule management optimization in the construction phase of green buildings. Through the discussion, it can be concluded that BIM utilization can be adopted from the pre-construction phase to the post-construction stage in the green building construction process. Besides these, the barriers to BIM utilization in the green building construction phase are also revealed in the discussion section, including the non-uniform data format, insufficient interactivity, ambiguous ownership, insufficient BIM training, and hesitation toward BIM adoption. Moreover, the challenges and future directions of BIM utilization in green building construction are identified. The findings of this study can facilitate construction personnel to be acquainted with BIM capabilities in the construction of green buildings to promote the utilization and optimization of BIM capabilities in the green building construction process.

Keywords: building information modeling; information technology; green building; sustainable building; construction

Citation: Cao, Y.; Kamaruzzaman, S.N.; Aziz, N.M. Green Building Construction: A Systematic Review of BIM Utilization. *Buildings* **2022**, *12*, 1205. https://doi.org/10.3390/buildings12081205

Academic Editors: Hongling Guo, Jia-Rui Lin and Yantao Yu

Received: 29 June 2022
Accepted: 8 August 2022
Published: 10 August 2022

Publisher's Note: MDPI stays neutral with regard to jurisdictional claims in published maps and institutional affiliations.

Copyright: © 2022 by the authors. Licensee MDPI, Basel, Switzerland. This article is an open access article distributed under the terms and conditions of the Creative Commons Attribution (CC BY) license (https://creativecommons.org/licenses/by/4.0/).

1. Introduction

As a multi-function method, Building Information Modeling (BIM) makes a significant contribution to the Architectural Engineering and Construction (AEC) industry. According to the British Standards Institution [1], BIM uses shared digital representations of building assets to promote the design, construction, and operational process and form a reliable basis for decision making. It is the process of generating and managing data about projects from the pre-construction phase to the post-construction phase [2]. In this process, the BIM that is three-dimensional, used in real time, and dynamic is adopted to improve the productivity and quality of projects in their lifecycle [2]. Despite no uniform definition of BIM across different countries and generations, there is a consensus in the AEC industry that BIM is not just a tool and software installed on a computer [3]; rather, it is a combination of software and process [3]. In summary, BIM is the multiple function method that integrates business process, digital representation, organization, and control of the process. It can provide three-dimensional (3D) modeling of the project, manage the project schedule throughout the whole lifecycle, provide a communication platform for all stakeholders, estimate and calculate project costs, detect clashes, and allow stakeholders to inspect and manage buildings throughout their building lifecycle [4].

As an effective information technology method in the AEC industry, BIM can provide a significant contribution to the construction of green buildings [5,6]. Green buildings are also known as healthy buildings. They are the buildings that can prompt positive influence and reduce the negative impact on the natural environment in the lifecycle of assets [7]. According to the Evaluation Standard for Green Building [8], green buildings contain one or more of the following features and standards:

1. Efficient utilization of resources and energy.
2. Utilization of renewable energy, such as wind and geothermal energy.
3. Adoption of pollution- and waste-reduction measures.
4. Utilization of non-toxic, ethical, sustainable, recycled, and re-used materials.
5. Quality indoor environment and comfortable residential experience.
6. Suitable for the local environment and climate.

Given the strict standards and requirements of green buildings, it is difficult for construction organizations to achieve the green buildings' requirements through traditional construction methods [9,10]. To overcome the obstacles of the conventional construction process, an increasing number of AEC participants recommended that BIM should be integrated into the construction of green buildings [4,11,12]. As an effective information technology tool, BIM was deemed the efficient solution to assist AEC corporates in overcoming the barriers in the construction process of green buildings [3,4,11–14]. Ghaffarianhoseini et al. [5] comprehensively summarized the benefits of BIM application: "these benefits range from its technical superiority, interoperability capabilities, early building information capture, use throughout the building lifecycle, integrated procurement, improved cost control mechanisms, reduced conflict and project team benefits." Moreover, BIM is the essential method for implementing full automation of information integration, collaboration and intellectual property issues, multi-party involvement, and collaboration [15].

Although the application of BIM in the AEC industry has had tremendous positive influences on the technology involved to provide effective and optimal parameters to facilitate users to achieve project requirements, the integration of BIM and green buildings in construction activities is still deficient [9,16–18]. The major hindrance to BIM utilization in green building construction was the unfamiliarity with the BIM capabilities of the construction organizations [19–23]. According to the statistics of Akhmetzhanova et al. [24] and Tatygulov et al. [25], 44% of respondents refused to adopt BIM because they were unfamiliar with BIM functions and had not received the appropriate training. Given the abovementioned content, it can be concluded that construction personnel generally suffer from a lack of familiarity with BIM within the AEC industry. To enhance familiarity with BIM capabilities that can be utilized in the green building construction process, this study was developed to perform a systematic review of BIM capabilities in the construction phase of green buildings.

Moreover, there are a few review articles that reviewed the BIM application in the construction phase of green buildings [3,26–29]. However, most of these review articles did not utilize a systematic review method in their research. According to Lu et al. [4], most reviews of BIM utilization in the construction phase are conducted by traditional and bibliometric review. Most traditional reviews lack the explicit retrieve and screen protocol of the literature in their studies [30,31]. Thus, the article-screening processes in traditional reviews are usually not transparent enough for the audience [31]. Moreover, no fixed and formal article search process guidance is identified in the traditional review, which has led to confusion in the article selection process in the traditional review to some extent [32,33].

From the perspective of a bibliometric review, this review is conducted through the quantitative analysis of bibliographic material (data) [34]. Although the bibliometric review can perform a more quantitative description and discussion of the research area, it has insufficient qualitative exploration and analysis of the particular study in this research [35]. Compared with the abovementioned review methods, the systematic review can provide a complete summary of the current literature relevant to the research questions and develop a reliable evaluation of future development directions [36,37]. Moreover, BIM software is evolving rapidly, and BIM features are changing and iterating continually. To fill the research gap of the insufficient systematic review of the BIM capabilities in green building construction, it is necessary to perform a contemporary systematic literature review to retrieve, summarize, discuss, and analyze the latest BIM capabilities that can be utilized during the construction stage of green buildings.

Given the abovementioned background content, this study was developed with the aim of performing a systematic literature review on BIM capabilities that can be utilized during the construction stage of green buildings. To achieve this aim, three objectives were developed:

1. Identify the BIM capabilities that can be utilized in the construction of green buildings.
2. Discuss and analyze the methods that BIM capabilities performed in green building construction.
3. Summarize the advantages, challenges, and future direction of BIM utilization in the construction of green buildings.

This study consists of the following sections. The introduction is presented in Section 1. The research methodology is illustrated in Section 2, in which the process of article retrieval and screening is demonstrated. Moreover, the search string and the inclusion and exclusion criteria for the study are also presented in Section 2. The results of the systematic review are shown in Section 3. Here, the reviewed BIM capabilities are categorized into four categories, including project quality improvement, lifecycle data storage and management, collaboration optimization, and planning and schedule management optimization. Moreover, the utilization methods of these BIM capabilities in green building construction are reviewed in Section 3. The discussion and analysis are conducted in Section 4, in which the authors discuss and analyze the BIM capabilities from the pre-construction phase to the post-construction phase of green buildings. Besides these, the future direction, advantages, and challenges of BIM utilization in green building construction are provided in Section 4. Section 5 is the conclusion, and also discusses the contributions and limitations of this study.

2. Research Methodology

This study aimed to perform a systematic literature review on BIM capabilities that can be utilized during the construction stage of green buildings. Given the precise scientific design insisted upon in the systematic review process, it can assist researchers in mitigating biases and random errors in the review process [38]. Moreover, a systematic review can facilitate authors to become familiar with the primary knowledge retrieved in the screened articles, and develop the research model through a robust approach, to explore the future research directions more precisely [39–41].

To achieve the abovementioned aim, the preferred reporting items for systematic reviews and meta-analyses (PRISMA) model [42] was used to conduct the systematic review. In this study, the PRISMA was conducted through four phases, as per the systematic literature review studies by Cho et al. [43] and Lee et al. [40]:

1. Determine the search database and keywords.
2. Develop the search strings based on the keywords, inclusion criteria, and exclusion criteria. Conduct the primary article screening through the search strings.
3. Conduct the qualitative screening of titles, keywords, and abstracts according to the inclusion criteria and exclusion criteria.
4. Perform the qualitative assessment and literature review of the full content of the remaining articles.

The process of the systematic literature review in this study is shown below.

Phase 1: To ensure the retrieved articles can meet the requirements of the systematic literature review, Web of Science (WoS) and Scopus were determined as the databases in this study. Articles that had not been peer reviewed were not permitted to be included in this study. The keywords were determined as follows: "Building Information Modeling", "Building Information Model", "BIM", "Green Building", "Sustainable Building", and "Construction".

Phase 2: In this phase, the search strings (as shown in Table 1) were developed to conduct the initial article search of this study. Moreover, the inclusion criteria and exclusion criteria (as shown in Table 2) were formulated for further qualitative screening. In this study, the retrieved articles included conference papers, articles, review articles, and proceedings papers that can be searched for through WoS and Scopus. Other types and database sources of articles were excluded in this process. Moreover, non-English articles were also excluded from the retrieval process. Through the initial article search and review, the overview of BIM capabilities in the construction process of green buildings was formed.

Table 1. Search string and initial search results.

Search Engine	Search String	Results
WoS	TS = (("building information modeling" OR "building information modelling" OR BIM) AND (green building OR sustainable building) AND (construction OR construct))	974
	Document Types: Articles or Proceedings Papers or Review Articles	969
	AND LANGUAGES: (ENGLISH)	965
Scopus	"TITLE-ABS-KEY (("building information modelling" OR "building information modeling" OR BIM) AND (green AND building OR sustainable AND building) AND (construction OR construct))	459
	AND (LIMIT-TO (DOCTYPE, "cp") OR LIMIT-TO (DOCTYPE, "ar") OR LIMIT-TO (DOCTYPE, "re") OR LIMIT-TO (DOCTYPE, "cr"))	445
	AND (LIMIT-TO (LANGUAGE, "English"))	437
Sum of the papers = 1402 Duplicates = 137 Invalid = 109 After removing duplicates and invalid papers = 1156 After title and keyword screening = 537 After abstract screening = 249 After review of full content of papers = 165 Total = 165		

The initial article search was conducted in April 2022. In this process, 1433 articles were retrieved from WoS and Scopus (974 in WoS, 459 in Scopus). Then, 137 duplicated articles and 109 invalid articles (articles that cannot be provided as an online version of the full content) were removed by the author. In the end, 1156 articles remained after this phase was complete. The remaining articles were carried over into the next step to conduct further qualitative screening of the titles, keywords, and abstracts.

The detailed search strings and initial search results are presented in Table 1, and the clear inclusion and exclusion criteria are presented in Table 2.

Phase 3: This phase involved performing the qualitative analysis based on the above-mentioned inclusion and exclusion criteria. In phase 3, the titles and keywords of the remaining articles were firstly screened manually according to the secondary inclusionary and exclusionary criteria. In this step, 619 articles were eliminated, and 537 articles remained. Then, abstract analysis was performed manually on the remaining 537 articles based on the secondary inclusionary and exclusionary criteria. In total, 288 articles were removed because their abstracts failed to meet the requirements of the secondary inclusion and exclusion criteria (mentioned in Table 2). After phase 3 was accomplished, 249 articles remained and were brought into the next phase.

Table 2. Inclusion criteria and exclusion criteria.

Primary Criteria		Secondary Criteria	
Inclusionary	**Exclusionary**	**Inclusionary**	**Exclusionary**
Journal articles that can be searched in Web of Science (WoS) or Scopus	Duplicated papers	Articles that contain BIM capabilities in the construction of green buildings	Articles that contain no BIM capabilities in the construction of green buildings
Conference paper and proceeding papers that are searchable through WoS or Scopus	Invalid articles (articles that cannot provide the online version of full-text content)	Articles that can support authors to accomplish research objectives	The articles that cannot provide support for authors to accomplish research objectives
Review articles that are searchable through WoS or Scopus			
Published in English	Non-English edited articles or papers		

Phase 4: The qualitative assessment was performed on the remaining 249 articles and a literature review of the full content based on the secondary inclusionary and exclusionary criteria was carried out (mentioned in Table 2). Moreover, these articles were also reviewed manually by the authors to identify whether they contain quality content on BIM utilization in green building construction. In this process, the inclusion and exclusion of an article relied on the subjective decisions of the authors without firm objective standards. In the end, 165 articles were included in the study, and were brought over to the next stage of the process (Section 3), namely, the systematic literature review.

The entire search and screening process of this study is presented in Figure 1.

Through the full-text review of the included studies, the BIM capabilities can be categorized according to their contribution areas (see in Table 3).

Table 3. Process of categorization determination.

Identify the BIM capabilities that can be utilized in the green building construction through full-text review.
Categorize the BIM capabilities according to their contribution areas.
Develop the classifications of BIM capabilities in green building construction.
Check for consistency by referring to other studies.
Verify the developed classifications in this study.

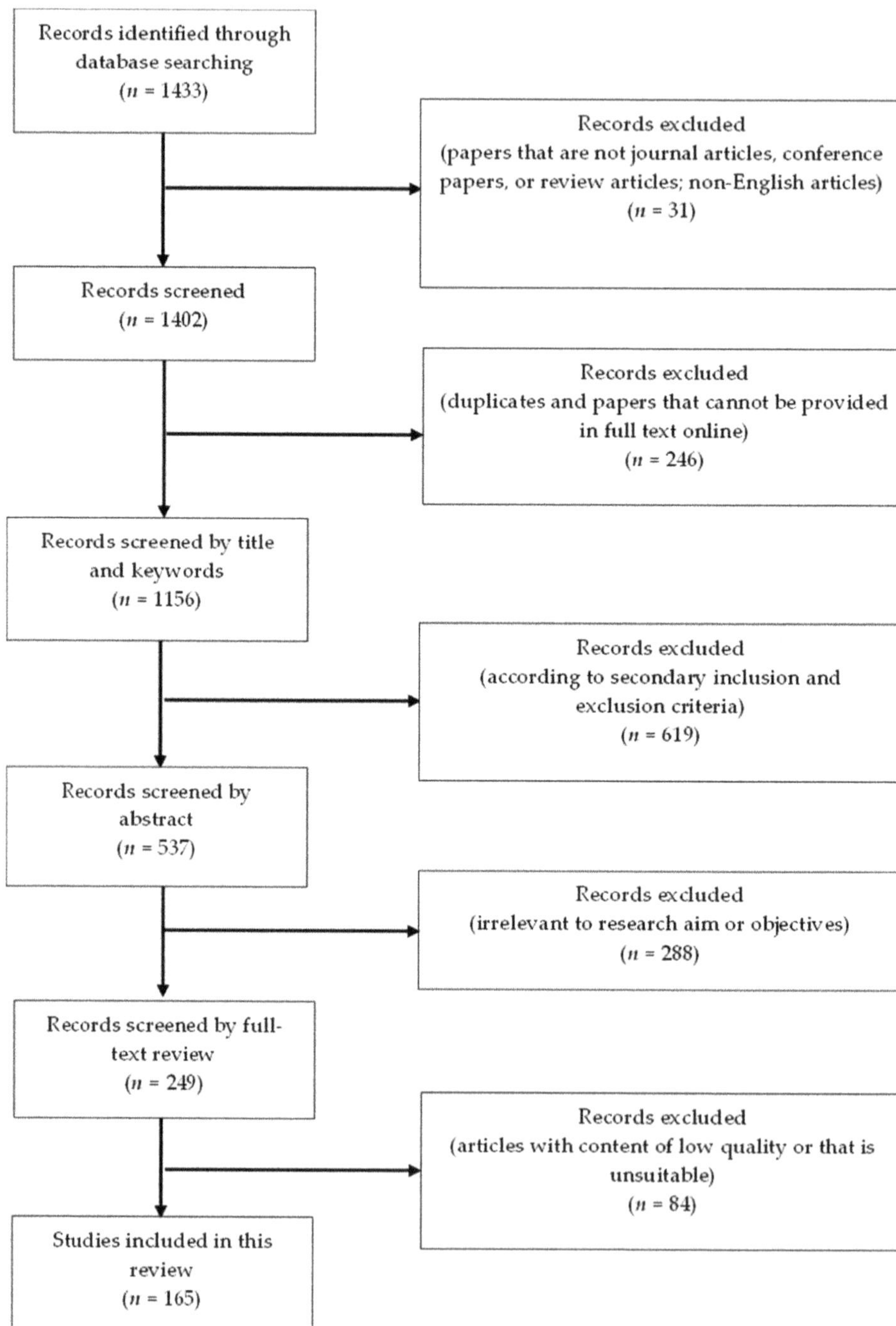

Figure 1. Flowchart of article screening process used in this study.

3. Results

3.1. Descriptive Analysis

After the search and screening carried out in Section 2, 165 articles were retrieved and included in this systematic review. In Section 3, these included articles were systematically reviewed and summarized by the authors.

The publication dates of the reviewed articles are presented in Figure 2. According to Figure 2, it can be identified that the earliest publication of the articles reviewed was 2010. From 2010 to 2015, the research on BIM capabilities in green building construction was still in the infancy stage. From 2010 to 2015, although the number of articles in this area generally shows moderate growth, the overall number of publications is still relatively few (from one to five).

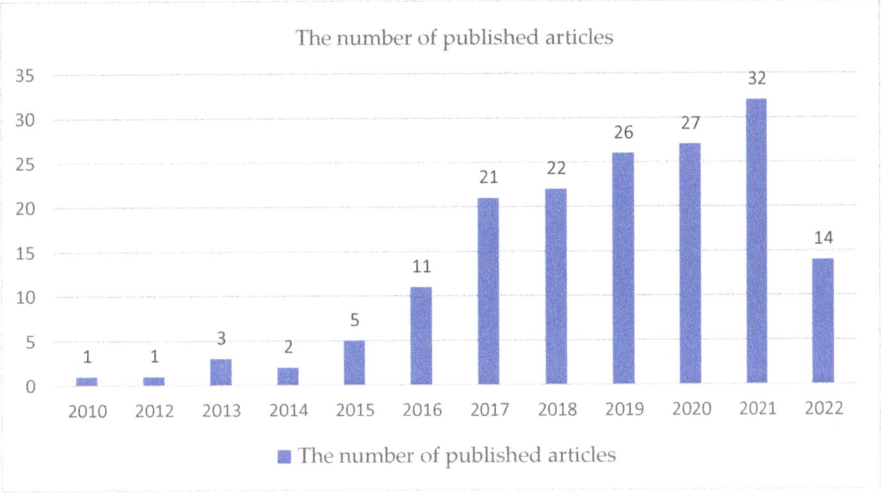

Figure 2. Number of reviewed articles per year.

Eleven of the reviewed articles were published in 2016, and the research on BIM utilization in green building construction was becoming more popular and attractive. From 2017 to 2021, BIM utilization in the construction phase of green buildings had become a prominent discipline and received wide and significant attention from researchers. The included articles published exceeded 20 per year and gradually increased in this period (from 21 in 2017 to 32 in 2021).

The number of included articles that were published in 2022 was 14. However, given that this study was performed between January 2022 and April 2022, it can be concluded that all reviewed articles were published in the first four months of 2022. Any articles published after April could not be reviewed because of the publication date limitations. Therefore, the decline in the number of reviewed articles in 2022 does not indicate that the BIM capabilities in green building construction are obsolescent. Moreover, among the reviewed articles in this study, 14 were published in the first four months of 2022. This phenomenon can also indicate indirectly that this field is still valued and vital in 2022.

From the perspective of publication journals, these 165 reviewed articles were taken from 64 journals and 12 conferences. According to the number of reviewed articles published in each journal, the rank of journals is presented below (due to the length of the study, only the top 10 journals are listed): *Automation in Construction* (25), *Sustainability* (11), *Procedia Engineering* (8), *Advanced Engineering Informatics* (6), *Buildings* (6), *Journal of Cleaner Production* (6), *Journal of Building Engineering* (6), *Renewable and Sustainable Energy Reviews* (5), *International Journal of Project Management* (4), and *Journal of Civil Engineering and*

Management (4). The ranking of journals according to the number of the reviewed articles being published is demonstrated in Figure 3 (Only the journals that ranked in the top ten for the number of articles in this study are included).

Figure 3. Number of reviewed articles published per journal.

3.2. Results Analysis

BIM is considered to be a potential technology in the construction stage of green buildings. To encourage the adoption and development of BIM in the AEC industry, the importance and potential of BIM implementation have been highlighted by many researchers [44]. Ghaffarianhoseini et al. [5] comprehensively summarized the benefits of BIM application: "these benefits range from its technical superiority, interoperability capabilities, early building information capture, use throughout the building lifecycle, integrated procurement, improved cost control mechanisms, reduced conflict and project team benefits."

In this study, 165 articles were included in the systematic review and analysis by the authors. Through the systematic review, the BIM capabilities are summarized and categorized according to their benefits and advantages during the construction phase of green buildings. In this study, the classification of BIM capabilities in the green building construction phase is as follows: project quality improvement, lifecycle data storage and management, collaboration optimization, and planning and schedule management optimization. Detailed information on the reviewed studies in terms of the four aspects mentioned above is presented in Appendix A.

3.2.1. Project Quality Improvement

BIM has made significant contributions to improving the project quality in the process of green building construction. As an object-orientated 3D model, BIM can integrate the knowledge of various disciplines to provide a quality platform for parametric modeling, spatial visualization, and asset-process simulation [45]. Based on the abovementioned functions, the situations of components can be demonstrated through BIM, which can facilitate the architects' and engineers' ability to conduct clash detection [46]. In green building construction, the clash detection in BIM can save up to 10% of the contract value and reduce the construction schedule by 7% [5]. In addition, the comparison between different design schemes can be conducted by BIM tools. It assists stakeholders in developing construction schemes with better efficiency and sustainability [17]. Moreover, BIM can provide the 3D demonstration at the initial stage of green building construction so that the clients can become familiar with the intent of the design in a timely manner [47]. This advantage helps the designers and civil engineers to effectively make changes in time to meet the clients' requirements. Based on the statistics from Noor et al. [48], the most efficient contributions from BIM are "better visualization compared to traditional CAD technology", "the ability of BIM in visualization", and "helps to ensure that the quality-related activities are being performed effectively", with mean values of 3.7500, 3.6471, and 3.9853, respectively. As a reference, a mean value between 3.5 and 4.49 can be defined as "much" in the mean scale formulated by Aguila et al. [49]. According to the questionnaire survey from Huang et al. [14], 87.8% (180/205) of respondents endorsed BIM facilitating designers' ability to integrate designs. Furthermore, 84.4% (173/205) of respondents agreed that the acceleration of model generation and modification could be achieved through BIM, which can improve working efficiency and quality [14].

For the construction organizations that adopted BIM in the green building construction process, the advantages of BIM can be extended from the pre-construction phase to the projects' final acceptance and retrofit phases [50]. From pre-construction to retrofit, BIM can assist stakeholders in performing construction activities' documentation, information real-time storage and exchange, monitoring and surveillance, emergency control, and asset demolition [5]. Moreover, BIM has a remarkable effect on the renovation of green buildings with the support of as-built data acquisition tools (such as laser scanning and infrared thermography) [51]. In the retrofit process, Najjar et al. [52] put forward that BIM utilization in lifecycle assessment can encourage architects to incorporate environmental criteria into their decision-making process. This characteristic enables the stakeholders from various aspects to consider ecological requirements reasonably when making decisions [52]. BIM also has practical functions in supporting the assessment of energy consumption and residential comfort in various renovation instructions [53]. In the interior retrofit case of the Diagnosis and Treatment Centre of the University Hospital of Jaén, through thermal simulation, daylight simulation, and energy analysis, the annual energy consumption in the renovated building dropped by 120.94 kWh/m^2 [54]. In the energy-efficient retrofit project of Xindian Central Public Retail Market in Taiwan, the air conditioner system load was reduced by 100 kWh due to the support of BIM [55].

Moreover, green buildings are required to meet the green building evaluation standards. Through the simulation and knowledge-storage functions in BIM, BIM-based green building assessment models were proposed to record the energy consumption during construction and to predict the energy performance of the building during the post-construction phase [56–59]. In the case study of Wu et al. [60], the researcher inputted the green building evaluation standard into the BIM to check the compliance of the green building item by item, and then determined the corresponding green building rating of the target projects.

3.2.2. Collaboration Optimization

Due to the strict quality and technology requirements of green building construction, the practical cooperation and collaboration of multiple stakeholders from various organizations in sustainable construction are required [10]. Collaboration is deemed one of the most critical features of BIM. Collaboration aims to achieve the best results in a cost-effective and timely manner by bringing together a variety of people and resources and using their collective knowledge and capabilities to accomplish tasks that would be difficult for an individual organization to perform [61]. Due to the massive scale and high complexity of green building construction, it is necessary for the various stakeholders in the lifecycle of projects to couple with other participants through project-specific collaborative relationships [13,62,63]. The practical cooperation between multi-disciplines has significant contributions to rework elimination, the reduction in clashes and misunderstandings, waste mitigation, and the definition of risks and uncertainties [64,65].

The collaboration between multi-disciplines and various stakeholders (or organizations) can be effectively conducted through BIM. As a collaboration and communication platform, BIM can develop a comprehensive shared operating environment intergraded by multiple discipline models [64,66,67]. Through the cooperation platform in BIM, the project information of their lifecycle is can more easily be updated, modified, inserted, and extracted by various stakeholders. These stakeholders pertain to multiple disciplines and organizations, and possess specific skill sets to fulfil BIM-related project requirements [68]. With the transparency of the BIM cooperation environment, the ownership of data through the lifecycle of projects is shared by various stakeholders [69]. To eliminate the barriers and ensure interoperability in collaboration, Industry Foundation Classes (IFC) are generally utilized as the standard file format specification [70]. In addition, the gbXML schema (Green Building XML) was formulated to enhance the transfer of building data from BIMs to engineering analysis software [26].

In BIM-utilized green building construction processes, the collaboration among various disciplines and stakeholders is basically achieved through the BIM-based construction network (BbCN) [64]. The BbCN contains team members from multiple organizations to conduct BIM-related activities on BIM-enabled projects [64]. Cao et al. [71] revealed that the enhancement of internal collaboration within the BbCN had been a particularly effective selling point for BIM. In the cooperation process in BbCN, some prerequisites need to be integrated, including the context, team, process, task, and actor [72]. With the assistance of effective management and transparent and shared information exchange, collaboration can be deemed a central element of success throughout the lifecycle of construction projects [73,74]. In addition to BbCN, Wang et al. [75] introduced the stake source system based on social network analysis (SNA), which can automatically recommend suitable stakeholders through SNA. Through this system, stakeholders can easily become familiar with others' responsibilities, work progress, and position.

In conclusion, BIM can effectively improve collaboration quality. As a digital representation tool and database inventory, all stakeholders can work on a sharing cooperation platform through the BIM application, which enhances the quality of the decision-making process [4]. The essential issues in the cooperation management process can be described as: "which building elements, from which trades, should be developed at what time and at what level?". This issue can be addressed by the Level of Detail (LOD) decision plan, which is conducted through BIM [76]. In state-owned assets projects and public–private partnership projects, satisfaction from the government is necessary for collaboration. According to the statistics from Zuhairi et al. [77], the most important driving factor in BIM implementation in Malaysia is "the advocation and enforcement in the implementation of BIM by the government" with relative importance indicators (RII) of 0.950. BIM implementation can effectively improve the government's satisfaction, thus improving the quality of cooperation. According to statistics from Huang et al. [14], 86.34% (177/205) of respondents believed that BIM played an essential role in the establishment of collaborative platforms. According to the questionnaire survey of the Collaboration Management (CM)-based BIM model developed by Lin and Yang [66], 92% of the respondents were satisfied with CM-based BIM creation work, and 86% of the respondents believed that it could enhance the management of model creation work in the collaboration process.

3.2.3. Lifecycle Data Storage and Management

In the process of green building construction, it is necessary for the construction organizations to obtain the data and information on corresponding green buildings that are produced in their design phase. If the information delivery is deficient, the construction schedule might be delayed.

BIM can be formulated as a multifunction repository that stores data throughout the project lifecycle [15,52,78]. The digital presentation and remarkable interoperability capabilities facilitate the exchange and revision of data by users throughout the entire green building construction process. This benefit can help stakeholders to capture comprehensive building information [4]. With the development of BIM, the international BIM information storage and exchange standard Industry Foundation Classes (IFC) were introduced into BIM [79]. This effectively eliminated the format barriers involved in collaboration using BIM [79].

BIM's knowledge storage and management function can effectively overcome the fragmented information issues that have developed in the green building construction process [80,81]. In the research of Solihin et al. [82], they integrate spatial operations into standardized SQL queries that are easy to query. This makes the data in BIM more accessible, and better assists stakeholders in making decisions. Lu et al. [4] maintained that one BIM model could contain information from multiple disciplines, which can continuously incorporate sustainability measures into the project throughout the design process. BIM can also make users familiar with the intricate relationship between stakeholders. Zheng et al. [83] propose a novel method based on the Stakeholder Value Network (SVN) that could quantify and visualize the value exchange between critical stakeholders when adopting BIM in sustainable constructions. This allows users to visualize and quantify the perceived value of BIM stakeholders [83].

Moreover, with the integration of BIM and third-party devices, BIM can achieve effective data collection and management in the construction phase of green buildings [84,85]. With the integration of BIM and GIS, BIM can assist the construction organization in obtaining information on the construction site and surrounding environment, including topography, terrain, soils, vegetation cover, road layout, and infrastructure layout [63,86–89]. Besides that, with the connection of BIM and the Internet, BIM can effectively retrieve and integrate weather conditions on green building construction sites, thus mitigating the natural hazard damage and ensuring the safety of construction personnel [90]. In addition to these, with the BIM, RFID, barcodes, 2D imaging, and photogrammetry, BIM can automatically capture and store the utilization situation, stock quantities, and input and output information of materials and equipment on the green building construction site [87,91–94].

3.2.4. Planning and Schedule Management Optimization

Quality scheduling and project management in green building construction can be efficiently guaranteed through BIM implementation. BIM can enhance the construction schedule management for stakeholders [95]. Not only can the resource requirements, equipment requirements, and excepted expenditure for the next step be obtained through BIM, but the percentage of progress, the number of expenses, and the deviation from the budget can also be predicted by BIM [96]. Moreover, real-time updates and quality visualization performance can be achieved by BIM to enhance planning activities [97,98]. The project duration and expenditure can be efficiently reduced through the project management of BIM. According to the summary from Ghaffarianhoseini et al. [5] in the aspect of project management, BIM can eliminate 40% of unforeseen modifications, provide cost estimates with a 3% error threshold, and reduce the generation time by up to 80%. In addition, Gao and Pishdad-Bozorgi [99] put forward that BIM can facilitate the integration of AEC knowledge, which has a significant advantage for the contractors and subcontractors of green building construction projects to enhance their management personnel training.

The use of BIM in the planning and schedule management of green building construction can be started at the pre-construction stage. Before the construction activities begin, the feasibility studies of green building construction plans can be verified through BIM simulation, to eliminate rework and waste for the subsequent activities [17]. In addition, Wang and Liu [100] highlighted that engineers and constructors could propose optimized methods and simulate their performance to conduct feasibility verification and trial and error if there are defects in previous plans. By evaluating the impact of construction activities on the surroundings, the corresponding environmental protection measures can be adopted to mitigate the negative influence of the construction [101]. In the site-planning process, BIM can compare various siting alternatives to determine the most suitable construction site layout with the most negligible impact on the surrounding environment [102]. In this process, BIM provides an appropriate framework for decision making by bringing together the necessary information at the right time, and clarifying details and existing conditions [102]. A meta-heuristic algorithm is used to optimize the construction site's layout after thoroughly considering all factors [88,103].

In multiple dimensions of BIM modeling, the stakeholders can utilize BIM to formulate schedules and conduct project management in the green building construction process. BIM is a multi-dimensionality tool. The BIM applications can be divided into BIM 3D, BIM 4D, BIM 5D, and BIM 6D according to their functions and application aspects [54,104–108]. To clarify, 3D means that BIM can provide detailed 3D model simulations of buildings [109–116]. BIM 4D integrates BIM 3D with the time dimension, so BIM 4D can simulate the green building construction process to support the schedule development and revision, constructability analysis, clash detection, and other functions [117–122]. BIM 5D is based on BIM 4D, with the addition of cost-related information [123,124]. Through BIM 5D, the construction organizations can effectively forecast and account for the expenditure of the green building construction project at different phases, and predict the return-on-investment (ROI) ratio [125–128]. BIM 6D is based on BIM 5D and adds sustainability management functions, which improves the sustainable efficiency and quality of the green building construction process [104,129].

As an advanced scheduling and modeling tool, BIM 4D can effectively conduct the integration of 3D visual modeling and project schedules. Compared with the regular Gantt chart, the construction schedules and sequence of tasks can be visually demonstrated in BIM, which helps stakeholders to become familiar with the green building construction sequences [130]. Through the integration of geometric information with the schedule and material information, Jupp [131] revealed the potential of BIM to identify work sequence errors and conflicts quickly. With the combination of the BIM management system, surveillance, barcode, and radio-frequency identification, materials that are transported, transferred, and utilized within the construction site can be automatically recorded and updated in the bill of materials (BOM) [132]. To eliminate the uncertainties in green building construction schedule plans, Yuan et al. [133] developed the Monte Carlo method (MCM) and BIM-based construction schedule early warning model (MCM-BIM-CSEWM) to address the logical relationships between construction activities and provide timely risk warnings. Irizarry et al. [134] also revealed that the supply chain is arranged in a better precise, efficient, and cost-effective method with the integration of BIM and geographic information systems (GIS). From the perspective of safety management and planning, through BIM 4D's simulation and visualization of the green building construction progress, BIM can conduct risk identification and safety training for management personnel and construction workers [135–139]. Moreover, through the BIM 4D hazard identification component established by Heidary et al. [140], BIM can assist green building construction managers in identifying and demonstrating the potential construction hazards in the early stages of green building projects [140].

In the BIM-generated schedule, the components and schedules of each sub-progress and each task can be contained and demonstrated in the entire construction schedule [141,142]. Based on the abovementioned functions, the intercomparison of construction schedules and the detection of recurring processes can be conducted expediently through BIM 4D [143]. Moreover, through the visualization in BIM, the construction progress can be visually simulated, including engineering design, field environment, projected material consumption, and machinery utilization [100]. This method enhances the predictability of construction and transforms the traditional pattern of construction plans [100]. By simulating the construction progress and potential construction accidents, personnel training and safety education can be provided by BIM utilization [135,136,144]. Nicał and Wodyński [129] put forward that the impact of construction activities on the surrounding environment can be simulated through BIM 6D, which is necessary to support green buildings to conform to specific green building evaluation standards.

In the process of scheduling and project management, some repetitive tasks can be conducted using the same or a similar method in the green building construction process [143], and BIM 4D can provide predefined process templates to execute the required tasks without wasting time or production [143]. To improve the generality of the template, the IFC was adopted to provide required object definitions in BIM 4D [145]. In addition to these, case-based reasoning (CBR) was utilized as an effective machine learning method in BIM [146]. The faults can be settled by utilizing or adjusting the previous solutions to tackle similar tasks [146]. In the process of CBR, new problems and malfunctions can be matched and solved by the most similar solution through the typical four phases of the CBR (retrieve, reuse, reserve, and modify), and then the new solution schemes can be retained for future similar disposal [147]. To improve the accuracy of the match between disputes and solutions, Sigalov and König [143] asserted that the graph indexing is settled in partial BIM software.

One of the prominent features of BIM is simulation. In traditional AEC tools, with the increasing size and complexity of green building construction processes, it is challenging to generate sufficient suitable design and construction schemes with the distraction of various undefined risks and uncertainties [148,149]. Therefore, the predictions about project progress and results are hard to keep accurately [148,149]. However, these barriers can be partially mitigated by the simulations of BIM. Based on virtual modeling, without the consumption of materials, reliable simulations and predictions of projects can be developed through the thorough consideration of factors and potential risks [150].

From the finance perspective, BIM can support green building construction teams to generate bills of quantities automatically, perform procurement plans and logistical layout, and conduct materials and equipment management. Given that the BOM can be updated in BIM in real time, BIM can reflect real-time expenditure. Moreover, the construction activities' feasibility research can be conducted with the support of a data repository, 3D visualization demonstration, and simulation in BIM [151]. In addition, BIM can provide quality procurement management and optimization for construction organizations. In the research of Vilas-Boas et al. [152], they proposed a four-dimensional BIM model to analyze and optimize procurement. This model provides procurement suggestions by comprehensively analyzing and comparing the following dimensions:

1. Product-based. Assess the properties, quality, and compliance of the purchased materials.
2. User-based. Assess whether the material accords with the green assessment criteria and whether it meets the requirements of stakeholders and participants.
3. Manufacturing-based. Test the operation status of the procured material, and check whether these materials have clashed with other parts.
4. Value-based. Calculate the value of each material and procurement link, and evaluate their cost performance. For stakeholders, value includes tangible and intangible benefits.

4. Discussion

Through the systematic review in Section 3, it can be determined that BIM capabilities have significant advantages in the construction phase of green buildings. BIM can develop apparent benefits to green building construction, including improving the quality of the projects, optimizing collaboration between different stakeholders, performing lifecycle data storage and management, and assisting construction organizations in conducting and optimizing their planning and schedule management in the process of constructing green buildings.

In this section, these abovementioned capabilities are discussed and analyzed by the authors. Through the discussion and analysis, it can be concluded that the implementation of BIM can be utilized in the green building pre-construction phase, construction phase, and post-construction phase. The detailed discussion and analysis processes are shown below.

4.1. BIM Capabilities in the Green Building Pre-Construction Phase

Given the enormous complexity, extensive information and uncertainty in pre-construction, the pre-construction phase of green buildings is always an area to which construction teams and subcontractors attach great importance [50,65,153,154]. According to the systematic review and discussion in this study, the BIM implementation in the pre-construction phase of green buildings can be adopted from the perspectives of information and knowledge delivery, feasibility study, construction team setup, construction plans and schedule formulation, construction cost estimation and budget formation, construction material supplement and transportation, and construction equipment management.

Given the extensive information and the complex structural construction of green buildings, it is vital for construction organizations to obtain detailed information about targeted green buildings before formal construction [9,155–157]. BIM can help the construction teams to transfer the green building model and related information generated during the design phase to the construction personnel in an error-free, omission-free manner [4,15,45,47,52,78–83]. For green building projects that do not adopt BIM in their design phase, BIM can also be utilized to automatically capture the related information to generate the green building information model by identifying CAD drawings, or integrating with photography, GIS, and 3D scanning [63,86–94].

In the aspect of feasibility studies, construction plans, and schedule formulation, through the simulation function of BIM, the consequences and impact of different green building construction plans can be simulated [5,17,48,99,100,117–122]. Moreover, it can assist the project managers in comparing different green building construction plans [102,103]. Through the BIM's comparison of the quality, estimated construction period, and environmental protection and resource-saving conditions of various alternatives, the most suitable green building construction plan can be selected [102,103]. In addition, based on green building project planning, construction site conditions, and feedback from subcontractors, BIM can assist the construction teams in developing the requirements and configurations of personnel, materials, and equipment [4,5,99,100,130,141–143,150]. Based on the abovementioned information, the logistics and transportation routes can also be automatically formulated through BIM [158–162]. Finally, through the information mentioned above, the estimated cost and budgets can be put forward by BIM [123,124,151].

Despite the various benefits of BIM utilization in the green building pre-construction phase, there are still some obvious challenges. Given that many formats can be adopted in BIM applications, the green building construction organizations might have format mismatch issues in the information delivery process from the design phase to construction [6,14]. Moreover, another hindrance is that project management personnel often refuse to adopt BIM in the construction process [22,77,163]. According to Akhmetzhanova et al. [24], 55% of companies refused to utilize BIM because clients or management do not support the adoption of the technology.

4.2. BIM Utilization in the Green Building Construction Phase

Compared with the BIM utilization at the pre-construction and post-construction phases, the most significant contributions of BIM capabilities are in the construction phase.

Through the discussion of reviewed BIM capabilities in Section 3, it can be determined that the BIM contributions in the construction phase can be summarized as below.

In the process of green building construction, 3D visual modeling is of significant importance to the construction organization [48,54,97,98,104–112,116]. It can demonstrate the entire green building project, the interior structures, and components in 3D [4,5,51,54,76,77,104–106]. Moreover, In BIM 4D, the stakeholders can become familiar with the conditions of buildings at different periods through visual 3D demonstrations [100,117–122,142,143,145,164]. From the perspective of safety management, BIM's visualization can assist construction teams in identifying the clashes and potential risks in a visual manner, so as to provide the corresponding safety training and avoid delay, construction waste production, and rework [5,14,46,99,133,135,136,138–140].

As a multi-function database, BIM can provide information and knowledge collection, storage, and management in the entire green building construction process, and integrate fragmented information in a unified format in the corresponding file [4,15,52,78–83,145]. Besides storing and categorizing the data that are generated in the green building construction process, BIM can automatically provide solutions to issues for stakeholders through CBR [143,146,147]. With the integration of BIM and third-party devices, the construction site's natural environment, climate, infrastructure, and the utilization conditions of human resources and materials are all available to stakeholders [5,63,85–94].

Moreover, simulation is an important characteristic of BIM utilization in green building construction. Simulation includes not only the simulation of the construction activities' impact on the surrounding environment, but also the prediction of construction processes and risks [4,5,10,17,46,100,101,117–122,129,131,133,141]. Given that green buildings are required to meet the green buildings' assessment standards, through the simulation function of BIM, the construction organization can perform comparisons of different construction schemes and select the scheme that can match the evaluation standards of green buildings in the most positive sense [52–55,102,103]. Based on the simulation and information management, the bills of quantity can be automatically generated by BIM to estimate the consumption of materials and the overall cost of the project [5,95,96,99,100,123–128,132,151,152].

BIM can provide an effective collaboration platform for all stakeholders to communicate and collaborate in the construction process of green buildings [10,13,26,62–67,69–71,73,74,122,165]. Through BIM, all project changes can be reflected in a timely manner, and decisions made by one stakeholder are immediately uploaded to the BIM platform and communicated to all other stakeholders [4,14,66,73,76,165]. Moreover, BIM can demonstrate the positions, responsibilities, and current status of all stakeholders, thus assisting stakeholders in obtaining an overview of other stakeholders' situations [64,72,75]. In addition, construction organizations can utilize BIM to communicate with design organizations and facilitate the management requirements of green buildings, and to develop their requirements about the corresponding green building projects [13,61,62,66–68,71]. Through the integration of BIM with the Internet of Things (IoT), the construction team can be assisted remotely by professionals worldwide to improve the projects' quality and solve existing issues [64,72,75].

Despite the significant advantages of BIM implementations in the green building construction phase, the challenges are still non-negligible. In the process of BIM utilization, many stakeholders cooperate using the same BIM platforms or files, which leads to the obscure copyright of the developed data in the green building construction process [165–170]. It is difficult for project managers to define the ownership of involved data [167,168]. Moreover, given the relative independence of the design and construction organization, some information required in the green building construction process might not be obtained in the delivered BIM files [6,14]. In the green building construction process, the main contractors might be required to cooperate with other subcontractors. In the context that the uniform regulatory framework is absent, the interactivity between the main contractors and subcontractors might be insufficient [24].

4.3. BIM Utilization in the Green Building Post-Construction Phase

The post-construction phase is the final phase of green building construction. It includes all of the final processes to hand over these green building projects to the building owner and facility management organizations [171]. Through the systematic review and discussion of the BIM utilization in green building construction, the BIM's contribution during the post-construction phase can be summarized as follows: final acceptance, information handover, and green building certificate inspection.

From the perspective of information handover and final acceptance, all building-related information and construction activities can be stored, categorized, and retrieved in BIM. This information is integrated into one single file in a given format (usually the IFC format) [4,15,45,47,52,78–83,172–174]. Moreover, through the 3D visualization demonstration of BIM, the entire 3D model of green buildings can be demonstrated for the final acceptance personnel to check [109–112,116]. Through BIM 4D, the specific situations of the target green building projects at different points in time can be provided [100,117–122,130].

The green buildings must meet the green building evaluation standards. Regardless of the various countries' green building evaluation standards, the construction organizations can adopt BIM's simulation function to examine these buildings [175,176]. Utilizing BIM to simulate and evaluate the energy consumption, environmental performance, material utilization, and residents' comfort of the green building can help to assess the compliance of the green building to the green building evaluation standards [13,59,177–181].

In the post-construction phase, there are still barriers that impede BIM utilization. In the process of information handover between green building construction organizations and green building facility management organizations, information omissions and mismatched formats are still the main challenges for both parties [6,14]. Moreover, given the high expenditure on BIM training, the facility management organizations might lack sufficient BIM operators. This issue can also disturb the handover of BIM documents [14,182,183].

5. Conclusions

Given the significant performance of BIM, there has been a dramatic increase in construction organizations that utilize BIM in the green building construction phase. Despite many studies exploring BIM utilization in green building construction, review articles in this area are relatively rare. To enhance the understanding of AEC practitioners in terms of the BIM capabilities in the green building construction phase and to bridge the abovementioned research gap, this study performed a systematic review of these BIM capabilities. Through the review of the retrieved articles, it can be summarized that the BIM implementations in green building construction are categorized into the following benefits: project quality improvement, lifecycle data storage and management, collaboration optimization, and planning and schedule management optimization. Moreover, through the discussion and analysis of the reviewed BIM capabilities, it can be concluded that BIM can make significant contributions in the pre-construction phase, construction phase, and post-construction phase of green building projects.

In spite of the tremendous abovementioned BIM benefits, there are still some obstacles when using BIM in the green building construction phase, including non-uniform data formats, insufficient interactivity, ambiguous ownership, insufficient BIM training, and BIM adoption hesitancy. Despite the abovementioned shortcomings of BIM at the present stage, through the comparison of the benefits and challenges of BIM capabilities in the green buildings' construction phase, it can be concluded that the BIM application still has significant potential benefits and improvements for green building construction. Through the systematic review, this study provided a comprehensive overview and understanding of BIM capabilities in the green building construction phase to promote and optimize BIM utilization in this area. Moreover, this study also pointed out the challenges and future direction of BIM capabilities in green buildings to encourage other researchers to overcome these issues.

In addition to the contributions mentioned above, this study also has some limitations. The limitations are presented below.

1. In this study, some reviewed BIM capabilities can be utilized in not only the construction phase of green buildings, but also in the design and facility management phase of other building types. This reduces the pertinence of the study to some extent. However, to provide a comprehensive systematic review and avoid the omissions of BIM capabilities in green building construction, these BIM capabilities are included in this study.
2. Due to the language skills limitations of the authors, only English articles were reviewed in this study. Non-English articles were excluded from the article screening process.
3. In this study, the majority of the reviewed BIM capabilities are on BIM utilization in the pre-construction phase and the construction of green buildings. Rarely are BIM functions reviewed that have been utilized in the post-construction phase of green building projects specifically. It is recommended that other researchers perform the corresponding studies to explore BIM utilization in the green building post-construction phase.

In conclusion, this study develops a comprehensive systematic review and discussion of BIM capabilities in the construction of green buildings. Given that the evolvement of BIM is rapid, the BIM capabilities are updated correspondingly with the development of internet technology. Thus, other researchers are welcome to further explore and review the BIM utilization in the construction of green buildings based on this study.

Author Contributions: Y.C. wrote the original draft preparation and conducted data curation and formal analysis; Y.C., S.N.K. and N.M.A. were responsible for the review and conceptualization; S.N.K. and N.M.A. were responsible for the supervision. All authors have read and agreed to the published version of the manuscript.

Funding: This research received no external funding.

Institutional Review Board Statement: Not applicable.

Informed Consent Statement: Not applicable.

Data Availability Statement: Not applicable.

Acknowledgments: I would first like to thank my supervisors, Syahrul Nizam Kamaruzzaman and Nur Mardhiyah Aziz. Your insightful supervision and constructive feedback pushed me to sharpen my thinking and brought my work to a higher level. Moreover, I would like to acknowledge Mingru Cao, Yan Zhang, Yuehua Ma, Runjia Chen, and Xiaoguang Lin. Your constructive suggestions and warm-hearted psychic support sustained me in completing this study.

Conflicts of Interest: The authors declare no conflict of interest.

Appendix A

Table A1. Reviewed studies in Section 3.2.1 (sorted by order of appearance).

Title of the Article	Publication Year
BIM in Off-Site Manufacturing for Buildings	2017
A Scientometric Review of Global BIM Research: Analysis and Visualization	2017
Building Information Modelling (BIM) Uptake: Clear Benefits, Understanding Its Implementation, Risks and Challenges	2017
Effect of BIM on Rework in Construction Projects in Singapore: Status Quo, Magnitude, Impact, and Strategies	2019
Applications of BIM: A Brief Review and Future Outline	2018

Table A1. Cont.

Title of the Article	Publication Year
Adoption of Building Information Modelling (Bim): Factors Contribution and Benefits	2018
Contribution and Obstacle Analysis of Applying BIM in Promoting Green Buildings	2021
BIM-Based Approach for Optimizing Life Cycle Costs of Sustainable Buildings	2018
Integration of BIM and LCA: Evaluating the Environmental Impacts of Building Materials at an Early Stage of Designing a Typical Office Building	2017
Measuring the Feasibility of Using of BIM Application to Facilitate GBI Assessment Process	2019
Sustainability and Energy Efficiency: BIM 6D. Study of the BIM Methodology Applied to Hospital Buildings. Value of Interior Lighting and Daylight in Energy Simulation	2020
Green BIM Assessment Applying for Energy Consumption and Comfort in the Traditional Public Market: A Case Study	2019
Integrating BIM-Based LCA and Building Sustainability Assessment	2020
Step-by-Step Implementation of BIM-LCA: A Case Study Analysis Associating Defined Construction Phases with Their Respective Environmental Impacts	2019
LCA and BIM: Visualization of Environmental Potentials in Building Construction at Early Design Stages	2018
Recommendations for Developing a BIM for the Purpose of LCA in Green Building Certifications	2020
Developing a Green Building Evaluation Standard for Interior Decoration: A Case Study of China	2019

Table A2. Reviewed studies in Section 3.2.2 (sorted by order of appearance).

Title of the Article	Publication Year
Critical Success Factors for Small Contractors to Conduct Green Building Construction Projects in Singapore: Identification and Comparison with Large Contractors	2020
Differing Perspectives on Collaboration in Construction	2012
Relationship Network Structure and Organizational Competitiveness: Evidence from BIM Implementation Practices in the Construction Industry	2018
BIM-Based Green Building Evaluation and Optimization: A Case Study	2021
Developing an Integrated BIM + GIS Web-Based Platform for a Mega Construction Project	2022
Collaboration Barriers in BIM-Based Construction Networks: A Conceptual Model	2019
BIM Tool Development Enhancing Collaborative Scheduling for Pre-Construction	2020
A Framework for Collaboration Management of BIM Model Creation in Architectural Projects	2018
Building Information Modelling in Construction: Insights from Collaboration and Change Management Perspectives	2018
Communications in Hybrid Arrangements: Case of Australian Construction Project Teams	2017

Table A2. *Cont.*

Title of the Article	Publication Year
Modelling Building Ownership Boundaries within BIM Environment: A Case Study in Victoria, Australia	2017
Interoperability Analysis of IFC-Based Data Exchange between Heterogeneous BIM Software	2018
Review of BIM's Application in Energy Simulation: Tools, Issues, and Solutions	2019
Identifying and Contextualizing the Motivations for BIM Implementation in Construction Projects: An Empirical Study in China	2017
Collaboration in BIM-Based Construction Networks: A Bibliometric-Qualitative Literature Review	2017
Sorting out the Essence of Owner–Contractor Collaboration in Capital Project Delivery	2015
The Conditions for Successful Automated Collaboration in Construction	2014
Collaborative Relationship Discovery in BIM Project Delivery: A Social Network Analysis Approach	2020
Building Information Modeling (BIM) for Green Buildings: A Critical Review and Future Directions	2017
Use of LoD Decision Plan in BIM-Projects	2017
Exploring the Barriers and Driving Factors in Implementing Building Information Modelling (BIM) in the Malaysian Construction Industry: A Preliminary Study	2014
Contribution and Obstacle Analysis of Applying BIM in Promoting Green Buildings	2021

Table A3. Reviewed studies in Section 3.2.3 (sorted by order of appearance).

Title of the Article	Publication Year
Transition from Building Information Modeling (BIM) to Integrated Digital Delivery (IDD) in Sustainable Building Management: A Knowledge Discovery Approach Based Review	2021
Enhancing a Building Information Model for an Existing Building with Data from a Sustainable Facility Management Database	2021
Integration of BIM and LCA: Evaluating the Environmental Impacts of Building Materials at an Early Stage of Designing a Typical Office Building	2017
Building Information Modeling (BIM) for Green Buildings: A Critical Review and Future Directions	2017
Comparative Analysis of Energy Performance Assessment for Green Buildings: China Green Building Rating System vs Other Major Certification Systems	2016
BIM-Based Performance Monitoring for Smart Building Management	2021
Blockchain-Enabled IoT-BIM Platform for Supply Chain Management in Modular Construction	
A Simplified Relational Database Schema for Transformation of BIM Data into a Query-Efficient and Spatially Enabled Database	2017
Quantifying and Visualizing Value Exchanges in Building Information Modeling (BIM) Projects	2019

Table A3. *Cont.*

Title of the Article	Publication Year
Application of ND BIM Integrated Knowledge-Based Building Management System (BIM-IKBMS) for Inspecting Post-Construction Energy Efficiency	2017
Improving Maintenance Performance by Developing an IFC BIM/RFID-Based Computer System	2021
3D Environmental Urban BIM Using LiDAR Data for Visualization on Google Earth	2022
Research Trend of the Application of Information Technologies in Construction and Demolition Waste Management	2020
Reducing Noise Pollution by Planning Construction Site Layout via a Multi-Objective Optimization Model	2019
Using BIM to Improve Building Energy Efficiency–A Scientometric and Systematic Review	2021
Developing an Integrated BIM + GIS Web-Based Platform for a Mega Construction Project	2022
Toward Sustainable Energy-Independent Buildings Using Internet of Things	2020
The Intelligent Use of RFID and BIM in Prefabricated, Prefinished, Volumetric Construction Work Flow	2020
Building Information Modeling (BIM)-Based Modular Integrated Construction Risk Management–Critical Survey and Future Needs	2020
An ICT-Enabled Product Service System for Reuse of Building Components	2019
Analysis of the Benefits, Challenges and Risks for the Integrated Use of BIM, RFID and WSN: A Mixed Method Research	2022

Table A4. Reviewed studies in Section 3.2.4 (sorted by order of appearance).

Title of the Article	Publication Year
A BIM-WMS Integrated Decision Support Tool for Supply Chain Management in Construction	2019
Research on Construction Schedule Management Based on BIM Technology	2017
Life Cycle Energy Efficiency in Building Structures: A Review of Current Developments and Future Outlooks Based on BIM Capabilities	2017
Real-Time Visualization of Building Information Models (BIM)	2015
Building Information Modelling (BIM) Uptake: Clear Benefits, Understanding Its Implementation, Risks and Challenges	2017
BIM-Enabled Facilities Operation and Maintenance: A Review	2019
Effect of BIM on Rework in Construction Projects in Singapore: Status Quo, Magnitude, Impact, and Strategies	2019
Research on the Project Management of BIM Project from the Perspective of Enterprise Strategy	2016
Integration of BIM and GIS in Sustainable Built Environment: A Review and Bibliometric Analysis	2019
Critical Success Factors for Implementing Building Information Modelling (BIM): A Longitudinal Review	2018
BIM-Based Applications of Metaheuristic Algorithms to Support the Decision-Making Process: Uses in the Planning of Construction Site Layout	2017

Table A4. *Cont.*

Title of the Article	Publication Year
Reducing Noise Pollution by Planning Construction Site Layout via a Multi-Objective Optimization Model	2019
Sustainability-Based Lifecycle Management for Bridge Infrastructure Using 6D BIM	2020
Building Performance Optimization Using CFD for 6D BIM Application—A Case Study	2021
Sustainability and Energy Efficiency: BIM 6D. Study of the BIM Methodology Applied to Hospital Buildings. Value of Interior Lighting and Daylight in Energy Simulation	2020
Evaluation of the Open Diversion Channel Capacity on Margatiga Dam Construction Project Using 6D BIM Analysis	2021
Integration of Aerobiological Information for Construction Engineering Based on LiDAR and BIM	2022
Permanent Magnet, Toroidal Winding Generator for 6D BIM Applications	2021
Research on PKIM Energy Construction Engineering Software System Based on Building BIM Technology	2022
Automated 3D Volumetric Reconstruction of Multiple-Room Building Interiors for as-Built BIM	2018
Green Construction Evaluation System Based on BIM Distributed Cloud Service	2021
Green Building Investment Control System Based on a Three-Dimensional Parametric Model of the Green Building	2021
Utilizing BIM and GIS for Representation and Visualization of 3D Cadastre	2019
A BIM Oriented Model to a 3D Indoor GIS for Space Management—A Requirement Analysis	2019
A Full Level-of-Detail Specification for 3D Building Models Combining Indoor and Outdoor Scenes	2018
Truss Construction of Green Fabricated Steel Structure Based on BIM Intelligent Technology	2021
Integrated EDM and 4D BIM-Based Decision Support System for Construction Projects Control	2022
Supporting Constructability Analysis Meetings with Immersive Virtual Reality-Based Collaborative BIM 4D Simulation	2018
Impacts of 4D BIM on Construction Project Performance	2021
The Effects of BIM Maturity Level on the 4D Simulation Performance: An Empirical Study	2021
BIM-Based Framework to Quantify Delays and Cost Overruns Due to Changes in Construction Projects	2022
4D Modelling Using Virtual Collaborative Planning and Scheduling	2021
Quantity Surveying and BIM 5D. Its Implementation and Analysis Based on a Case Study Approach in Spain	2021
Implementing 5D BIM on Construction Projects: Contractor Perspectives from the UK Construction Sector	2020
Machine Learning-Integrated 5D BIM Informatics: Building Materials Costs Data Classification and Prototype Development	2022
Cash Flow System Development Framework within Integrated Project Delivery (IPD) Using BIM Tools	2021

Table A4. *Cont.*

Title of the Article	Publication Year
A BIM-database-integrated system for construction cost estimation	2021
Application of BIM Technology in Construction Cost Management of Building Engineering	2021
Enhancing Facility Management through BIM 6D	2016
The Adoption of 4D BIM in the UK Construction Industry: An Innovation Diffusion Approach	2017
4D BIM for Environmental Planning and Management	2017
Integrating BIM and GIS to Improve the Visual Monitoring of Construction Supply Chain Management	2013
Improving Effectiveness of Safety Training at Construction Worksite Using 3D BIM Simulation	2020
Information Technology and Safety: Integrating Empirical Safety Risk Data with Building Information Modeling, Sensing, and Visualization Technologies	2016
An Automated Safety Risk Recognition Mechanism for Underground Construction at the Pre-Construction Stage Based on BIM	2018
A Research Framework of Mitigating Construction Accidents in High-Rise Building Projects via Integrating Building Information Modeling with Emerging Digital Technologies	2021
A Research Framework of Mitigating Construction Accidents in High-Rise Building Projects via Integrating Building Information Modeling with Emerging Digital Technologies	2021
Using BIM as a Tool to Teach Construction Safety	2017
Semi-Automatic Construction Hazard Identification Method Using 4D BIM	2021
BIM-Based Framework for Automatic Scheduling of Facility Maintenance Work Orders	2018
Investigating Benefits and Criticisms of BIM for Construction Scheduling in SMEs: An Italian Case Study	2018
Recognition of Process Patterns for BIM-Based Construction Schedules	2017
BIM-Based Augmented Reality Inspection and Maintenance of Fire Safety Equipment	2020
Automated Schedule and Progress Updating of IFC-Based 4D BIMs	2017
Retrieving Similar Cases for Construction Project Risk Management Using Natural Language Processing Techniques	2017
BIM-Based Risk Identification System in Tunnel Construction	2016
Construction Planning, Programming and Control	2013
Knowledge-Based Schedule Generation and Evaluation	2010
BIM-Integrated Construction Operation Simulation for Just-In-Time Production Management	2016
Informetric Analysis and Review of Literature on the Role of BIM in Sustainable Construction	2019
Outlining a New Collaborative Business Model as a Result of the Green Building Information Modelling Impact in the AEC Supply Chain	2019

References

1. *ISO 19650-1:2018*; Organization and Digitization of Information about Buildings and Civil Engineering Works, Including Building Information Modelling (BIM)—Information Management Using Building Information Modelling. British Standards Institution: London, UK, 2018.
2. National Bureau of Statistics of China. *NBS National BIM Report 2019*; National Bureau of Statistics of China: Beijing, China, 2019.
3. Wong, J.K.W.; Zhou, J. Enhancing environmental sustainability over building life cycles through green BIM: A review. *Autom. Constr.* **2015**, *57*, 156–165. [CrossRef]
4. Lu, Y.; Wu, Z.; Chang, R.; Li, Y. Building Information Modeling (BIM) for green buildings: A critical review and future directions. *Autom. Constr.* **2017**, *83*, 134–148. [CrossRef]
5. Ghaffarianhoseini, A.; Tookey, J.; Ghaffarianhoseini, A.; Naismith, N.; Azhar, S.; Efimova, O.; Raahemifar, K. Building Information Modelling (BIM) uptake: Clear benefits, understanding its implementation, risks and challenges. *Renew. Sustain. Energy Rev.* **2017**, *75*, 1046–1053. [CrossRef]
6. Raouf, A.M.I.; Al-Ghamdi, S.G. Building information modelling and green buildings: Challenges and opportunities. *Arch. Eng. Des. Manag.* **2019**, *15*, 1–28. [CrossRef]
7. US Green Building Council. The Definition of Green Building. Available online: https://www.usgbc.org/articles/what-green-building (accessed on 8 April 2022).
8. *GB/T 50378-2014*; Evaluation Standard for Green Building. Ministry of Housing and Urban-Rural Development: Beijing, China, 2018.
9. Hwang, B.-G.; Shan, M.; Lye, J.-M. Adoption of sustainable construction for small contractors: Major barriers and best solutions. *Clean Technol. Environ. Policy* **2018**, *20*, 2223–2237. [CrossRef]
10. Shan, M.; Liu, W.-Q.; Hwang, B.-G.; Lye, J.-M. Critical success factors for small contractors to conduct green building construction projects in Singapore: Identification and comparison with large contractors. *Environ. Sci. Pollut. Res.* **2020**, *27*, 8310–8322. [CrossRef]
11. Mohanta, A.; Das, S. Causal Analysis of Slow BIM Adoption in Eastern India with a Special Focus on Green Building Sector. *J. Inst. Eng. Ser. A* **2021**, *103*, 319–337. [CrossRef]
12. Sarkar, R.; Narang, K.; Daalia, A.; Gautam, V.; Nathani, U.; Shaw, R. Incorporation of BIM Based Modeling in Sustainable Development of Green Building from Stakeholders Perspective. In *Ecosystem-Based Disaster and Climate Resilience*; Mukherjee, M., Shaw, R., Eds.; Disaster and Risk Research: GADRI Book Series; Springer: Singapore, 2021; pp. 307–323. ISBN 978-981-164-814-4.
13. Guo, K.; Li, Q.; Zhang, L.; Wu, X. BIM-based green building evaluation and optimization: A case study. *J. Clean. Prod.* **2021**, *320*, 128824. [CrossRef]
14. Huang, B.; Lei, J.; Ren, F.; Chen, Y.; Zhao, Q.; Li, S.; Lin, Y. Contribution and obstacle analysis of applying BIM in promoting green buildings. *J. Clean. Prod.* **2021**, *278*, 123946. [CrossRef]
15. Liu, Z.; Lu, Y.; Shen, M.; Peh, L.C. Transition from building information modeling (BIM) to integrated digital delivery (IDD) in sustainable building management: A knowledge discovery approach based review. *J. Clean. Prod.* **2021**, *291*, 125223. [CrossRef]
16. Kang, T.W.; Choi, H.S. BIM-based Data Mining Method considering Data Integration and Function Extension. *KSCE J. Civ. Eng.* **2018**, *22*, 1523–1534. [CrossRef]
17. Hwang, B.-G.; Zhao, X.; Yang, K.W. Effect of BIM on Rework in Construction Projects in Singapore: Status Quo, Magnitude, Impact, and Strategies. *J. Constr. Eng. Manag.* **2019**, *145*, 04018125. [CrossRef]
18. Smits, W.; van Buiten, M.; Hartmann, T. Yield-to-BIM: Impacts of BIM maturity on project performance. *Build. Res. Inf.* **2017**, *45*, 336–346. [CrossRef]
19. Ascione, F.; Bianco, N.; De Stasio, C.; Mauro, G.M.; Vanoli, G.P. Simulation-based model predictive control by the multi-objective optimization of building energy performance and thermal comfort. *Energy Build.* **2016**, *111*, 131–144. [CrossRef]
20. Chan, D.W.; Olawumi, T.O.; Ho, A.M. Perceived benefits of and barriers to Building Information Modelling (BIM) implementation in construction: The case of Hong Kong. *J. Build. Eng.* **2019**, *25*, 100764. [CrossRef]
21. Sant'Anna, D.; Dos Santos, P.; Vianna, N.; Romero, M. Indoor environmental quality perception and users' satisfaction of conventional and green buildings in Brazil. *Sustain. Cities Soc.* **2018**, *43*, 95–110. [CrossRef]
22. Zhang, L.; Chu, Z.; He, Q.; Zhai, P. Investigating the Constraints to Buidling Information Modeling (BIM) Applications for Sustainable Building Projects: A Case of China. *Sustainability* **2019**, *11*, 1896. [CrossRef]
23. Zhang, L.; Chu, Z.; Song, H. Understanding the Relation between BIM Application Behavior and Sustainable Construction: A Case Study in China. *Sustainability* **2019**, *12*, 306. [CrossRef]
24. Akhmetzhanova, B.; Nadeem, A.; Hossain, A.; Kim, J.R. Clash Detection Using Building Information Modeling (BIM) Technology in the Republic of Kazakhstan. *Buildings* **2022**, *12*, 102. [CrossRef]
25. Tatygulov, A.; Gizatulina, A.S.; Zhamankulov, A. Level of BIM Development and Applying in Design and Engineering Survey Companies in the Republic of Kazakhstan. Research Results. *Bull. Natl. Eng. Acad. Repub. Kaz.* **2020**, *4*, 100–106. [CrossRef]
26. Kamel, E.; Memari, A.M. Review of BIM's application in energy simulation: Tools, issues, and solutions. *Autom. Constr.* **2019**, *97*, 164–180. [CrossRef]
27. Kylili, A.; Fokaides, P.A. Policy trends for the sustainability assessment of construction materials: A review. *Sustain. Cities Soc.* **2017**, *35*, 280–288. [CrossRef]

28. Liu, Z.; Lu, Y.; Peh, L.C. A Review and Scientometric Analysis of Global Building Information Modeling (BIM) Research in the Architecture, Engineering and Construction (AEC) Industry. *Buildings* **2019**, *9*, 210. [CrossRef]
29. Rooshdi, R.R.R.M.; Ismail, N.A.A.; Sahamir, S.R.; Marhani, M.A. Integrative Assessment Framework of Building Information Modelling (BIM) and Sustainable Design for Green Highway Construction: A Review. *Chem. Eng. Trans.* **2021**, *89*, 55–60. [CrossRef]
30. Byrne, J.A. Improving the peer review of narrative literature reviews. *Res. Integr. Peer Rev.* **2016**, *1*, 12. [CrossRef]
31. Sutton, A.; Clowes, M.; Preston, L.; Booth, A. Meeting the review family: Exploring review types and associated information retrieval requirements. *Health Inf. Libr. J.* **2019**, *36*, 202–222. [CrossRef]
32. Gasparyan, A.Y.; Ayvazyan, L.; Blackmore, H.; Kitas, G. Writing a narrative biomedical review: Considerations for authors, peer reviewers, and editors. *Rheumatol. Int.* **2011**, *31*, 1409–1417. [CrossRef]
33. Khangura, S.; Konnyu, K.; Cushman, R.; Grimshaw, J.; Moher, D. Evidence summaries: The evolution of a rapid review approach. *Syst. Rev.* **2012**, *1*, 10. [CrossRef]
34. Merigó, J.M.; Yang, J.-B. A bibliometric analysis of operations research and management science. *Omega* **2017**, *73*, 37–48. [CrossRef]
35. Brika, S.K.M.; Algamdi, A.; Chergui, K.; Musa, A.A.; Zouaghi, R. Quality of Higher Education: A Bibliometric Review Study. *Front. Educ.* **2021**, *6*, 666087. [CrossRef]
36. Gopalakrishnan, S.; Ganeshkumar, P. Systematic reviews and meta-analysis: Understanding the best evidence in primary healthcare. *J. Fam. Med. Prim. Care* **2013**, *2*, 9. [CrossRef]
37. Jahan, N.; Naveed, S.; Zeshan, M.; Tahir, M.A. How to Conduct a Systematic Review: A Narrative Literature Review. *Cureus* **2016**, *8*, e864. [CrossRef] [PubMed]
38. Cook, D.J.; Mulrow, C.D.; Haynes, R.B. Systematic Reviews: Synthesis of Best Evidence for Clinical Decisions. *Ann. Intern. Med.* **1997**, *126*, 376. [CrossRef]
39. Colenberg, S.; Jylhä, T.; Arkesteijn, M. The relationship between interior office space and employee health and well-being—A literature review. *Build. Res. Inf.* **2021**, *49*, 352–366. [CrossRef]
40. Lee, K.-T.; Im, J.-B.; Park, S.-J.; Kim, J.-H. Conceptual Framework to Support Personalized Indoor Space Design Decision-Making: A Systematic Literature Review. *Buildings* **2022**, *12*, 716. [CrossRef]
41. Hoang, G.T.T.; Dupont, L.; Camargo, M. Application of Decision-Making Methods in Smart City Projects: A Systematic Literature Review. *Smart Cities* **2019**, *2*, 433–452. [CrossRef]
42. Page, M.J.; McKenzie, J.E.; Bossuyt, P.M.; Boutron, I.; Hoffmann, T.C.; Mulrow, C.D.; Shamseer, L.; Tetzlaff, J.M.; Akl, E.A.; Brennan, S.E.; et al. The PRISMA 2020 Statement: An Updated Guideline for Reporting Systematic Reviews. *Syst. Rev.* **2021**, *10*, 89. [CrossRef]
43. Cho, S.; Lee, K.-T.; Choi, Y.I.; Jung, S.J.; Park, S.-J.; Bae, S.; Kim, J. Networking human biomarker and hazardous chemical elements from building materials: Systematic literature review and in vivo test. *Build. Environ.* **2021**, *192*, 107603. [CrossRef]
44. Al-Ashmori, Y.Y.; Othman, I.; Rahmawati, Y.; Amran, Y.H.M.; Sabah, S.H.A.; Rafindadi, A.D.; Mikić, M. BIM benefits and its influence on the BIM implementation in Malaysia. *Ain Shams Eng. J.* **2020**, *11*, 1013–1019. [CrossRef]
45. Abanda, F.; Tah, J.; Cheung, F. BIM in off-site manufacturing for buildings. *J. Build. Eng.* **2017**, *14*, 89–102. [CrossRef]
46. Zhao, X. A scientometric review of global BIM research: Analysis and visualization. *Autom. Constr.* **2017**, *80*, 37–47. [CrossRef]
47. Pezeshki, Z.; Ivari, S.A.S. Applications of BIM: A Brief Review and Future Outline. *Arch. Comput. Methods Eng.* **2018**, *25*, 273–312. [CrossRef]
48. Noor, S.M.; Junaidi, S.R.; Ramly, M.K.A. Adoption of Building Information Modelling (Bim): Factors Contribution and Benefits. *J. Inf. Syst. Technol. Manag.* **2018**, *3*, 47–63.
49. Aguila, G.M.; De Castro, E.L.; Dotong, C.I.; Laguador, J.M. Employability of Computer Engineering Graduates from 2013 to 2015 in One Private Higher Education Institution in the Philippines. *Asia Pac. J. Educ. Arts Sci.* **2016**, *3*, 48–54.
50. Marzouk, M.; Azab, S.; Metawie, M. BIM-based approach for optimizing life cycle costs of sustainable buildings. *J. Clean. Prod.* **2018**, *188*, 217–226. [CrossRef]
51. Sanhudo, L.; Ramos, N.M.M.; Martins, J.P.; Almeida, R.M.S.F.; Barreira, E.; Simões, M.L.; Cardoso, V. Building information modeling for energy retrofitting—A review. *Renew. Sustain. Energy Rev.* **2018**, *89*, 249–260. [CrossRef]
52. Najjar, M.; Figueiredo, K.; Palumbo, M.; Haddad, A. Integration of BIM and LCA: Evaluating the environmental impacts of building materials at an early stage of designing a typical office building. *J. Build. Eng.* **2017**, *14*, 115–126. [CrossRef]
53. Solla, M.; Ismail, L.H.; Shaarani, A.S.M.; Milad, A. Measuring the Feasibility of Using of BIM Application to Facilitate GBI Assessment Process. *J. Build. Eng.* **2019**, *25*, 100821. [CrossRef]
54. Montiel-Santiago, F.; Hermoso-Orzáez, M.; Terrados-Cepeda, J. Sustainability and Energy Efficiency: BIM 6D. Study of the BIM Methodology Applied to Hospital Buildings. Value of Interior Lighting and Daylight in Energy Simulation. *Sustainability* **2020**, *12*, 5731. [CrossRef]
55. Lin, P.-H.; Chang, C.-C.; Lin, Y.-H.; Lin, W.-L. Green BIM Assessment Applying for Energy Consumption and Comfort in the Traditional Public Market: A Case Study. *Sustainability* **2019**, *11*, 4636. [CrossRef]
56. Carvalho, J.P.; Alecrim, I.; Bragança, L.; Mateus, R. Integrating BIM-Based LCA and building sustainability assessment. *Sustainability* **2020**, *12*, 7468. [CrossRef]

57. Di Bari, R.; Jorgji, O.; Horn, R.; Gantner, J.; Ebertshäuser, S. Step-by-step implementation of BIM-LCA: A case study analysis associating defined construction phases with their respective environmental impacts. *IOP Conf. Ser. Earth Environ. Sci.* **2019**, *323*, 012105. [CrossRef]
58. Röck, M.; Hollberg, A.; Habert, G.; Passer, A. LCA and BIM: Visualization of environmental potentials in building construction at early design stages. *Build. Environ.* **2018**, *140*, 153–161. [CrossRef]
59. Veselka, J.; Nehasilová, M.; Dvořáková, K.; Ryklová, P.; Volf, M.; Růžička, J.; Lupíšek, A. Recommendations for Developing a BIM for the Purpose of LCA in Green Building Certifications. *Sustainability* **2020**, *12*, 6151. [CrossRef]
60. Wu, Z.; Li, H.; Feng, Y.; Luo, X.; Chen, Q. Developing a green building evaluation standard for interior decoration: A case study of China. *Build. Environ.* **2019**, *152*, 50–58. [CrossRef]
61. Hughes, D.; Williams, T.; Ren, Z. Differing perspectives on collaboration in construction. *Constr. Innov.* **2012**, *12*, 355–368. [CrossRef]
62. Cao, D.; Li, H.; Wang, G.; Luo, X.; Tan, D. Relationship Network Structure and Organizational Competitiveness: Evidence from BIM Implementation Practices in the Construction Industry. *J. Manag. Eng.* **2018**, *34*, 04018005. [CrossRef]
63. Zhao, L.; Mbachu, J.; Liu, Z. Developing an Integrated BIM+GIS Web-Based Platform for a Mega Construction Project. *KSCE J. Civ. Eng.* **2022**, *26*, 1505–1521. [CrossRef]
64. Oraee, M.; Hosseini, M.R.; Edwards, D.J.; Li, H.; Papadonikolaki, E.; Cao, D. Collaboration barriers in BIM-based construction networks: A conceptual model. *Int. J. Proj. Manag.* **2019**, *37*, 839–854. [CrossRef]
65. Tallgren, M.V.; Roupé, M.; Johansson, M.; Bosch-Sijtsema, P. BIM tool development enhancing collaborative scheduling for pre-construction. *J. Inf. Technol. Constr.* **2020**, *25*, 374–397. [CrossRef]
66. Lin, Y.-C.; Yang, H.-H. A Framework for Collaboration Management of BIM Model Creation in Architectural Projects. *J. Asian Arch. Build. Eng.* **2018**, *17*, 39–46. [CrossRef]
67. Matthews, J.; Love, P.E.D.; Mewburn, J.; Stobaus, C.; Ramanayaka, C. Building information modelling in construction: Insights from collaboration and change management perspectives. *Prod. Plan. Control* **2018**, *29*, 202–216. [CrossRef]
68. Hosseini, M.R.; Zavadskas, E.K.; Xia, B.; Chileshe, N.; Mills, A. Communications in Hybrid Arrangements: Case of Australian Construction Project Teams. *Eng. Econ.* **2017**, *28*, 290–300. [CrossRef]
69. Atazadeh, B.; Kalantari, M.; Rajabifard, A.; Ho, S. Modelling building ownership boundaries within BIM environment: A case study in Victoria, Australia. *Comput. Environ. Urban Syst.* **2017**, *61*, 24–38. [CrossRef]
70. Lai, H.; Deng, X. Interoperability analysis of IFC-based data exchange between heterogeneous BIM software. *J. Civ. Eng. Manag.* **2018**, *24*, 537–555. [CrossRef]
71. Cao, D.; Li, H.; Wang, G.; Huang, T. Identifying and contextualising the motivations for BIM implementation in construction projects: An empirical study in China. *Int. J. Proj. Manag.* **2017**, *35*, 658–669. [CrossRef]
72. Oraee, M.; Hosseini, M.R.; Papadonikolaki, E.; Palliyaguru, R.; Arashpour, M. Collaboration in BIM-based construction networks: A bibliometric-qualitative literature review. *Int. J. Proj. Manag.* **2017**, *35*, 1288–1301. [CrossRef]
73. Suprapto, M.; Bakker, H.L.M.; Mooi, H.G.; Moree, W. Sorting out the essence of owner–contractor collaboration in capital project delivery. *Int. J. Proj. Manag.* **2015**, *33*, 664–683. [CrossRef]
74. van Gassel, F.; Láscaris-Comneno, T.; Maas, G. The conditions for successful automated collaboration in construction. *Autom. Constr.* **2014**, *39*, 85–92. [CrossRef]
75. Wang, Y.; Thangasamy, V.K.; Hou, Z.; Tiong, R.L.; Zhang, L. Collaborative relationship discovery in BIM project delivery: A social network analysis approach. *Autom. Constr.* **2020**, *114*, 103147. [CrossRef]
76. Grytting, I.; Svalestuen, F.; Lohne, J.; Sommerseth, H.; Augdal, S.; Lædre, O. Use of LoD Decision Plan in BIM-projects. *Procedia Eng.* **2017**, *196*, 407–414. [CrossRef]
77. Zakari, Z.; Ali, N.M.A.; Haron, A.T.; Ponting, A.M.; Hamid, Z.A. Exploring the Barriers and Driving Factors in Implementing Building Information Modelling (BIM) in the Malaysian Construction Industry: A Preliminary Study. *J. Inst. Eng. Malays.* **2014**, *75*, 1. [CrossRef]
78. Loeh, R.; Everett, J.; Riddell, W.; Cleary, D. Enhancing a Building Information Model for an Existing Building with Data from a Sustainable Facility Management Database. *Sustainability* **2021**, *13*, 7014. [CrossRef]
79. He, Y.; Ding, Y. Comparative Analysis of Energy Performance Assessment for Green Buildings: China Green Building Rating System vs. Other Major Certification Systems. *HVAC* **2016**, *46*, 79–86.
80. Edirisinghe, R.; Woo, J. BIM-based performance monitoring for smart building management. *Facilities* **2021**, *39*, 19–35. [CrossRef]
81. Li, X.; Lu, W.; Xue, F.; Wu, L.; Zhao, R.; Lou, J.; Xu, J. Blockchain-Enabled IoT-BIM Platform for Supply Chain Management in Modular Construction. *J. Constr. Eng. Manag.* **2022**, *148*, 04021195. [CrossRef]
82. Solihin, W.; Eastman, C.; Lee, Y.-C.; Yang, D.-H. A simplified relational database schema for transformation of BIM data into a query-efficient and spatially enabled database. *Autom. Constr.* **2017**, *84*, 367–383. [CrossRef]
83. Zheng, X.; Lu, Y.; Li, Y.; Le, Y.; Xiao, J. Quantifying and visualizing value exchanges in building information modeling (BIM) projects. *Autom. Constr.* **2019**, *99*, 91–108. [CrossRef]
84. GhaffarianHoseini, A.; Zhang, T.; Nwadigo, O.; GhaffarianHoseini, A.; Naismith, N.; Tookey, J.; Raahemifar, K. Application of nD BIM Integrated Knowledge-based Building Management System (BIM-IKBMS) for inspecting post-construction energy efficiency. *Renew. Sustain. Energy Rev.* **2017**, *72*, 935–949. [CrossRef]

85. Kameli, M.; Hosseinalipour, M.; Sardroud, J.M.; Ahmed, S.M.; Behruyan, M. Improving maintenance performance by developing an IFC BIM/RFID-based computer system. *J. Ambient Intell. Humaniz. Comput.* **2020**, *12*, 3055–3074. [CrossRef]
86. Fernández-Alvarado, J.; Fernández-Rodríguez, S. 3D environmental urban BIM using LiDAR data for visualisation on Google Earth. *Autom. Constr.* **2022**, *138*, 104251. [CrossRef]
87. Li, C.Z.; Zhao, Y.; Xiao, B.; Yu, B.; Tam, V.W.; Chen, Z.; Ya, Y. Research trend of the application of information technologies in construction and demolition waste management. *J. Clean. Prod.* **2020**, *263*, 121458. [CrossRef]
88. Ning, X.; Qi, J.; Wu, C.; Wang, W. Reducing noise pollution by planning construction site layout via a multi-objective optimization model. *J. Clean. Prod.* **2019**, *222*, 218–230. [CrossRef]
89. Pereira, V.; Santos, J.; Leite, F.; Escórcio, P. Using BIM to improve building energy efficiency–A scientometric and systematic review. *Energy Build.* **2021**, *250*, 111292. [CrossRef]
90. Motlagh, N.H.; Khatibi, A.; Aslani, A. Toward Sustainable Energy-Independent Buildings Using Internet of Things. *Energies* **2020**, *13*, 5954. [CrossRef]
91. Abbott, E.L.; Chua, D.K. The Intelligent Use of RFID and BIM in Prefabricated, Prefinished, Volumetric Construction Work Flow. In Proceedings of the MATEC Web of Conferences, Cape Town, South Africa, 24–26 September 2018; EDP Sciences: Les Ulis, France, 2020; Volume 312, p. 04005.
92. Darko, A.; Chan, A.P.; Yang, Y.; Tetteh, M.O. Building information modeling (BIM)-based modular integrated construction risk management—Critical survey and future needs. *Comput. Ind.* **2020**, *123*, 103327. [CrossRef]
93. Ness, D.; Xing, K.; Kim, K.; Jenkins, A. An ICT-enabled Product Service System for Reuse of Building Components. *IFAC-PapersOnLine* **2019**, *52*, 761–766. [CrossRef]
94. Seyis, S.; Sönmez, A.M. Analysis of the benefits, challenges and risks for the integrated use of BIM, RFID and WSN: A mixed method research. *Constr. Innov.* 2022; *ahead of print*. [CrossRef]
95. Chen, P.-H.; Nguyen, T.C. A BIM-WMS integrated decision support tool for supply chain management in construction. *Autom. Constr.* **2019**, *98*, 289–301. [CrossRef]
96. Li, X.; Xu, J.; Zhang, Q. Research on Construction Schedule Management Based on BIM Technology. *Procedia Eng.* **2017**, *174*, 657–667. [CrossRef]
97. Eleftheriadis, S.; Mumovic, D.; Greening, P. Life cycle energy efficiency in building structures: A review of current developments and future outlooks based on BIM capabilities. *Renew. Sustain. Energy Rev.* **2017**, *67*, 811–825. [CrossRef]
98. Johansson, M.; Roupé, M.; Bosch-Sijtsema, P. Real-time visualization of building information models (BIM). *Autom. Constr.* **2015**, *54*, 69–82. [CrossRef]
99. Gao, X.; Pishdad-Bozorgi, P. BIM-enabled facilities operation and maintenance: A review. *Adv. Eng. Inform.* **2019**, *39*, 227–247. [CrossRef]
100. Wang, Y.; Liu, J. Research on the Project Management of BIM Project from the Perspective of Enterprise Strategy. In Proceedings of the 2016 International Conference on Economy, Management and Education Technology, Chongqing, China, 28–29 May 2016; Atlantis Press: Amsterdam, The Netherlands, 2016.
101. Wang, H.; Pan, Y.; Luo, X. Integration of BIM and GIS in sustainable built environment: A review and bibliometric analysis. *Autom. Constr.* **2019**, *103*, 41–52. [CrossRef]
102. Antwi-Afari, M.; Li, H.; Pärn, E.; Edwards, D. Critical success factors for implementing building information modelling (BIM): A longitudinal review. *Autom. Constr.* **2018**, *91*, 100–110. [CrossRef]
103. Amiri, R.; Sardroud, J.M.; de Soto, B.G. BIM-based Applications of Metaheuristic Algorithms to Support the Decision-making Process: Uses in the Planning of Construction Site Layout. *Procedia Eng.* **2017**, *196*, 558–564. [CrossRef]
104. Kaewunruen, S.; Sresakoolchai, J.; Zhou, Z. Sustainability-based lifecycle management for bridge infrastructure using 6D BIM. *Sustainability* **2020**, *12*, 2436. [CrossRef]
105. Le, H.T.; Nguyen, T.T. Building Performance Optimization Using CFD for 6D BIM Application—A Case Study. In Proceedings of the AIP Conference, Ho Chi Minh, Vietnam, 16 March 2021; AIP Publishing LLC: Melville, NY, USA, 2021; Volume 2420, p. 020003.
106. Hermawan, F.D.; Monica, S. Evaluation of the Open Diversion Channel Capacity on Margatiga Dam Construction Project Using 6D BIM Analysis. *IOP Conf. Ser. Earth Environ. Sci.* **2021**, *930*, 012045. [CrossRef]
107. Quevedo-Martínez, E.; Cortés-Pérez, J.P.; Coloma, J.F.; Fernández-Alvarado, J.F.; García, M.; Fernández-Rodríguez, S. Integration of Aerobiological Information for Construction Engineering Based on LiDAR and BIM. *Remote Sens.* **2022**, *14*, 618. [CrossRef]
108. Tien, L.H. Design Criteria for Axial Flux, Permanent Magnet, Toroidal Winding Generator for 6D BIM Applications. In Proceedings of the AIP Conference, Ho Chi Minh, Vietnam, 16 March 2021; AIP Publishing LLC: Melville, NY, USA, 2021; Volume 2420, p. 020004.
109. Fu, Y. Research on PKIM Energy Construction Engineering Software System Based on Building BIM Technology. *Wirel. Commun. Mob. Comput.* **2022**, *2022*, 2546708. [CrossRef]
110. Jung, J.; Stachniss, C.; Ju, S.; Heo, J. Automated 3D volumetric reconstruction of multiple-room building interiors for as-built BIM. *Adv. Eng. Inform.* **2018**, *38*, 811–825. [CrossRef]
111. Li, Y.; Gao, X.; Liu, X.; Zhang, R.; Wu, Y. Green Construction Evaluation System Based on BIM Distributed Cloud Service. *IOP Conf. Ser. Earth Environ. Sci.* **2021**, *760*, 012055. [CrossRef]

112. Monastyreva, D.; Astafieva, N. Green building investment control system based on a three-dimensional parametric model of the green building. *E3S Web Conf.* **2021**, *258*, 09079. [CrossRef]
113. Sun, J.; Mi, S.; Olsson, P.-O.; Paulsson, J.; Harrie, L. Utilizing BIM and GIS for Representation and Visualization of 3D Cadastre. *ISPRS Int. J. Geo-Inf.* **2019**, *8*, 503. [CrossRef]
114. Syed Mustorpha, S.N.A.; Wan Mohd, W.M.N. A Bim Oriented Model to a 3d Indoor GIS for Space Management-a Requirement Analysis. *IOP Conf. Ser. Earth Environ. Sci.* **2019**, *385*, 012046. [CrossRef]
115. Tang, L.; Li, L.; Ying, S.; Lei, Y. A Full Level-of-Detail Specification for 3D Building Models Combining Indoor and Outdoor Scenes. *ISPRS Int. J. Geo-Inf.* **2018**, *7*, 419. [CrossRef]
116. Xu, J. Truss construction of green fabricated steel structure based on BIM intelligent technology. *Int. J. Crit. Infrastruct.* **2021**, *17*, 54. [CrossRef]
117. Ayman, H.M.; Mahfouz, S.Y.; Alhady, A. Integrated EDM and 4D BIM-Based Decision Support System for Construction Projects Control. *Buildings* **2022**, *12*, 315. [CrossRef]
118. Boton, C. Supporting constructability analysis meetings with Immersive Virtual Reality-based collaborative BIM 4D simulation. *Autom. Constr.* **2018**, *96*, 1–15. [CrossRef]
119. Crowther, J.; Ajayi, S.O. Impacts of 4D BIM on construction project performance. *Int. J. Constr. Manag.* **2021**, *21*, 724–737. [CrossRef]
120. Haji, M.D.; Taghaddos, H.; Sebt, M.; Chokan, F.; Zavari, M. The Effects of BIM Maturity Level on the 4D Simulation Performance: An Empirical Study. *Int. J. Eng.* **2021**, *34*, 606–614.
121. Honnappa, D.; Padala, S.P.S. BIM-based framework to quantify delays and cost overruns due to changes in construction projects. *Asian J. Civ. Eng.* **2022**, *23*, 707–725. [CrossRef]
122. Tallgren, M.V.; Roupé, M.; Johansson, M. 4D modelling using virtual collaborative planning and scheduling. *J. Inf. Technol. Constr.* **2021**, *26*, 763–782. [CrossRef]
123. Aragó, A.B.; Hernando, J.R.; Saez, F.J.L.; Bertran, J.C. Quantity surveying and BIM 5D. Its implementation and analysis based on a case study approach in Spain. *J. Build. Eng.* **2021**, *44*, 103234. [CrossRef]
124. Moses, T.; Heesom, D.; Oloke, D. Implementing 5D BIM on construction projects: Contractor perspectives from the UK construction sector. *J. Eng. Des. Technol.* **2020**, *18*, 1867–1888. [CrossRef]
125. Banihashemi, S.; Khalili, S.; Sheikhkhoshkar, M.; Fazeli, A. Machine learning-integrated 5D BIM informatics: Building materials costs data classification and prototype development. *Innov. Infrastruct. Solut.* **2022**, *7*, 215. [CrossRef]
126. Elghaish, F.; Abrishami, S.; Abu Samra, S.; Gaterell, M.; Hosseini, M.R.; Wise, R.J. Cash flow system development framework within integrated project delivery (IPD) using BIM tools. *Int. J. Constr. Manag.* **2021**, *21*, 555–570. [CrossRef]
127. Le, H.T.T.; Likhitruangsilp, V.; Yabuki, N. A BIM-Database-Integrated System for Construction Cost Estimation. *Asean Eng. J.* **2021**, *11*, 45–59. [CrossRef]
128. Yang, J. Application of BIM Technology in Construction Cost Management of Building Engineering. *J. Phys. Conf. Ser.* **2021**, *2037*, 012046. [CrossRef]
129. Nicał, A.K.; Wodyński, W. Enhancing Facility Management through BIM 6D. *Procedia Eng.* **2016**, *164*, 299–306. [CrossRef]
130. Gledson, B.J.; Greenwood, D. The adoption of 4D BIM in the UK construction industry: An innovation diffusion approach. *Eng. Constr. Arch. Manag.* **2017**, *24*, 950–967. [CrossRef]
131. Jupp, J. 4D BIM for Environmental Planning and Management. *Procedia Eng.* **2017**, *180*, 190–201. [CrossRef]
132. Yu, Q.; Li, K.; Luo, H. A BIM-based Dynamic Model for Site Material Supply. *Procedia Eng.* **2016**, *164*, 526–533. [CrossRef]
133. Yuan, Z.; Wang, Y.; Sun, C. Construction schedule early warning from the perspective of probability and visualization. *J. Intell. Fuzzy Syst.* **2017**, *32*, 877–888. [CrossRef]
134. Irizarry, J.; Karan, E.P.; Jalaei, F. Integrating BIM and GIS to improve the visual monitoring of construction supply chain management. *Autom. Constr.* **2013**, *31*, 241–254. [CrossRef]
135. Ahn, S.; Kim, T.; Park, Y.-J.; Kim, J.-M. Improving Effectiveness of Safety Training at Construction Worksite Using 3D BIM Simulation. *Adv. Civ. Eng.* **2020**, *2020*, 2473138. [CrossRef]
136. Hallowell, M.R.; Hardison, D.; Desvignes, M. Information Technology and Safety: Integrating Empirical Safety Risk Data with Building Information Modeling, Sensing, and Visualization Technologies. *Constr. Innov.* **2016**, *16*, 323–347. [CrossRef]
137. Li, M.; Yu, H.; Liu, P. An automated safety risk recognition mechanism for underground construction at the pre-construction stage based on BIM. *Autom. Constr.* **2018**, *91*, 284–292. [CrossRef]
138. Manzoor, B.; Othman, I.; Pomares, J.C.; Chong, H.-Y. A Research Framework of Mitigating Construction Accidents in High-Rise Building Projects via Integrating Building Information Modeling with Emerging Digital Technologies. *Appl. Sci.* **2021**, *11*, 8359. [CrossRef]
139. Soemardi, B.W.; Erwin, R.G. Using BIM as a Tool to Teach Construction Safety. *MATEC Web Conf.* **2017**, *138*, 05007. [CrossRef]
140. Heidary, M.S.; Mousavi, M.; Alvanchi, A.; Barati, K.; Karimi, H. Semi-Automatic Construction Hazard Identification Method Using 4D BIM. In Proceedings of the International Symposium on Automation and Robotics in Construction, Dubai, United Arab Emirates, 2–4 November 2021; IAARC: Lyon, France, 2021; Volume 38, pp. 590–597.
141. Chen, W.; Chen, K.; Cheng, J.C.; Wang, Q.; Gan, V.J. BIM-based framework for automatic scheduling of facility maintenance work orders. *Autom. Constr.* **2018**, *91*, 15–30. [CrossRef]

142. Malacarne, G.; Toller, G.; Marcher, C.; Riedl, M.; Matt, D.T. Investigating benefits and criticisms of BIM for construction scheduling in SMES: An Italian case study. *Int. J. Sustain. Dev. Plan.* **2018**, *13*, 139–150. [CrossRef]
143. Sigalov, K.; König, M. Recognition of process patterns for BIM-based construction schedules. *Adv. Eng. Inform.* **2017**, *33*, 456–472. [CrossRef]
144. Chen, Y.-J.; Lai, Y.-S.; Lin, Y.-H. BIM-based augmented reality inspection and maintenance of fire safety equipment. *Autom. Constr.* **2020**, *110*, 103341. [CrossRef]
145. Hamledari, H.; McCabe, B.; Davari, S.; Shahi, A. Automated Schedule and Progress Updating of IFC-Based 4D BIMs. *J. Comput. Civ. Eng.* **2017**, *31*, 04017012. [CrossRef]
146. Zou, Y.; Kiviniemi, A.; Jones, S.W. Retrieving similar cases for construction project risk management using Natural Language Processing techniques. *Autom. Constr.* **2017**, *80*, 66–76. [CrossRef]
147. Zhang, L.; Wu, X.; Ding, L.; Skibniewski, M.; Lu, Y. Bim-Based Risk Identification System in Tunnel Construction. *J. Civ. Eng. Manag.* **2016**, *22*, 529–539. [CrossRef]
148. Cooke, B.; Williams, P. *Construction Planning, Programming and Control*; John Wiley & Sons: Hoboken, NJ, USA, 2013.
149. Mikulakova, E.; König, M.; Tauscher, E.; Beucke, K. Knowledge-based schedule generation and evaluation. *Adv. Eng. Inform.* **2010**, *24*, 389–403. [CrossRef]
150. Jeong, W.; Chang, S.; Son, J.; Yi, J.-S. BIM-Integrated Construction Operation Simulation for Just-In-Time Production Management. *Sustainability* **2016**, *8*, 1106. [CrossRef]
151. Santos, R.; Costa, A.A.; Silvestre, J.D.; Pyl, L. Informetric analysis and review of literature on the role of BIM in sustainable construction. *Autom. Constr.* **2019**, *103*, 221–234. [CrossRef]
152. Vilas-Boas, J.; Mirnoori, V.; Razy, A.; Silva, A. Outlining a New Collaborative Business Model as a Result of the Green Building Information Modelling Impact in the AEC Supply Chain. In *IFIP Advances in Information and Communication Technology, Proceedings of the Collaborative Networks and Digital Transformation, Turin, Italy, 23–25 September 2019*; Camarinha-Matos, L.M., Afsarmanesh, H., Antonelli, D., Eds.; Springer International: Cham, Switzerland, 2019; Volume 568, pp. 405–417. ISBN 978-3-030-28463-3.
153. Hamid, A.B.A.; Embi, M.R. Review on Application of Building Information Modelling in Interior Design Industry. *MATEC Web Conf.* **2016**, *66*, 3. [CrossRef]
154. Lau, S.E.N.; Zakaria, R.; Aminudin, E.; Saar, C.C.; Yusof, A.; Wahid, C.M.F.H.C. A Review of Application Building Information Modeling (BIM) during Pre-Construction Stage: Retrospective and Future Directions. In Proceedings of the IOP Conference Series: Earth and Environmental Science, Ho Chi Minh, Vietnam, 17–19 April 2018; IOP: Bristol, UK, 2018; Volume 143, p. 012050.
155. Aloise-Young, P.A.; Ross, E.C.; Dickmann, E.M.; Cross, J.E.; Zimmerle, D.; Nobe, M.C. Overcoming barriers to direct current power: Lessons learned from four commercial building case studies. *Energy Effic.* **2020**, *14*, 10. [CrossRef]
156. Holloway, S.; Parrish, K. The Contractor's Role in the Sustainable Construction Industry. *Proc. Inst. Civ. Eng. Eng. Sustain.* **2015**, *168*, 53–60. [CrossRef]
157. Karji, A.; Namian, M.; Tafazzoli, M. Identifying the Key Barriers to Promote Sustainable Construction in the United States: A Principal Component Analysis. *Sustainability* **2020**, *12*, 5088. [CrossRef]
158. Deng, H.; Tian, M.; Ou, Z.; Deng, Y. Obstacle-Aware Rescue Routing on Construction Site Based on BIM and Computer Vision. In Proceedings of the ICCREM 2021: Challenges of the Construction Industry under the Pandemic, Beijing, China, 16–17 October 2021; pp. 331–337.
159. Elmalı, Ö.; Bayram, S. Adoption of BIM Concept in the Turkish AEC Industry. *Iran J. Sci. Technol. Trans. Civ. Eng.* **2022**, *46*, 435–452. [CrossRef]
160. Fahad, M.; Bus, N. Geolocation in the Semantic BIM. In Proceedings of the 2019 IEEE International Conference on Engineering, Technology and Innovation (ICE/ITMC), Valbonne Sophia-Antipolis, France, 17–19 June 2019; IEEE: Piscataway, NJ, USA, 2019; pp. 1–7.
161. Frías, E.; Díaz-Vilariño, L.; Balado, J.; Lorenzo, H. From BIM to Scan Planning and Optimization for Construction Control. *Remote Sens.* **2019**, *11*, 1963. [CrossRef]
162. Pérez, C.T.; Costa, D.B. Increasing production efficiency through the reduction of transportation activities and time using 4D BIM simulations. *Eng. Constr. Arch. Manag.* **2021**, *28*, 2222–2247. [CrossRef]
163. Dixit, M.K.; Venkatraj, V.; Ostadalimakhmalbaf, M.; Pariafsai, F.; Lavy, S. Integration of Facility Management and Building Information Modeling (BIM): A Review of Key Issues and Challenges. *Facilities* **2019**, *37*, 455–483. [CrossRef]
164. Chen, S.-Y. A green building information modelling approach: Building energy performance analysis and design optimization. *MATEC Web Conf.* **2018**, *169*, 01004. [CrossRef]
165. Atazadeh, B.; Mirkalaei, L.H.; Olfat, H.; Rajabifard, A.; Shojaei, D. Integration of cadastral survey data into building information models. *Geo-Spat. Inf. Sci.* **2021**, *24*, 387–402. [CrossRef]
166. Alreshidi, E.; Mourshed, M.; Rezgui, Y. Requirements for cloud-based BIM governance solutions to facilitate team collaboration in construction projects. *Requir. Eng.* **2018**, *23*, 1–31. [CrossRef]
167. Ardani, J.A.; Utomo, C.; Rahmawati, Y. Model Ownership and Intellectual Property Rights for Collaborative Sustainability on Building Information Modeling. *Buildings* **2021**, *11*, 346. [CrossRef]
168. Baharom, M.H.; Abdullah Habib, S.N.H.; Ismail, S. Building Information Modelling (BIM): Contractual Issues of Intellectual Property Rights (IPR) in Construction Projects. *Int. J. Sustain. Constr. Eng. Technol.* **2021**, *12*, 170–178. [CrossRef]

169. Beach, T.; Petri, I.; Rezgui, Y.; Rana, O. Management of Collaborative BIM Data by Federating Distributed BIM Models. *J. Comput. Civ. Eng.* **2017**, *31*, 04017009. [CrossRef]
170. Hosseini, M.R.; Roelvink, R.; Papadonikolaki, E.; Edwards, D.J.; Pärn, E. Integrating BIM into Facility Management: Typology Matrix of Information Handover Requirements. *Int. J. Build. Pathol. Adapt.* **2018**, *36*, 2–14. [CrossRef]
171. Salem, D.; Bakr, A.; El Sayad, Z. Post-construction stages cost management: Sustainable design approach. *Alex. Eng. J.* **2018**, *57*, 3429–3435. [CrossRef]
172. Andriamamonjy, A.; Saelens, D.; Klein, R. An automated IFC-based workflow for building energy performance simulation with Modelica. *Autom. Constr.* **2018**, *91*, 166–181. [CrossRef]
173. Porsani, G.B.; Del Valle de Lersundi, K.; Gutiérrez, A.S.-O.; Bandera, C.F. Interoperability between Building Information Modelling (BIM) and Building Energy Model (BEM). *Appl. Sci.* **2021**, *11*, 2167. [CrossRef]
174. Mirahadi, F.; McCabe, B.; Shahi, A. IFC-centric performance-based evaluation of building evacuations using fire dynamics simulation and agent-based modeling. *Autom. Constr.* **2019**, *101*, 1–16. [CrossRef]
175. Ansah, M.K.; Chen, X.; Yang, H.; Lu, L.; Lam, P.T. A review and outlook for integrated BIM application in green building assessment. *Sustain. Cities Soc.* **2019**, *48*, 101576. [CrossRef]
176. Solla, M.; Elmesh, A.; Memon, Z.A.; Ismail, L.H.; Al Kazee, M.F.; Latif, Q.B.A.I.; Yusoff, N.I.; Alosta, M.; Milad, A. Analysis of BIM-Based Digitising of Green Building Index (GBI): Assessment Method. *Buildings* **2022**, *12*, 429. [CrossRef]
177. Atabay, S.; Gurgun, A.P.; Koc, K. Incorporating BIM and Green Building in Engineering Education: Assessment of a School Building for LEED Certification. *Pract. Period. Struct. Des. Constr.* **2020**, *25*, 04020040. [CrossRef]
178. Khoshdelnezamiha, G.; Liew, S.C.; Bong, V.N.S.; Ong, D.E.L. A BIM-Based Automated Assessment Tool for Green Building Index. *IOP Conf. Ser. Mater. Sci. Eng.* **2020**, *943*, 012059. [CrossRef]
179. Lim, Y.-W.; Seghier, T.E.; Ahmad, M.H.; Leng, P.C.; Yasir, A.M.; Rahman, N.A.; Chan, W.L.; Syed Mahdzar, S.S. Green Building Design and Assessment with Computational BIM: The Workflow and Case Study. In *Building Information Modelling (BIM) in Design, Construction and Operations IV*; WIT Press: Santiago de Compostela, Spain, 2021; pp. 3–13.
180. Olawumi, T.O.; Chan, D.W.M. Green-building information modelling (Green-BIM) assessment framework for evaluating sustainability performance of building projects: A case of Nigeria. *Arch. Eng. Des. Manag.* **2021**, *17*, 458–477. [CrossRef]
181. Seghier, T.E.; Khosakitchalert, C.; Lim, Y.-W. A BIM-Based Method to Automate Material and Resources Assessment for the Green Building Index (GBI) Criteria. In *Lecture Notes in Civil Engineering, Proceedings of 2021 4th International Conference on Civil Engineering and Architecture, Seoul, Korea, 10–12 July 2021*; Kang, T., Lee, Y., Eds.; Springer Nature: Singapore, 2022; Volume 201, pp. 527–536, ISBN 9789811669316.
182. Ilter, D.; Ergen, E. BIM for building refurbishment and maintenance: Current status and research directions. *Struct. Surv.* **2015**, *33*, 228–256. [CrossRef]
183. Pishdad-Bozorgi, P.; Gao, X.; Eastman, C.; Self, A.P. Planning and developing facility management-enabled building information model (FM-enabled BIM). *Autom. Constr.* **2018**, *87*, 22–38. [CrossRef]

buildings

MDPI

Article

What Drives the Intelligent Construction Development in China?

Xiaoli Yan [1,*], Yingxue Zhou [2], Tao Li [1,3] and Feifei Zhu [1]

1 School of Management Studies, Shanghai University of Engineering Science, Shanghai 201620, China
2 SIVA DETAO School of Design, Shanghai Institute of Visual Arts, Shanghai 201620, China
3 Fujian Post & Telecom Planning and Designing Institute Co., Ltd., Fuzhou 350001, China
* Correspondence: yanxiaoli821@sues.edu.cn or yanxiaoli821@163.com

Abstract: Intelligent construction (IC) integrates intelligent technologies with the construction industry to improve efficiency and sustainability. IC development involves many driving factors, but only the critical factors play essential roles. Thus, it is necessary to identify these key factors to understand and promote IC development thoroughly. Although there are many studies on IC-related technologies, a focus on identifying the driving factors of IC is lacking. We aimed to identify the key driving factors for IC development, analyze the relationship between the key factors and IC, and then produce general laws to guide IC by conducting an empirical study in China. We employed a five-stage research design and proposed the following general laws of how the key factors drive the development of IC: (1) initially, there exits the opportunity that drives companies to generate IC; (2) subsequently, the planning and pressure of a firm strategy, structure, and rivalry further drive companies to try to develop IC; (3) afterward, government policy vigorously promotes IC practices of the participating companies and accelerates the development of IC; and (4) finally, the market forces begin to play a leading role, and companies spontaneously carry out IC activities when the policy effect reaches a certain level. The findings indicate that policies to promote IC development should be consistent with its development stage, and the key driving factors of different stages should be paid attention to. Although the context of this study is China, the findings can provide references for IC's development globally.

Keywords: intelligent construction; driving factors; driving force theory; Porter Diamond Model; grey relation analysis

Citation: Yan, X.; Zhou, Y.; Li, T.; Zhu, F. What Drives the Intelligent Construction Development in China? *Buildings* **2022**, *12*, 1250. https://doi.org/10.3390/buildings12081250

Academic Editors: Hongling Guo, Jia-Rui Lin and Yantao Yu

Received: 27 June 2022
Accepted: 10 August 2022
Published: 15 August 2022

Publisher's Note: MDPI stays neutral with regard to jurisdictional claims in published maps and institutional affiliations.

Copyright: © 2022 by the authors. Licensee MDPI, Basel, Switzerland. This article is an open access article distributed under the terms and conditions of the Creative Commons Attribution (CC BY) license (https://creativecommons.org/licenses/by/4.0/).

1. Introduction

The global construction industry is booming, prompting both an increase in construction projects and a demand for intellectual development [1]. Luckily, the application of emerging technologies has responded to the intellectual development demand. These emerging technologies in the construction industry usually include the four main kinds of business digitalization; computer-integrated design; data acquisition, optimization, and predictive analytics; and robotics and automation [2]. With the increasing application of these emerging technologies to the construction industry, the intellectual development of this industry is inevitable [3,4].

At present, the discussion of "the process or product of the construction using emerging technologies" is mostly limited to "digital construction" [5] or "smart construction" [6] or "Construction 4.0" [7,8] in developed countries. IC is an innovative development model that combines emerging technologies with the construction industry under the background of the new technological revolution [9]. The words "digital construction", "smart construction", "Construction 4.0", and "intelligent construction" have similar connotations. That is, emerging intelligent technologies are used in the construction industry to improve quality, save costs, reduce pollution, and improve the efficiency of the desired processes, further promoting industrial upgrading [9]. IC is the key to transforming and upgrading the construction industry. In July 2020, the "Guiding Opinions on Promoting the Coordinated Development of Intelligent Construction and Building Industrialization" jointly issued by 13 Chinese government ministries and commissions proposed increasing the application of IC in all aspects of construction to form an IC industry (the guiding opinions on promoting the coordinated development of intelligent construction and building industrialization. http://www.gov.cn/zhengce/zhengceku/2020-07/28/content_5530762.htm, accessed on 8 August 2022).

In recent years, applying emerging technologies in the construction industry has significantly promoted IC development. Correspondingly, lots of research focuses on IC-related technologies' applications in construction, such as BIM [10–12], the Internet of Things [13,14], 3D scanning and printing [15,16], computer vision [17], and intelligent equipment [18–20]; and the improvements in safety, quality, scheduling, etc.

The above articles have studied the application of various emerging technologies in IC development from different perspectives. They have made significant contributions to promoting IC development, but it is not enough to only focus on IC technology to promote IC development. IC development is driven by many factors, and some critical factors play an essential role in its development, such as government policies [9] and a labor shortage [21]. These factors and their driving effects may also vary with IC development stages. However, few studies have systematically investigated the driving factors of IC development. Therefore, clarifying the critical driving factors at different stages and grasping the fundamental laws that promote IC development will help to explain IC thoroughly and provide a theoretical basis for guiding IC development in the future. For example, it could help the government and companies clarify the current focuses of IC development and provide a basis for the policy formation and practice acceleration. Due to the importance of research on driving factors of IC development and the lack of existing research, there is an urgent need to study the driving factors to improve the construction industry's performance. Thus, this paper aims to identify the driving factors for IC development, determine the key factors, analyze the relationship between them with IC, and finally, explore the general laws for driving IC development and provide recommendations to promote it.

1.1. Driving Force Theory (DFT)

The concept of "driving force" in physics is mainly used to describe the effect of the force generated in the driving process of a vehicle [22]. In business, the "driving force" refers to the force that makes a company move toward the target direction to a specific state under the internal and external driving environment to achieve a particular goal. Driving factors influence the driving process [23,24].

With the intersection of disciplines, the basic concept of driving force has also been extended to the fields of management science [25,26], sociology [27], economics [28], and environmental science [29]. For example, Chen et al. [25] examined the driving force of co-evolutionary dynamics between multistage overseas merger and acquisition (M&A) integration and knowledge network reconfiguration. Pichlak M. [26] contributed to showing the driving force for technological eco-innovation development. Qin Z. et al. [29] investigated the driving forces of agricultural intensification.

In addition, some scholars have researched explicitly from the perspective of driving factors, such as the driving factors of innovation systems [30,31], the driving factors of land use in development zones [32], and the driving factors of the industrialization of new buildings [33]. In summary, many scholars have researched the driving issues in various disciplines from the aspects of the driving force, driving environment, and driving factors, which provide good references for this paper.

Driving force theory provides a good reference and inspiration for this paper. Industrial philosophy theory suggests that any industry's generation, growth, and maturity are driven by many factors [34]. IC is no exception. Many factors continue to interact in IC development, which restructure IC's industrial, supply, value, and innovation chains [9]. For IC development, the driving force is the power or impact exerted or imposed on it by the driving factors such as policy, market, competition among enterprises, and opportunities. The power and impact may have different magnitudes, called driving strengths. Therefore, identifying these driving factors and their relationships with IC based on the driving force theory and making targeted improvements are significant to better promoting IC development.

1.2. Driving Factors of IC Development

The literature review showed very few studies on IC driving factors. However, some research indirectly reflects the driving forces and factors of IC. For example, Ding [9] proposed that IC's application could improve resource utilization efficiency, responded to customer needs, and met the requirements of sustainable development. In addition, some research performed some analyses related to IC development, including government policies [35–37], technology [38–40], labor and professionals [21,41], etc. In this article, we identify these as the driving factors of IC. Therefore, the initial list of IC driving factors taken from the literature is shown in Table 1.

Table 1. Initial list of driving factors for the development of IC from previous studies.

Range of Driving Factors	Literature Reference
Perfection and matching of laws, regulations, and standard systems	Liu et al. [35], Mao & Zhang [36], Okpala et al. [37]
Government support and incentive policies	Liu et al. [35], Ding [9], Yue & Li [42], Zhou & Wu [43]
Demonstration projects	Zhang et al. [44], Lin et al. [45], Yang et al. [46], Fan Q.et al. [47]
The dilemma of traditional construction methods	Ding [9], Memari et al. [48], Zhou et al. [49];
Market and consumer demand	Ding [9]
Corporate Strategy	Ding [9], Mao & Zhang [36].
The intelligent technology application system	Okpala et al. [38], Ogunrinde et al. [39], Shi et al. [40]
Support for the industrial system	Mao & Zhang [36]
Talent training system	Kim et al. [21], Heravi & Eslamdoost [41], Liu et al. [50]

The studies above mentioned some factors that drive IC development from different angles, but they were not systematic or complete. Based on driving force theory, this article innovatively and systematically studies the driving factors that promote IC development and provides suggestions.

We arranged this paper as follows: the second section introduces the research method, the third section presents the research results, the fourth section provides the discussion and suggestions, and finally, the conclusions are reached.

2. Research Methodology

Many factors drive IC development, but only a few driving factors play a critical role. In addition, IC development is a gradual process, and its degree of development is related to the driving strength. Generally, when the driving strength is low, the company will be unwilling to carry out activities related to IC, and it is difficult for IC to develop. On the other hand, as the driving strength gradually increases, companies become more active, and IC gradually grows and matures. Moreover, the driving factors that play crucial roles are not the same with different driving strength levels. Therefore, it is imperative to clarify the critical driving factors at various levels, grasp the focuses of IC development in different stages, and drive its development.

Based on the research purpose and the above analysis, we designed the five-stage, comprehensive approach shown in Figure 1. First, we identified and determined the driving factors through a literature review and an expert symposium in stages one and two; secondly, we introduced Porter Diamond Model (PDM) to classify the driving factors based on its advantages in stage three; thirdly, we carried out the investigation in stage four; and finally, we determined the relationships between the factors and IC development based on grey relation analysis (GRA) according to their use in solving problems in stage five.

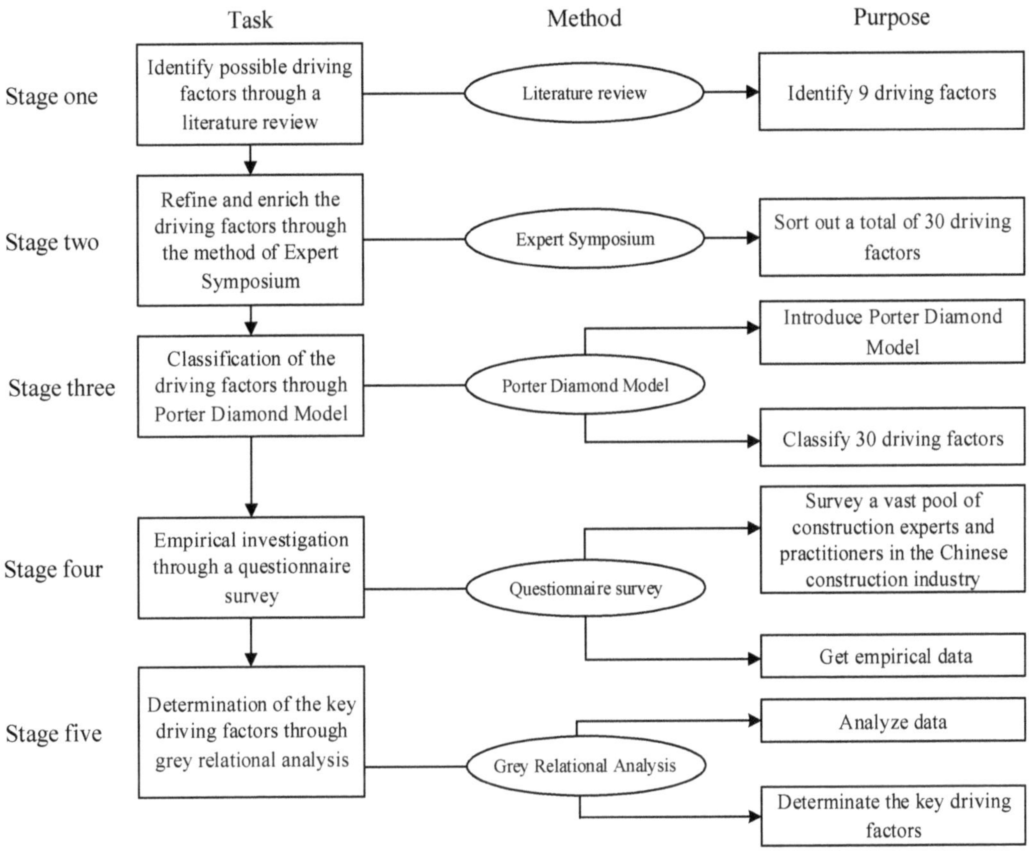

Figure 1. Five-stage research protocol.

2.1. *Stage One—Identifying Possible Driving Factors through a Literature Review*

This stage involved a literature review for identifying driving factors affecting IC development. First, we searched for the relevant literature published between 2010 and 2021 across various databases with the keywords "intelligent construction", "smart construction", "digital construction", "construction industry development", "emerging technologies in construction", etc., and we roughly searched for literature related to driving factors. Eighteen papers were selected. Next, we closely examined the literature to identify the possible factors. A list of nine driving factors for IC development drawn from the literature review is presented in Table 1. These nine factors formed an initial list of factors.

2.2. *Stage Two—Refining and Enriching the Driving Factors through an Expert Symposium*

The objective of this stage was to identify the driving factors that specifically facilitated IC implementation in China. After the preliminary identification, we held an expert symposium to refine and optimize the driving factors identified. First, we set the criteria for selecting experts to ensure the quality of the symposium. These experts were to be representative and authoritative. They worked in all aspects of construction activities along the construction lifecycle, and had rich knowledge and practical experience in the construction industry. In addition, they were to have a certain influence in the industry and have professional insights into Chinese IC development status. Secondly, we invited experts under the set standards through various channels, explained the purpose and significance of this research to them in advance, and won their support. Among the invited experts, six were from design, construction, supervision, and maintenance companies; four were government officials; and five were scholars involved in IC. All invited experts had more than ten years of experience. Then, we held a symposium to discuss IC driving factors in China.

The symposium adopted a hybrid method and was divided into two stages. The first stage used the brainstorming method. The reason for choosing to brainstorm is that these experts were able to discuss in an unrestricted atmosphere, think positively, inspire each other, brainstorm ideas, and fully express their opinions [51]. With the brainstorming method, we can produce high-quality, creative results. In the second stage, the content of the first stage was fully discussed. Finally, a consensus was reached, and a more consistent conclusion was obtained.

The main task of the symposium included disassembling, enriching, and supplementing the driving factors. Some factors were subdivided and disassembled. For example, the role of the government was subdivided into "financial subsidies and tax incentives for IC technology"; "rewards for IC projects"; "mandatory standards in the approval of planning and design schemes" to align with China's situation. Some factors were classified. For example, "government support and incentive policies" and "demonstration projects" were both government-led policy incentives in China and were classified into one category to avoid omission during the optimization. In addition, some factors were enriched and supplemented. For example, some other factors not covered by the nine initial factors were added, such as the reduction of pollution, green development, and less labor that China has advocated for in recent years. Finally, we identified 30 driving factors through the expert symposium, as shown in Table 2.

Table 2. Driving factors of IC development.

Code	Driving Factors
1	The necessity for gradual improvement of the construction industry's industrial structure
2	Innovation and reform of upgrading of the construction industry
3	Severe pollution from construction solid waste
4	Enormous noise pollution during construction
5	Severe air pollution during construction
6	The necessity to improve the existing construction technology
7	Possible economic benefits of IC
8	Willingness to transform the construction process management
9	Decision-makers' expectations for IC benefits
10	Decision-makers' requirements for the construction period and quality
11	Level of decision-makers' awareness of sustainable development
12	Long-term strategic goals for decision-makers
13	Company's resource investment in IC
14	Anticipation of potential market opportunities in the future
15	Competitiveness of related companies
16	Promotion and application of IC technology
17	Financial guarantee for the research and development of IC technology
18	IC technical staffing
19	Building materials and energy consumption
20	Lack of labor force
21	Consumers' needs and preferences for IC
22	Consumers' awareness and understanding of IC
23	Market access system for IC
24	Low labor productivity in the construction industry
25	Continuous increase in labor costs
26	Increase in the number of prefabricated component factories

Table 2. *Cont.*

Code	Driving Factors
27	Information technology serves industry development
28	Financial subsidies and tax incentives for IC technology
29	Rewards for IC projects
30	Mandatory standards in the approval of planning and design schemes

2.3. Stage Three—Classifying Driving Factors through PDM

The objective of this stage was to categorize 30 factors based on the perspective of industry development, which was conducive to a structured questionnaire survey and the analysis below.

PDM, as a theoretical model, explains why the industry is competitive and is widely used in research on competitive advantages in industry [52–55]. It states that four elements determine the development of a specific industry: factor conditions; demand conditions; related and supporting industries; and firm strategy, structure, and rivalry. In addition, there are two variables: opportunity and government, both of which are closely related to the development of the entire industry and affect four significant factors (see Figure 2). Therefore, PDM is also used for the analysis of the characteristics of factors [27,56,57].

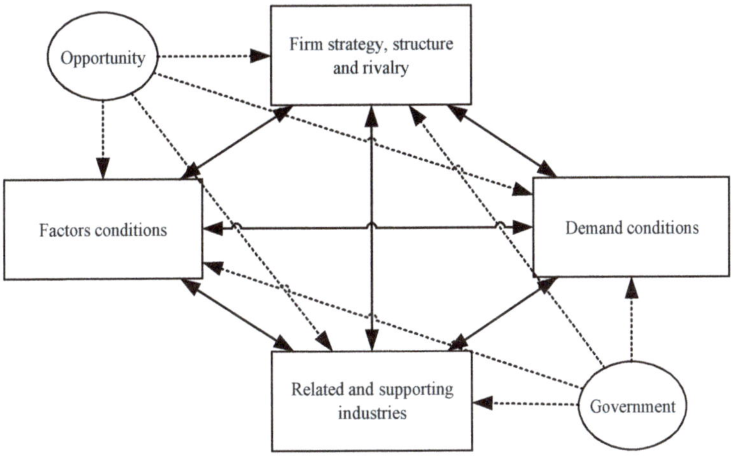

Figure 2. Porter Diamond Model (source: Porter, M. E. [52]).

Based on PDM, the four elements and two variables in IC development were identified as follows: (1) Opportunity refers to the construction industry urgently needing to transform and upgrade the status quo, and IC has the advantages of high efficiency and sustainability; (2) the firm strategy, structure, and rivalry refer to the corporate strategy and organizational structure of the company formulated for the development of IC. In addition, competition in the same industry affects some factors, such as the decision-making and market share of the company; (3) the factor conditions mainly refer to the primary conditions required for IC, including natural resources, infrastructure, human resources, etc.; (4) the demand conditions refer to the market's demand for IC; (5) the related and supporting industries refer to the synergy between upstream and downstream companies of IC; and (6) government actions and government policies form the governmental roles: participating in IC projects, leading the development of IC projects, promoting market demand, promoting industrial development, etc.

Based on the definitions of the four elements and two variables in PDM, the 30 driving factors were classified in Table 3.

Table 3. Classification of driving factors for the development of IC.

Category	Driving Factors
Opportunity	The necessity for gradual improvement of the construction industry's industrial structure of (OT1) Upgrading innovation and reform of the construction industry (OT2) Severe pollution from construction solid waste (OT3) Enormous noise pollution during construction (OT4) Severe air pollution during construction (OT5) The necessity to improve the existing construction technology (OT6) Possible economic benefits of IC mode (OT7)
Firm strategy, structure, and rivalry	Willingness to transform the construction process management (FS1) Decision-makers' expectations for the benefits of IC (FS2) Decision-makers' requirements for the construction period and quality (FS3) Level of decision-makers awareness of sustainable development (FS4) Long-term strategic goals for decision-makers (FS5) Company's resource investment in IC (FS6) Anticipation of potential market opportunities in the future (FS7) Competitiveness of related companies (FS8)
Factors conditions	Promotion and application of IC technology (FC1)Financial guarantee for the research and development of IC technology (FC2)IC technical staffing (FC3) Building materials and energy consumption (FC4) Lack of labor force (FC5)
Demand conditions	Consumers' needs and preferences for IC (DC1) Consumers' awareness and understanding of IC (DC2) Market access system for IC (DC3)
Related and supporting industries	Low labor productivity in the construction industry (RS1) Continuous increase in labor costs (RS2) Increase in the number of prefabricated component factories (RS3) Information technology serves industry development (RS4)
Government	Financial subsidies and tax incentives for IC technology (GM1) Rewards for IC projects (GM2) Mandatory standards in the approval of planning and design schemes (GM3)

2.4. Stage Four—Empirical Investigation through the Questionnaire

We developed a questionnaire in this stage to obtain empirical data to determine the critical factors by surveying a vast pool of construction experts and practitioners. Excluding the initial respondent information, the questionnaire was divided into six parts according to the driving factors, and the 30 questions corresponded to the 30 driving factors. These respondents were required to rate the strengths of the identified IC driving factors using a five-point Likert scale with options ranging from "1" to "5", with "1" being the weakest rating and "5" being the highest rating. At the beginning of the questionnaire, we explained the definition and scope of IC to ensure understanding by the respondents.

This research adopted a simple random sampling method to conduct surveys using relevant personnel in high, medium, and low positions from organizations related to IC development, such as construction companies, research institutions, government departments, and IC technology developers, scattered in different cities in China. We distributed the questionnaires online via a Web-based platform. At the end of the survey, 150 questionnaires were returned, and 132 had valid data.

As is shown in Table 4, 100% of the respondents were Chinese. The findings also show that 10.61% of respondents were senior management staff, 45.45% were middle-level management staff, and 43.94% were low-level management staff in their organizations. Further analysis of their work experience revealed that 78.03% of the respondents possessed a minimum of 5 years of experience in the construction industry, and 16.67% had more than twenty years of experience. In addition, 92.42% of the respondents acknowledged they understood IC above a moderate level. We designed this question to obtain the respondents' subjective personal views on IC to judge the respondents' background. Besides the respondents' information, we got empirical data to analyze in stage five.

Table 4. Respondent's characteristics.

Category	Characteristic	Frequency	Percentage (%)
Nationality	China	132	100
Organization	Construction companies	44	33.33
	Research institutions	67	50.76
	IC technology development company	3	2.27
	Construction Industry Association	0	0
	Government departments	5	3.79
	Other	13	9.85
Position level	Senior management	14	10.61
	Middle-level management	60	45.45
	Low-level physical operators	58	43.94
Working years	Less than five years	29	21.97
	5–10 years	31	23.48
	10–20 years	50	37.88
	More than 20 years	22	16.67
Level of understanding of IC	Thoroughly understand	8	6.06
	Understand	56	42.42
	Moderately understand	58	43.94
	Slightly understand	9	6.82
	Does not understand	1	0.76

2.5. Stage Five—Determine the Critical Driving Factors through Grey Relation Analysis (GRA)

Not all 30 driving factors described above play a key role in IC development. Identifying the key driving factors in various stages of IC development is beneficial for policy formulation and corporate strategic planning in IC development.

GRA is specially applied in fuzzy problems with uncertain relations [58–60]. It is an analysis method that measures the importance of factors by using the order of the relevance degree influenced by other factors. For example, it is used in measuring sensitive factors for landslides and concrete structures' durability [61,62], travel modes and their influence factors [63], and sources of risk for abnormal driving on expressways in a port city [64]. This research was carried out the survey data by grey relation analysis to measure the driving factors and the relationships between the critical driving factors and IC development.

First, we accumulated the evaluation results under different factors' driving strengths. Therefore, we obtained five sets of cumulative data from the lowest to the highest driving strength. We performed the grey correlation analysis as follows:

The reference sequence X_0 or the development status of IC, also known as the parent sequence, was established.

The comparison series $x_i(k)$ using the 30 driving factor indicators, in which $i = 1, 2, \ldots, 30$, $k = 1, 2, \ldots, 5$, was set.

Data collection was initialized and calculated, as shown in Formula (1):

$$X_i(k) = \frac{x_i(k)}{x_i(1)} \quad (1)$$

The correlation coefficient between the reference sequence and the comparison series was calculated—that is, the driving effect of each driving factor on the development of

IC under different driving strengths. The correlation coefficient $\xi_{i(k)}$ was calculated with Formula (2), where the resolution factor is generally taken as $\rho = 0.5$ [3,65]:

$$\xi_i(k) = \frac{\min_i \min_k |X_0(k) - X_i(k)| + \rho \max_i \max_k |X_0(k) - X_i(k)|}{|X_0(k) - X_i(k)| + \rho \max_i \max_k |X_0(k) - X_i(k)|} \quad (2)$$

The grey relationship degrees were arranged from small to large. The calculated grey relationship degree is a relative weighted value. When the value is large, it indicates that the factor is essential, and the designer should focus on it. In contrast, the smaller the value is, the less critical the factor, which can be temporarily considered an unimportant reference under cost.

3. Results

The comparison of driving factors considering five levels of driving strength was calculated through GRA, and the results are shown in Table 5.

Table 5. Driving effects of the driving factors for IC development.

Code	Category	Driving Factors	Driving Strength				
			5 (Highest)	4	3	2	1 (Lowest)
1	Opportunity	OT1	0.183	0.496	0.784	0.958	1
2		OT2	0.173	0.593	0.791	0.972	1
3		OT3	0.203	0.426	0.701	0.885	1
4		OT4	0.208	0.412	0.654	0.891	1
5		OT5	0.201	0.412	0.686	0.899	1
6		OT6	0.191	0.478	0.782	0.961	1
7		OT7	0.251	0.543	0.837	0.961	1
8	Firm strategy, structure, and rivalry	FS1	0.156	0.537	0.832	0.973	1
9		FS2	0.199	0.600	0.849	0.921	1
10		FS3	0.248	0.539	0.877	0.948	1
11		FS4	0.236	0.512	0.791	0.953	1
12		FS5	0.188	0.527	0.852	0.929	1
13		FS6	0.232	0.448	0.671	0.889	1
14		FS7	0.244	0.590	0.866	0.946	1
15		FS8	0.226	0.501	0.793	0.855	1
16	Factors conditions	FC1	0.242	0.506	0.847	0.949	1
17		FC2	0.219	0.489	0.721	0.952	1
18		FC3	0.169	0.489	0.729	0.939	1
19		FC4	0.144	0.452	0.783	0.931	1
20		FC5	0.319	0.661	0.831	0.946	1
21	Demand conditions	DC1	0.240	0.483	0.700	0.90	1
22		DC2	0.174	0.456	0.700	0.85	1
23		DC3	0.221	0.496	0.781	0.877	1
24	Government	GM1	0.248	0.631	0.815	0.923	1
25		GM2	0.218	0.658	0.796	0.908	1
26		GM3	0.246	0.614	0.773	0.901	1
27	Related and supporting industries	RS1	0.342	0.531	0.831	0.959	1
28		RS2	0.371	0.672	0.860	0.963	1
29		RS3	0.212	0.532	0.805	0.920	1
30		RS4	0.241	0.530	0.819	0.936	1

According to the GRA method, the index value in each column in Table 5 represents the closeness of the correlation between the factor and the driving strength, which is a relative value and reflects the degree of the driving effect. As at level 1, each factor is not closely related to the driving degree, this column was set as the reference sequence, and the value was set to one during the calculation process. From level 2 to level 5, the close relationship changes. Therefore, we can compare the values of each column vertically.

Table 5 shows that the different driving factors have different driving effects under various driving strength levels. For example, at level 2 of driving strength, the 30 driving factors have different driving effects. Among them, FS1 ranks number 1 with the value of 0.973, and OT2 ranks number 2 with the value of 0.972. Therefore, we can obtain the critical driving factors under different driving strengths. To achieve this aim, we sorted and analyzed the top ten driving factors with the most apparent driving effects under different driving strength levels, as shown in Figure 3.

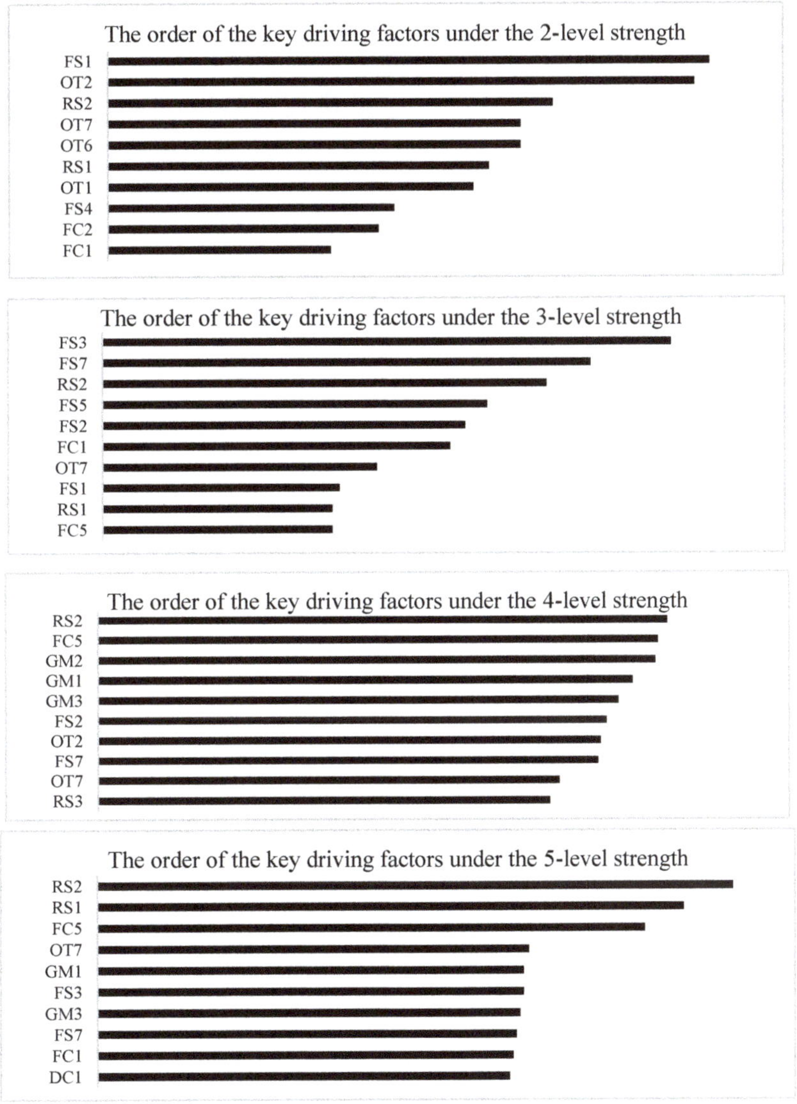

Figure 3. Ranking of critical driving factors under different drive strengths.

We draw the following conclusions about the relationships between the critical driving factors and IC development based on Figure 3:

(1) Under level 2 driving strength, four key driving factors belong to the category of opportunity; two key driving factors belong to the category of firm strategy, structure, and rivalry; two key driving factors belong to the category of factor conditions; and two key driving factors belong to the category of related and supporting industries. Among them, "the willingness to transform the construction process management" and "upgrading innovation and reform of construction industry" have the most significant impacts.

(2) Under level 3 driving strength, five key driving factors belong to the category of firm strategy, structure, and rivalry; two key driving factors belong to the category of factor conditions; two key driving factors belong to the category of related and supporting industries; and one key driving factor belongs to the category of opportunity. Among them, decision-makers' requirements on the construction period and quality have the greatest impact.

(3) Under level 4 driving strength, three key driving factors belong to the category of government; two key driving factors belong to the category of firm strategy, structure, and rivalry; two key driving factors belong to the category of opportunity; two key driving factors belong to the category of related and supporting industries; and one key driving factor belongs to the category of factor conditions. Among them, the continuous increase in labor costs, the lack of a labor force, and the rewards for IC projects have the most significant impacts.

(4) Under level 5 driving strength, the numbers of key driving factors under the element categories are balanced. The continuous increase in labor costs has the most significant impact and has opened a gap for other key driving factors.

4. Discussion

4.1. The General Laws for Driving IC Development

We found a certain regularity for various driving factors driving IC development. That is, the stage of IC development impacts driving strength levels, and the critical driving factors that play leading roles change with the stage, as shown in Figure 4.

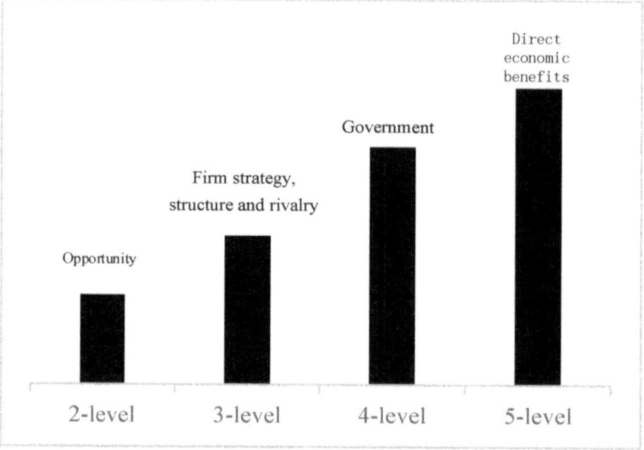

Figure 4. Key factors that play leading roles at different driving strengths.

(1) To drive IC development, the driving factors of opportunity are essential to achieve driving strength of level 2. Level 2 is a low driving strength level which can only trigger the initial motivation for the company to engage in IC. As an advanced construction mode, IC can solve the development difficulties of the traditional construction mode. Therefore, IC presents a significant opportunity for companies in the construction industry and is involved in the industry's development trends.

(2) When the company has the initial motivation, the factors of firm strategy, structure, and rivalry are critical to reaching level 3 driving strength. Upon reaching level 3, the company will transform the motivation for IC into practice. The company needs to recognize and respond to industry trends and transform its motivations into corporate practices. This is mainly affected by the firm's strategy, structure, and rivalries, which include factors such as the company's expectations for future potential market opportunities, the expectations of decision-makers for IC benefits, the willingness to transform the construction process, the long-term strategy for company development, and cooperation and competition with peer companies.

(3) When the company attempts to implement IC practices further, the factors of government are essential for achieving level 4 driving strength. Driven at this level, the company is genuinely implementing IC practices in a wait-and-see state of practice. The government plays a key role. The survey data and the IC status quo in China show that most IC practices depend on government policies. On the one hand, government policies force all companies to develop IC. On the other hand, they provide preferential subsidies and policy support to encourage companies to develop IC. The government's policies offer guarantees for companies with specific motives.

(4) When a company develops IC on a large scale, the market will play a huge role. The pursuit of interests by the company drives IC development. The direct economic factors driving the company to conduct IC activities spontaneously are essential indicators of achieving level 5 drive strength. For example, it is difficult to achieve the quality, cost, and construction period goals of construction products with traditional construction methods, resulting in a decline in corporate profits, reduced competitiveness, and even corporate decline. Therefore, companies will spontaneously practice IC based on its advantages for improving efficiency and quality, resource-friendliness, and sustainable development under the pressure of pursuing profits.

IC development under the influence of driving factors is a gradual and regular process. First, there must be an opportunity to drive the company's willingness to conduct IC. Second, the corporate strategy, structural adjustment, and competition in the industry drive companies to try more to develop IC. Furthermore, the support of the government vigorously promotes IC practices. When the government guides this promotion to a certain level, the market begins to play a leading role, so companies spontaneously carry out IC activities because economic interests drive them. This analysis is in line with the general state of China's IC development.

4.2. Suggestions for IC Development

At present, opportunities already exist for IC development in Chinese companies. On the one hand, most companies are aware of the development prospects of IC. A few of them have raised IC to the level of corporate strategy. On the other hand, although labor costs and shortages have become irreversible trends, the current impact on profits is still within the company's bearable range. However, Chinese construction companies have always been confined to traditional production methods and lack industrialization and digital development.

IC implementation requires substantial input and investment, but whether IC can generate more profits for companies remains to be further investigated. Therefore, some companies still doubt IC and are in a wait-and-see state. Under the above environmental characteristics, government policies are currently the most critical driving factor. Therefore, stimulating companies' vitality and promoting them to transform their wishes into IC practices will be the most direct and effective way to drive China's IC development.

There are significant differences in various regions of China regarding IC development. Due to diverse and complex regional development conditions, construction companies have apparent differences in development planning, innovation capabilities, willingness, and strength to develop IC. As the Chinese government has issued corresponding guidelines, some regional governments have actively formulated implementation opinions to respond.

However, some regions are not sensitive to these policies and trends, which leads to uneven regional development. Local governments and companies should adjust local policies, perceive the regional IC development status, seize opportunities, and clarify the current key driving factors. The government must play important guiding and supporting roles, and companies should formulate strategic plans.

Based on the above analysis, this paper proposes the following policy strategies:

(1) Understand the general laws driving IC development. It is necessary to understand the key driving factors at different stages of IC development, employ the driving role for its maximum utility, and vigorously promote the IC development process.

(2) Emphasize the role of government policies. The government must ensure and encourage relevant entities to practice IC, develop and improve the market, establish a long-term force to promote IC development, and make corresponding policy adjustments according to IC development status to form a virtuous environment.

(3) Implement various IC development strategies. Different regions must formulate suitable IC development plans and policy choices according to their characteristics and foundations. That is, they should take different implementation priorities according to their conditions.

5. Conclusions

IC is the key to adapting to the intellectual development trends of the global construction industry. The advancement of IC involves multiple factors and is inseparable from the driving forces of various driving factors. However, a focus on identifying the driving factors and the relationships between these factors and IC development was lacking. In this manuscript, we designed a five-stage method to compensate for this research gap. We obtained the key driving factors and arranged them, and we summarized and compared the key factors of levels 2–5. On this basis, the general laws that drive IC development were outlined and confirmed in combination with the current situation in China. Finally, we put forward relevant policy recommendations.

This paper contributes to identifying and determining the critical driving factors for IC development and clarifying the general laws of the relationship among them. It will help the government and companies understand the current focus of IC development and provide a basis for the policy formation and practice acceleration. This paper achieves the following:

(1) Identification of driving factors from the literature, which were refined and enriched through an expert symposium, and categorized based on PDM;

(2) Analysis of the key driving factors of IC development in China through the GRA method and clarification of the relationships between the key driving factors and the stages of IC development under different driving strength levels;

(3) Summarization of the general laws of the driving effect on IC development;

(4) Recommendations for IC development based on the general law, combined with the analysis of the current overall development status and regional development status of IC in China.

Even though the analysis of driving factors was performed from the Chinese perspective, some basic principles apply to many other countries, especially some developing countries, as the external environment of IC development is similar. Moreover, IC development all of over the world also follows some similar general rules. Therefore, the global construction industry can leverage the outcomes of this study.

At the same time, this manuscript did limit its range of applicability with its focus. Differences exist in IC development in the various countries of the world, and even different regions in the same country may have significant differences. Therefore, countries and regions should adapt to local conditions, understand their IC development status, identify the current critical driving factors, and take different measures to deal with them. Future research can investigate the driving factors of specific countries or regions of the world and conduct comparative studies.

Author Contributions: Conceptualization, X.Y.; writing—original draft preparation, X.Y., Y.Z. and T.L.; writing—review and editing, X.Y., Y.Z. and T.L.; investigation, T.L. and F.Z.; Data collection, T.L. and F.Z.; Data analysis, T.L. and F.Z.; project administration, X.Y.; funding acquisition, X.Y. All authors have read and agreed to the published version of the manuscript.

Funding: This research and The APC was funded by the Key Project of 2020 Scientific and Innovative Action Plan of Shanghai Science and Technology Commission: "Research on the Key Issues and Countermeasures of the Transformation of Traditional Industries Driven by Digital Technology-The Origin, Architecture and Realization of the Intelligent Construction Mode of the Construction Industry", grant number 20692101300.

Data Availability Statement: Some or all data, models, or code generated or used during the study are available from the corresponding author by request.

Acknowledgments: The authors are also very thankful for the anonymous referees and editors whose suggestions and comments helped to improve the manuscript's quality.

Conflicts of Interest: The authors declare that they have no conflict of interest.

References

1. Guo, P.; Tian, W.; Li, H.; Zhang, G.; Li, J. Global characteristics and trends of research on construction dust: Based on bibliometric and visualized analysis. *Environ. Sci. Pollut. Res.* **2020**, *27*, 37773–37789. [CrossRef] [PubMed]
2. Qi, B.; Razkenari, M.; Costin, A.; Kibert, C.; Fu, M. A systematic review of emerging technologies in industrialized construction. *J. Build. Eng.* **2021**, *39*, 102265. [CrossRef]
3. Chen, C.; Chuang, M. Integrating the Kano model into a robust design approach to enhance customer satisfaction with product design. *Int. J. Prod. Econ.* **2008**, *114*, 667–681. [CrossRef]
4. Giel, B.K.; Issa, R.R.A. Return on investment analysis of using building information modeling in construction. *J. Comput. Civ. Eng.* **2013**, *27*, 511–521. [CrossRef]
5. Arunothayan, A.R.; Nematollahi, B.; Ranade, R.; Bong, S.H.; Sanjayan, J. Development of 3D-printable ultra-high performance fiber-reinforced concrete for digital construction. *Constr. Build. Mater.* **2020**, *257*, 119546. [CrossRef]
6. Štefanič, M.; Stankovski, V. A review of technologies and applications for smart construction. *Proc. Inst. Civ. Eng.-Civ. Eng.* **2018**, *172*, 83–87. [CrossRef]
7. Perrier, N.; Bled, A.; Bourgault, M.; Cousin, N.; Danjou, C.; Pellerin, R.; Roland, T. Construction 4.0: A survey of research trends. *J. Inf. Technol. Constr.* **2020**, *25*, 416–437. [CrossRef]
8. Sawhney, A.; Riley, M.; Irizarry, J.; Pérez, C.T. A proposed framework for Construction 4.0 based on a review of Literature. In Proceedings of the 56th Annual Associated Schools of Construction (ASC) International Conference_ASC 2020, Liverpool, UK, 14–18 April 2020.
9. Ding, L.Y. *Digital Construction Introduction*; China Building Industry Press: Beijing, China, 2019.
10. Zaid, N.-U.B.; Hamzah, N.; Khoiry, M.A. Review building information modelling for infrastructure: Benefits for constructor. *J. Comput. Theor. Nanosci.* **2020**, *17*, 620–628. [CrossRef]
11. Jia, H.; Dong, S.; Fu, S. Application of BIM technology in intelligent construction and installation of prefabricated buildings. *Constr. Technol.* **2018**, *23*, 40–43.
12. Pezeshki, Z.; Ivari, S.A.S. Applications of BIM: A brief review and future outline. *Arch. Comput. Methods Eng. State Art Rev.* **2018**, *25*, 273–312. [CrossRef]
13. Zhong, R.Y.; Peng, Y.; Xue, F.; Fang, J.; Zou, W.; Luo, H.; Ng, S.T.; Lu, W.; Shen, G.Q.P.; Huang, G.Q. Prefabricated construction enabled by the Internet-of-Things. *Autom. Constr.* **2017**, *76*, 59–70. [CrossRef]
14. Lu, W.; Huang, G.Q.; Li, H. Scenarios for applying RFID technology in construction project management. *Autom. Constr.* **2011**, *20*, 101–106. [CrossRef]
15. Besklubova, S.; Skibniewski, M.J.; Zhang, X. Factors affecting 3D printing technology adaptation in construction. *J. Constr. Eng. Manag.* **2021**, *147*, 04021026. [CrossRef]
16. Wu, S.L.; Deng, H.L.; Chen, K.J.; Zhu, M.Y.; Huang, D.H.; Fu, S.Y. Visual monitoring technology of the tunnel 3D laser scanning and engineering applications. *Adv. Mater. Res.* **2013**, *779*, 463–468. [CrossRef]
17. Seo, J.; Han, S.; Lee, S.; Kim, H. Computer vision techniques for construction safety and health monitoring. *Adv. Eng. Inform.* **2015**, *29*, 239–251. [CrossRef]
18. Kim, S.; Peavy, M.; Huang, P.-C.; Kim, K. Development of BIM-integrated construction robot task planning and simulation system. *Autom. Constr.* **2021**, *127*, 103720. [CrossRef]
19. Pan, M.; Pan, W. Determinants of Adoption of Robotics in Precast Concrete Production for Buildings. *J. Manag. Eng.* **2019**, *35*, 05019007. [CrossRef]
20. Zhou, Z.; Irizarry, J.; Lu, Y. A Multidimensional Framework for Unmanned Aerial System Applications in Construction Project Management. *J. Manag. Eng.* **2018**, *34*, 04018004. [CrossRef]

21. Kim, S.; Chang, S.; Castro-Lacouture, D. Dynamic modeling for analyzing impacts of skilled labor shortage on construction project management. *J. Manag. Eng.* **2020**, *36*, 04019035. [CrossRef]
22. Rajamani, R. *Vehicle Dynamics and Control*; Springer Science: Berlin/Heidelberg, Germany, 2006.
23. Ghauri, P.; Wang, F. *Key Factors in the International Market Driving Process: The Role of Internal Competencies and External Network Actors*; The Annual Conference of the Consortium of International Marketing Research: Essex, UK, 2008. [CrossRef]
24. Li, W.; Zhong, X.; Wu, X.; Ye, N. Grey Correlation analysis on the economic factors of regional logistics in Guangdong province. *E3S Web Conf.* **2021**, *275*, 01060. [CrossRef]
25. Chen, F.; Zhu, J.; Wang, W. Driving force of industrial technology innovation: Coevolution of multistage overseas M&A integration and knowledge network reconfiguration. *J. Bus. Ind. Mark.* **2021**, *36*, 1344–1357. [CrossRef]
26. Pichlak, M. The Drivers of Technological Eco-Innovation—Dynamic Capabilities and Leadership. *Sustainability* **2021**, *13*, 5354. [CrossRef]
27. Li, X.; Huang, S.; Chen, Q. Analyzing the driving and dragging force in China's inter-provincial migration flows. *Int. J. Mod. Phys. C* **2019**, *30*, 1940015. [CrossRef]
28. Li, X.; Li, G. Research on the Driving Force of the Regional Economy to the Development of Ocean Port Shipping Based on Multiple Regression Analysis. *J. Coast. Res.* **2020**, *111*, 168–171. [CrossRef]
29. Qin, Z.; Storozum, M.; Liu, H.; Zhang, X.; Kidder, T.R. Investigating environmental changes as the driving force of agricultural intensification in the lower reaches of the Yellow River: A case study at the Sanyangzhuang site. *Quat. Int.* **2019**, *521*, 25–34. [CrossRef]
30. Brancati, E.; Brancati, R.; Guarascio, D.; Zanfei, A. Innovation drivers of external competitiveness in the great recession. *Small Bus. Econ.* **2021**, *58*, 1497–1516. [CrossRef]
31. Pascual-Fernández, P.; Santos-Vijande, M.L.; López-Sánchez, J.; Molina, A. Key drivers of innovation capability in hotels: Implications on performance. *Int. J. Hosp. Manag.* **2021**, *94*, 102825. [CrossRef]
32. Cui, J.; Kong, X.; Chen, J.; Sun, J.; Zhu, Y. Spatially explicit evaluation and driving factor identification of land use conflict in Yangtze River Economic Belt. *Land* **2021**, *10*, 43. [CrossRef]
33. Xiahou, X.; Yuan, J.; Liu, Y.; Tang, Y.; Li, Q. Exploring the Driving Factors of Construction Industrialization Development in China. *Int. J. Environ. Res. Public Health* **2018**, *15*, 442. [CrossRef]
34. Ford, H. My Philosophy of Industry. George G. Harrap & Co. Ltd.: London, UK, 1929.
35. Liu, Z.; Liu, S.; Zhao, Y.; Du, X. Development status and future trends of intelligent construction technology. *Archit. Technol.* **2019**, *7*, 772–779.
36. Mao, C.; Zhang, L. Core industry selection for the smart construction industry Chain. *J. Eng. Manag.* **2021**, *1*, 19–24.
37. Okpala, I.; Nnaji, C.; Awolusi, I. Emerging construction technologies: State of standard and regulation implementation. In Proceedings of the ASCE International Conference on Computing in Civil Engineering 2019: Data, Sensing, and Analytics, Atlanta, GA, USA, 17–19 June 2019; pp. 153–161. [CrossRef]
38. Okpala, I.; Nnaji, C.; Karakhan, A.A. Utilizing Emerging Technologies for Construction Safety Risk Mitigation. *Pract. Period. Struct. Des. Constr.* **2020**, *25*, 04020002. [CrossRef]
39. Ogunrinde, O.; Nnaji, C.; Amirkhanian, A. Application of emerging technologies for highway construction quality management: A review. In Proceedings of the Construction Research Congress 2020: Project Management and Controls, Materials, and Contracts, Tempe, AZ, USA, 8–10 March 2020; pp. 1030–1039. [CrossRef]
40. Shi, F.; Wang, Q.; Wang, Y. Research on top-level redesign of smart construction system based on case study. In Proceedings of the ICCREM 2019: Innovative Construction Project Management and Construction Industrialization, Banff, AB, Canada, 21–24 May 2019; pp. 117–124. [CrossRef]
41. Heravi, G.; Eslamdoost, E. Applying artificial neural networks for measuring and predicting construction-labor productivity. *J. Constr. Eng. Manag.* **2015**, *141*, 04015032. [CrossRef]
42. Yue, Y.; Li, Q. Research on the development trend of China's construction industry modernization: The case of Jiangsu, China. In Proceedings of the ICCREM 2016: BIM Application and Off-Site Construction, Edmonton, AB, Canada, 29 September–1 October 2017; pp. 162–168. [CrossRef]
43. Zhou, S.; Wu, S. A Study on the development trend of china's information of construction industry: A perspective of public policy analysis. In Proceedings of the ICCREM 2017: International Conference on Construction and Real Estate Management 2017, Guangzhou, China, 10–12 November 2017; pp. 395–404. [CrossRef]
44. Zhang, S.; Pan, F.; Wang, C.; Sun, Y.; Wang, H. BIM-based collaboration platform for the management of epc projects in hydropower engineering. *J. Constr. Eng. Manag.* **2017**, *143*, 04017087. [CrossRef]
45. Lin, M.; Wang, Q.; Wang, M.; Li, J. Exploration and practice of intelligent construction in island tunnel project of Hong Kong-Zhuhai-Macao Bridge. *Sci. Technol. Prog. Policy* **2018**, *24*, 81–85.
46. Yang, Z.; Zeng, B.; Wang, X. Study on BIM+ intelligent construction for Beijing urban sub-center district. *Archit. Technol.* **2018**, *9*, 990–992.
47. Fan, Q.; Lu, Y.; Zhou, S.; Yang, N.; Lin, E.; Li, G. Research and practice on intelligent construction technology system of Jinsha River hydropower projects. *J. Hydraul. Eng.* **2019**, *50*, 294–304.
48. Memari, A.M.; Huelman, P.H.; Iulo, L.D.; Laquatra, J.; Martin, C.; McCoy, A.; Nahmens, I.; Williamson, T. Residential Building Construction: State-of-the-Art Review. *J. Arch. Eng.* **2014**, *20*, B4014005. [CrossRef]

49. Zhou, Z.; Goh, Y.M.; Shen, L. Overview and Analysis of Ontology Studies Supporting Development of the Construction Industry. *J. Comput. Civ. Eng.* **2016**, *30*, 04016026. [CrossRef]
50. Liu, S.; Luo, H.; Sun, J.; Ding, L. Thoughts on the design of practical teaching plan for undergraduate intelligent construction program. *Res. High. Educ. Eng.* **2020**, *1*, 20–24.
51. Rossiter, J.R.; Lilien, G.L. New "brainstorming" principles. *Aust. J. Manag.* **1994**, *19*, 61–72. [CrossRef]
52. Porter, M.E. The competitive advantage of nations. *Harv. Bus. Rev.* **1990**, *68*, 73–93.
53. Deng, F.; Liu, G.; Jin, Z. Factors formulating the competitiveness of the Chinese construction industry: Empirical investigation. *J. Manag. Eng.* **2013**, *29*, 435–445. [CrossRef]
54. Liu, J.; Li, G. Research on the development of 3D printing construction industry based on diamond model. In *ICCREM 2018: Innovative Technology and Intelligent Construction*; American Society of Civil Engineers: Reston, VA, USA, 2018; pp. 164–176. [CrossRef]
55. Niu, Y.; Deng, X.; Zhao, X.; Zhang, N. Hexagonal diamond model for international competitive advantages of high-speed railway industry. *J. Manag. Eng.* **2020**, *36*, 04020001. [CrossRef]
56. Cho, D.-S.; Moon, H.-C.; Kim, M.-Y. Does one size fit all? A dual double diamond approach to country-specific advantages. *Asian Bus. Manag.* **2009**, *8*, 83–102. [CrossRef]
57. Saban, E.; Hande, U. Examining the competitive structure of Turkish tourism industry in comparison with Diamond Model. *Procedia-Soc. Behav. Sci.* **2012**, *62*, 620–627.
58. Wang, E.X.; Wu, C.Y. A study on the satisfaction of inbound tourism service quality based on grey correlation analysis. *Tour. Trib.* **2008**, *11*, 30–34.
59. Li, Y.; Zhang, X.; Yao, T.; Sake, A.; Peng, N. The developing trends and driving factors of environmental information disclosure in China. *J. Environ. Manag.* **2021**, *288*, 112386. [CrossRef]
60. Ma, Y.; Du, G.; Zheng, S.; Shi, W. Grey correlation analysis of influencing factors on logistics transportation development in Guizhou province. *J. Phys. Conf. Ser.* **2021**, *1774*, 012025. [CrossRef]
61. Wang, Y.; Yin, K.L.; An, G.F. Grey correlation analysis of sensitive factors of landslide. *Rock Soil Mech.-Wuhan* **2004**, *25*, 91–93.
62. Liu, J.X.; Liang, B.L. Grey correlation analysis of sensitive factors of concrete structures durability. *Key Eng. Mater.* **2009**, *400–402*, 471–476. [CrossRef]
63. Fang, S.; Yao, X.; Zhang, J.; Han, M. Grey Correlation Analysis on Travel Modes and their Influence Factors. *Procedia Eng.* **2017**, *174*, 347–352. [CrossRef]
64. Wang, X.; Gan, Y.; Lian, M.; Bi, B.; Tang, Y. Identification of risk sources of abnormal driving vehicles of expressway in port city. *J. Coast. Res.* **2020**, *104*, 317–321. [CrossRef]
65. Wu, L.; Gao, H.; Wang, K.-C.; Yang, C.-H. A Green-IKE Inference System Based on Grey Neural Network Model for Humanized Sustainable Feeling Assessment about Products. *Math. Probl. Eng.* **2020**, *2020*, 6391463. [CrossRef]

buildings

MDPI

Article

Integrated Schematic Design Method for Shear Wall Structures: A Practical Application of Generative Adversarial Networks

Yifan Fei [1], Wenjie Liao [2,*], Shen Zhang [3], Pengfei Yin [3], Bo Han [4], Pengju Zhao [1], Xingyu Chen [1] and Xinzheng Lu [2]

1 Beijing Engineering Research Center of Steel and Concrete Composite Structures, Tsinghua University, Beijing 100084, China
2 Key Laboratory of Civil Engineering Safety and Durability of Ministry of Education, Tsinghua University, Beijing 100084, China
3 Central-South Architectural Design Institute Co., Ltd., Wuhan 430071, China
4 Beijing Institute of Architectural Design Co., Ltd., Beijing 100045, China
* Correspondence: liaowj17@tsinghua.org.cn

Abstract: The intelligent design method based on generative adversarial networks (GANs) represents an emerging structural design paradigm where design rules are not artificially defined but are directly learned from existing design data. GAN-based methods have exhibited promising potential compared to conventional methods in the schematic design phase of reinforced concrete (RC) shear wall structures. However, for the following reasons, it is challenging to apply GAN-based approaches in the industry and to integrate them into the structural design process. (1) The data form of GAN-based methods is heterogeneous from that of the widely used computer-aided design (CAD) methods, and (2) GAN-based methods have high requirements on the hardware and software environment of the user's computer. As a result, this study proposes an integrated schematic design method for RC shear wall structures, providing a workable GAN application strategy. Specifically, (1) a preprocessing method of architectural CAD drawings is proposed to connect the GAN with the upstream architectural design; (2) a user-friendly cloud design platform is built to reduce the requirements of the user's local computer environment; and (3) a heterogeneous data transformation method and a parametric modeling procedure are proposed to automatically establish a structural analysis model based on GAN's design, facilitating downstream detailed design tasks. The proposed method makes it possible for the entire schematic design phase of RC shear wall structures to be intelligent and automated. A case study reveals that the proposed method has a heterogeneous data transformation accuracy of 97.3% and is capable of generating shear wall layout designs similar to the designs of a competent engineer, with 225 times higher efficiency.

Keywords: intelligent structural design; generative adversarial networks; parametric modeling; reinforced concrete shear wall structures; schematic design

Citation: Fei, Y.; Liao, W.; Zhang, S.; Yin, P.; Han, B.; Zhao, P.; Chen, X.; Lu, X. Integrated Schematic Design Method for Shear Wall Structures: A Practical Application of Generative Adversarial Networks. *Buildings* **2022**, *12*, 1295. https://doi.org/10.3390/buildings12091295

Academic Editor: André Furtado

Received: 14 July 2022
Accepted: 22 August 2022
Published: 24 August 2022

Publisher's Note: MDPI stays neutral with regard to jurisdictional claims in published maps and institutional affiliations.

Copyright: © 2022 by the authors. Licensee MDPI, Basel, Switzerland. This article is an open access article distributed under the terms and conditions of the Creative Commons Attribution (CC BY) license (https://creativecommons.org/licenses/by/4.0/).

1. Introduction

Intelligent structural design is an essential aspect of the fourth industrial revolution in the architecture, engineering, and construction (AEC) sector [1–3]. A reinforced concrete (RC) shear wall structure is an effective lateral force-resistant structural system commonly employed in high-rise residential buildings and is an important research object in intelligent structural design [4,5]. Schematic design is the first step in the structural design of RC shear-wall structures, which mainly involves the spatial layout of the primary force-transmitting components, including shear walls and beams. It is an essential basis for subsequent detailed design tasks and significantly impacts the final design outcomes [6].

Currently, the schematic design is usually manually completed by experienced engineers, resulting in low design efficiency and high labor costs. Existing intelligent schematic design methods can generally be separated into rule-based and learning-based methods [7,8]. Rule-based methods rely significantly on user-defined design rules, which tend to be less effective for complex real-world problems. Additionally, the length of time they require (usually several hours to dozens of hours) hinders their application in the industry. In contrast, learning-based methods do not require artificially defined explicit design rules but can automatically discover and master design laws from existing design data. Moreover, they have the advantage of extremely high design efficiency in the application stage [8,9]. As a typical representative of learning-based methods, generative adversarial networks (GAN)-based methods have recently made substantial strides in intelligent structural design. Existing studies have shown that GAN-based methods can effectively learn from existing design data and efficiently complete structural designs. The overall performance of the structures designed by GANs is close to that of structures designed by engineers [10–14].

However, several obstacles prevent existing GAN-based methods from being effectively applied in the industry. (1) GANs are based on computer vision techniques, and their inputs are in the form of pixel images. Consequently, GANs cannot directly perform structural designs based on the architectural computer-aided design (CAD) drawings commonly used in the industry. (2) GANs have a high requirement in terms of the computer environment. In terms of software, a deep learning framework and dependent libraries are needed. In terms of hardware, a graphics processing unit (GPU) is needed to achieve high design efficiency. (3) The outputs of GAN-based methods are also pixel images, where structural design-related information is unstructured data, making it challenging to establish the structural analysis model required for subsequent detailed design tasks.

This study focuses on the above research gaps and proposes a systematic solution, i.e., an integrated schematic design method based on GAN, as shown in Figure 1. First, a preprocessing method for architectural CAD drawings is proposed. Second, a cloud design platform is built based on the concept of software as a service (SaaS). Third, a high-precision data transformation method is proposed for transforming pixel images into structured data. Subsequently, a parametric modeling procedure is constructed to establish the structural analysis model. The proposed method can be easily embedded in the existing structural design process and can automatically complete the schematic design task traditionally manually finished by engineers. It should be noted that, at present, the structural designs are mainly stored in the form of 2D CAD drawings in China, and mainstream building information modeling software (e.g., Revit) supports exporting 3D models into 2D drawings. Therefore, this study takes 2D CAD drawings as the input of the structural design workflow.

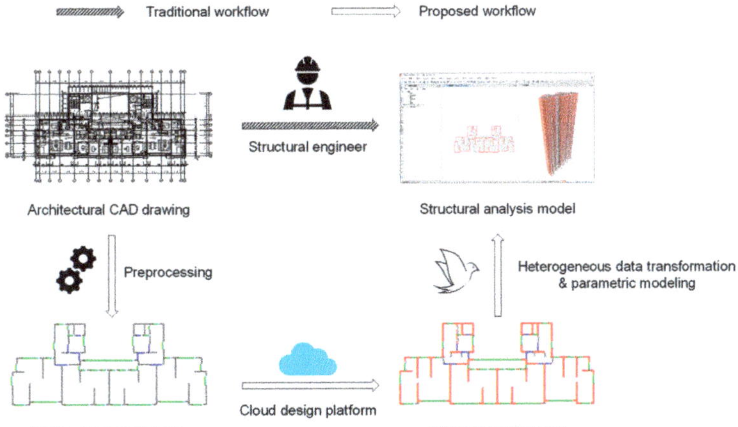

Figure 1. Traditional and proposed workflows of schematic design.

The remainder of this study is organized as follows. Section 2 is the literature review. Section 3 presents the framework of the integrated schematic design method. Section 4 introduces the preprocessing method of architectural CAD drawings. Section 5 introduces the intelligent design method based on GANs. Section 6 introduces the heterogeneous data transformation method and the parametric modeling procedure. A typical case study using the proposed method is presented in Section 7. Finally, conclusions are drawn in Section 8.

2. Literature Review

2.1. Learning-Based Structural Design Method

In recent years, machine learning has been extensively applied in the AEC sector [2]. As a novel paradigm, the machine learning-based structural design method has attracted substantial attention [7–9]. Compared with traditional rule-based methods, it can automatically discover and master design rules from existing design data without artificially defining them. Additionally, once the machine learning model is trained, it has the advantage of extremely high design efficiency. For example, Almasabha et al. [15] used several machine learning algorithms in the design of shear links for steel buildings; Zheng et al. [16] adopted artificial neural networks to speed up the topological design of shell structures; and Chang and Cheng [17] applied graph neural networks in the structural optimization of framed structures.

More recently, breakthroughs have been made in structural design methods using computer vision techniques, particularly GANs [18]. A GAN consists of a generator and a discriminator, where the generator strives to generate real-looking designs to fool the discriminator, and the discriminator tries to discriminate between real and fake designs. In a game between the two, the generator can learn to generate realistic designs after Nash equilibrium is reached. Liao et al. [10] and Pizarro et al. [11] effectively applied GANs to the shear wall layout design. Liao et al. [12] further proposed a "fused-text-image-to-image" GAN to consider the influence of design conditions on an intelligent structural design. Zhao et al. [13] expanded the applicability of GANs to the beam–slab system of shear wall residential buildings. Liao et al. [10] and Zhao et al. [13] evaluated the structural design performance of GANs using the intersection over union (IoU) of model-generated and engineer-designed structural pixel images. However, this evaluation method measures unstructured pixel-by-pixel consistency, which is not equivalent to the structural layout consistency on which the schematic design task focuses. Meanwhile, the performance of solely data-driven GANs depends on the quality and quantity of the training data, which limits their applications [10,12]. Consequently, Lu et al. [14] further embedded physical mechanisms into GANs and proposed a physics-enhanced GAN for the shear wall layout design. The physics-enhanced GAN features better interpretability, and its performance is less affected by training data. However, the inputs and outputs of the above method are still in the form of pixel images, limiting its embedment in the existing structural design process.

2.2. Parametric Modeling

Parametric modeling is a crucial tool for automated structural design, which can significantly improve design efficiency [19] and potentially benefit design creativity [20]. Existing studies have offered various parametric design systems that can automatically search for optimal solutions by combining optimization algorithms with parametric models [21–24]. However, these methods require structured design data as input and are difficult to apply to the unstructured design data obtained by GAN-based methods.

2.3. Transformation between Pixel Image and Structured Design Data

The input and output of the GAN-based method are unstructured pixel images, but structured design data are commonly used in the structural design process. In practical applications, it is necessary to convert the structured design data (architectural CAD drawing) into an architectural pixel image (GAN's input) and then convert the structural pixel image (GAN's output) into structured design data (structural analysis model). Pizarro and Massone [25] proposed a method to extract the polygons of wall contours from architectural CAD drawings, but the error rate was around 15%, requiring manual inspection and correction. To establish the structural analysis model from the structural pixel image, Lu et al. [14] proposed a vectorization method for pixel images of shear walls, but the accuracy was unsatisfactory, resulting in errors and missing elements frequently. Therefore, there is still a lack of high-precision preprocessing and heterogeneous data transformation methods for GAN-based methods.

3. Framework

The proposed integrated schematic design method based on GAN for RC shear wall structures is shown in Figure 2. It can complete structural design and establish structural analysis models according to the architectural CAD drawings and design conditions within 10 min, accomplishing the intelligent and automated design of RC shear wall structures. The proposed method includes the following modules.

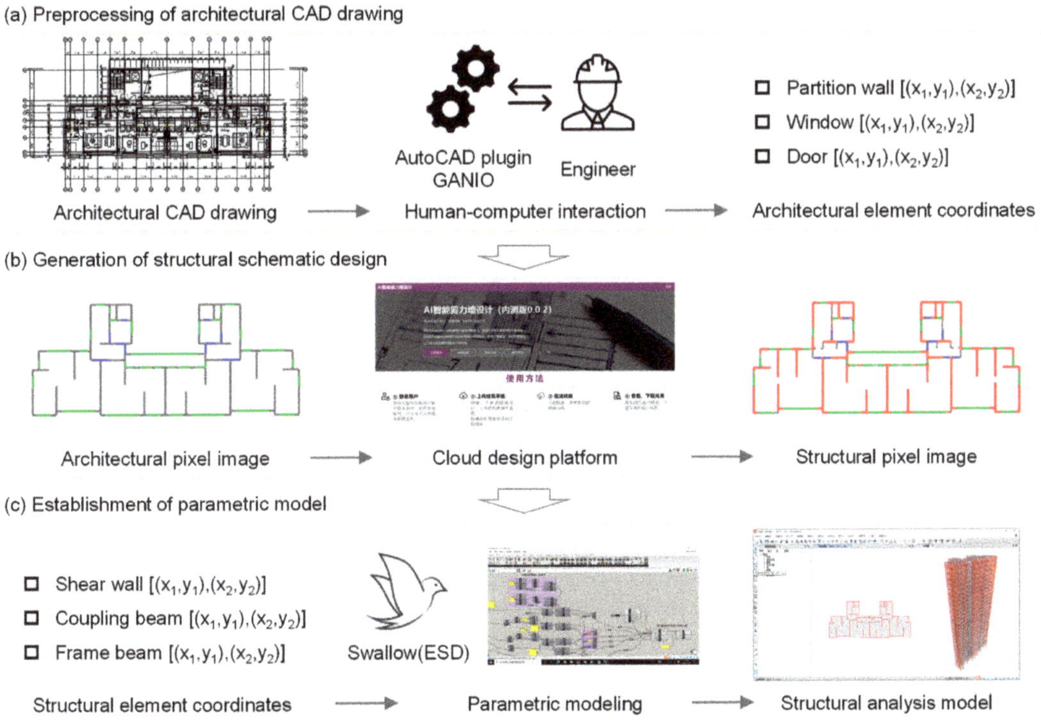

Figure 2. Key modules of the proposed integrated schematic design method.

(1) Preprocessing of architectural CAD drawings: Figure 2a shows the extraction of architectural elements using the AutoCAD plugin GANIO developed based on the AutoCAD application programming interface (API) using C# [26]. GANIO can automatically identify and extract essential architectural elements (i.e., partition walls, doors, and windows) and output their coordinates. Engineers can also check and adjust the extraction results through human–computer interaction. Subsequently, the architectural pixel image can be generated based on the architectural element coordinates. This process requires approximately 5 min.

(2) Generation of structural schematic design: Figure 2b shows the cloud design platform developed based on SaaS, which can swiftly generate a schematic design of the shear

wall structure. After the architectural pixel image is uploaded, the cloud platform inputs it into the pre-trained GAN deployed on the cloud server. The GAN generates the corresponding structural pixel image in seconds and outputs it to the cloud platform for users to download. This process requires approximately 1 min.

(3) Establishment of structural analysis model: Figure 2c shows the automatic modeling from the pixel image to the structural analysis model. First, identify and extract the key structural elements in the structural pixel image and obtain their coordinates. Next, utilize the parametric modeling software Swallow (ESD) [27], developed based on the Grasshopper API, to import structural element coordinates and establish a parametric model according to a predetermined modeling procedure. Finally, export the parametric model to ETABS for structural analysis. This process requires approximately 2 min.

It should be noted that the floor area affects the time consumption of the preprocessing of architectural CAD drawings and the establishment of a structural analysis model. Their time consumption mentioned above is based on a common RC shear wall structure with a floor area of around 500 m^2. The time consumption of the generation of structural schematic design is affected by the hardware performance and bandwidth of the cloud server. Its time consumption mentioned above is based on a common cloud server equipped with one Intel ® Xeon ® E5-2682 v4 CPU (two cores, 2.5 GHz), one NVIDIA P4 GPU (8 GB), and a bandwidth of 1 Mbps.

4. Preprocessing of Architectural CAD Drawings

Architectural CAD drawings contain numerous elements, as shown in Figure 2a. However, the elements related to structural design are sparse, mainly including three categories: partition walls (where shear walls can be positioned), doors, and windows (where shear walls cannot be positioned) [10,14]. To enable deep neural networks to extract the key features of architectural design and avoid the influence of irrelevant data, Liao et al. [10] proposed an architectural design representation method using semantic pixel images, extracting the key elements in the architectural CAD drawing and representing their categories with different colors in the RGB pixel image. However, manually completing these operations is inefficient, prone to errors, and unrealistic for industrial applications. Therefore, this study develops a CAD plugin, GANIO, based on the AutoCAD API [26], which can automatically extract and output the axis coordinates of critical elements. The coordinates are also used in Section 6.1 for the automatic identification and extraction of shear walls in semantic structural pixel images.

Specifically, the user interface of the GANIO plugin, depicted in Figure 3a, has three major functions: parameter setup, axis extraction, and coordinate export. The first step involves setting up six parameters. The first three are the maximum wall thickness, minimum wall thickness, and minimum wall length. These parameters are the thresholds for determining whether an element is a partition wall. The remaining three parameters are the layer names of the partition wall, door, and window. GANIO extracts the corresponding elements from a specified layer. The second step is to select the target elements and click on the "Extraction" button. GANIO locates key elements by matching parallel lines, calculates the coordinates of their axes, and draws the axes on a new layer. Engineers can check and adjust the extracted axes using an AutoCAD user interface. The third step is to select the extracted axes and click on the "Export" button. The axis coordinates of the key architectural elements are exported in a readable text format. Finally, according to the coordinates and categories of the key elements, Python-OpenCV is used to represent the key elements as RGB pixel images, as shown in Figure 3b. Distinct categories of elements are represented by different colors: the partition wall is gray (RGB = (132, 132, 132)), the door is blue (RGB = (0, 0, 255)), and the window is green (RGB = (0, 255, 0)).

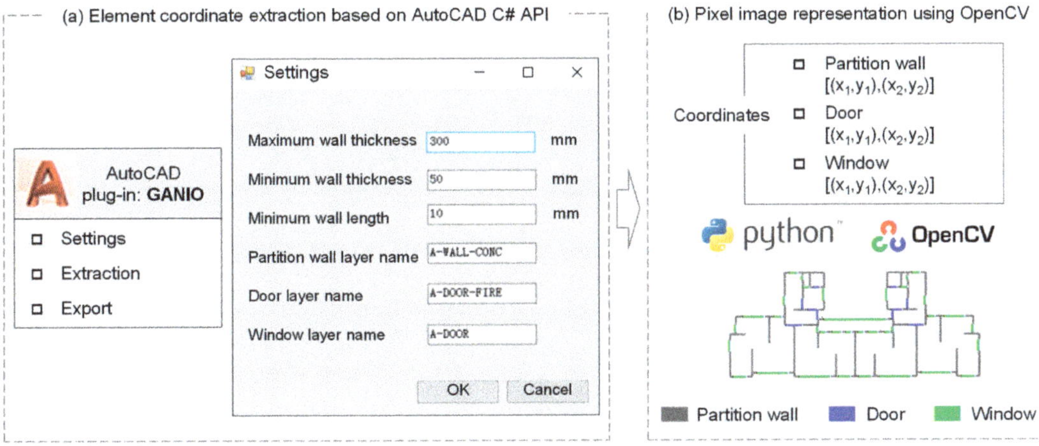

Figure 3. Preprocessing of architectural CAD drawings (the user interface of GANIO is in Chinese, and the figure is translated into English for convenient reading).

5. Intelligent Structural Design Based on GANs

5.1. Physics-Enhanced GAN

Experience and mechanics are two indispensable aspects of structural design. This study adopts the physics-enhanced GAN proposed by Lu et al. [14] (referred to as StructGAN-PHY) to generate the structural schematic design. The architecture of a conventional data-driven GAN is shown in Figure 4a (referred to as StructGAN), which only comprises a generator and a discriminator [10]. The architecture of StructGAN-PHY is shown in Figure 4b. Apart from a generator and a discriminator, StructGAN-PHY also comprises a physics evaluator. The generator generates a structural design according to the architectural design and design conditions. The discriminator judges whether the generated structural design is real or fake and forms an image loss $L_{\text{G-img}}$, which is fed back to the generator to improve the image quality of its designs. Meanwhile, the physics evaluator predicts the physical performance of the generated structural design considering the design conditions and forms a physics loss $L_{\text{G-PHY}}$, which is fed back to the generator to improve the physical performance of its designs. The loss functions of the generator and discriminator are shown in Equations (1) and (2), respectively. The generator, discriminator, and physics evaluator work together in the training stage until the model performance is stabilized and the Nash equilibrium is reached.

$$L_{\text{G}} = \omega_{\text{img}} L_{\text{G-img}} + \omega_{\text{PHY}} L_{\text{G-PHY}}, \tag{1}$$

$$L_{\text{D}} = L_{\text{D-GAN}}, \tag{2}$$

where $L_{\text{G-img}}$ is the image loss, as shown in Equation (3); $L_{\text{G-PHY}}$ is the physics loss predicted by the physics evaluator; ω_{img} and ω_{PHY} are the weights of $L_{\text{G-img}}$ and $L_{\text{G-PHY}}$, respectively [14]; is the discriminator loss [28].

$$L_{\text{G-img}} = L_{\text{G-GAN}} + \lambda_{\text{FM}} (L_{\text{G-FM}} + L_{\text{G-VGG}}), \tag{3}$$

where $L_{\text{G-GAN}}$, $L_{\text{G-FM}}$, and $L_{\text{G-VGG}}$ are different types of image losses and λ_{FM} is the weight [10,28].

The physics evaluator is a surrogate model based on neural networks, which can output the physics loss of a generated structural design corresponding to its physical performance. For details, please refer to Lu et al. [14]. For RC shear wall structures, the inter-story drift under earthquakes is a critical indicator that reflects their physical performance, which can be evaluated by P_{drift} as shown in Equation (4). The physical loss predicted by the physical evaluator is an approximation to P_{drift}.

$$P_{\text{drift}} = \begin{cases} 1 - \frac{d_{\max}}{d_{\text{limit}}}, d_{\max} \le d_{\text{limit}} \\ \left(\frac{d_{\max}}{d_{\text{limit}}} - 1\right)^{0.5}, d_{\max} > d_{\text{limit}} \end{cases}, \quad (4)$$

where $d_{\max}$ is the maximum inter-story drift; $d_{\text{limit}} = 0.001$ is the drift limit specified by the Chinese design code [29].

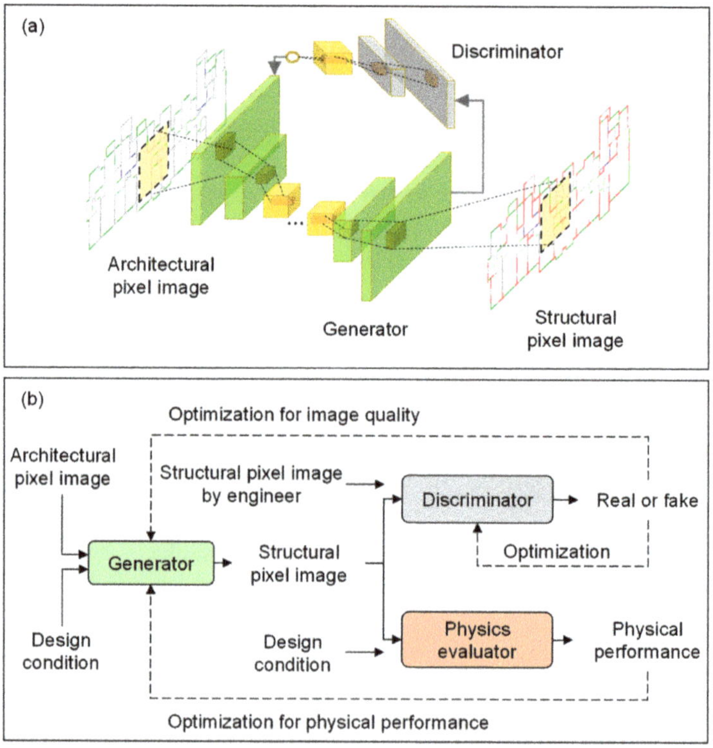

Figure 4. Intelligent structural design based on GAN: (**a**) data-driven GAN (StructGAN); (**b**) physics-enhanced GAN (StructGAN-PHY).

5.2. Dataset

This study collects 159 sets of architectural and structural CAD drawings and their design conditions from 10 top architectural design institutes in China. The collected CAD drawings have been used in real-world construction projects. Before that, they had been comprehensively optimized by the engineers to guarantee that all design code requirements were fulfilled. Therefore, the CAD drawings have a high design quality. The preprocessing method described in Section 4 is adopted to extract the coordinates of partition walls, doors, and windows from architectural CAD drawings and obtain corresponding architectural pixel images. Similarly, the coordinates of the shear walls are extracted from the structural CAD drawings, and the corresponding structural pixel images are obtained. One hundred thirty-five sets of architectural pixel images and their corresponding structural pixel images are used as the training set. Then, the training set is enlarged four times through data augmentation (flipping and mirroring). The remaining 24 sets of architectural pixel images are used as the test set, and their corresponding structural pixel images are not visible to the GAN. Typical architectural and structural pixel images and their design conditions (seismic intensity and structural height) are shown in Figure 5.

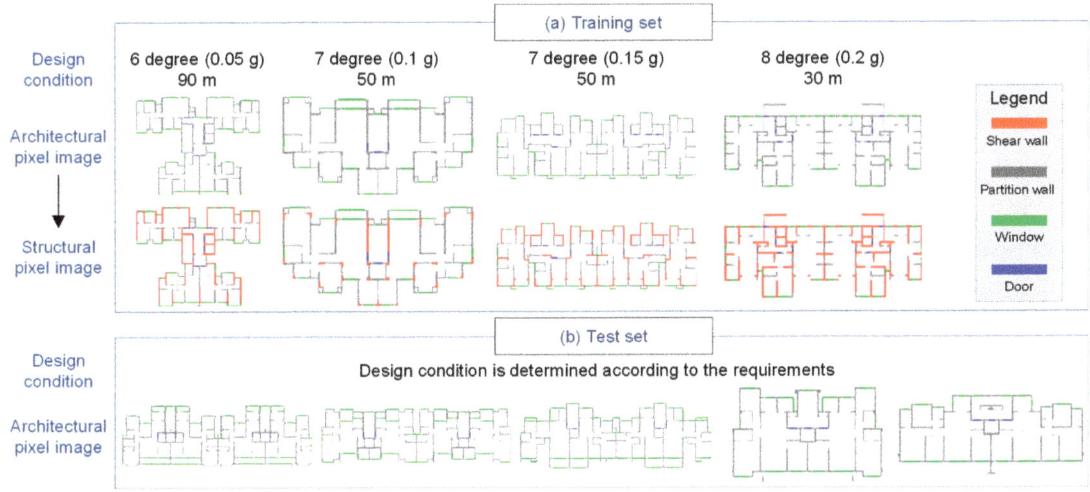

Figure 5. Typical datasets for StructGAN-PHY training [14].

Although the dataset is small in comparison to other deep learning tasks, it is effective for the training of StructGAN-PHY for the following reasons. (1) The focus of this research is on common shear wall structures in residential buildings. StructGAN-PHY can effectively learn general design rules from a relatively small dataset because the structures to be designed are in similar forms. (2) Key structural design elements are extracted from CAD drawings and processed into semantic pixel images. It is easier for StructGAN-PHY to learn from the preprocessed semantic images, and therefore fewer data are needed. (3) The incorporation of the physics mechanism reduces the model's reliance on data even further. (4) Data augmentation is used to increase the size of the training set.

5.3. Cloud Design Platform

Using StructGAN-PHY as the core algorithm, this study develops a cloud design platform based on the concept of SaaS, as illustrated in Figure 6. The cloud platform provides software services for users, with minimum requirements for users' local hardware and software, and the design process is straightforward and efficient. The client provides a human–computer interaction interface, including project creation, file upload, project design, and result download functions. The server is used to handle client requests and manage user data. All computing and design processes are performed on a GPU-powered server.

(1) Client: Figure 6a shows the homepage of the cloud platform, which has a login entry, manual, technical support, version history, and introduction to the technical details of the core algorithm. Figure 6b shows the window for creating a new project, including inputting the project name, uploading the architectural pixel image, selecting the design conditions (i.e., seismic intensity and structural height), and inputting the scale (unit: mm/pixel). The seismic intensity can be selected among 6 degrees (0.05 g), 7 degrees (0.10 g), 7 degrees (0.15 g), 8 degrees (0.20 g), 8 degrees (0.30 g), and 9 degrees (0.40 g). The numbers in brackets represent the seismic design acceleration with an exceedance probability of 10% in 50 years. The structural height can be selected as <40, 40–60, 60–80, 80–100, and >100 m. Figure 6c shows the project list. The initial status of a project is "to be converted". Clicking the "Convert" button calls the pre-trained StructGAN-PHY deployed on the server for the design. The obtained design result

is a structural pixel image (Figure 6d), where red (RGB = (255, 0, 0)) represents the shear wall layout. This pixel image can be downloaded by the user for subsequent parametric modeling.

(2) Server: The Python, Flask, and Nginx environments are set up on a Windows system. The website front-end is developed based on HTML and CSS, where Flask-login is adopted for the login interface management. The database for user data management adopts PyMySql, and data can be delivered by Flask-sqlalchemy. The website backend is developed based on Python, where the PyTorch deep learning framework and its dependent libraries are installed to run the pre-trained StructGAN-PHY model. Furthermore, Nginx builds network services, connecting the client and the server.

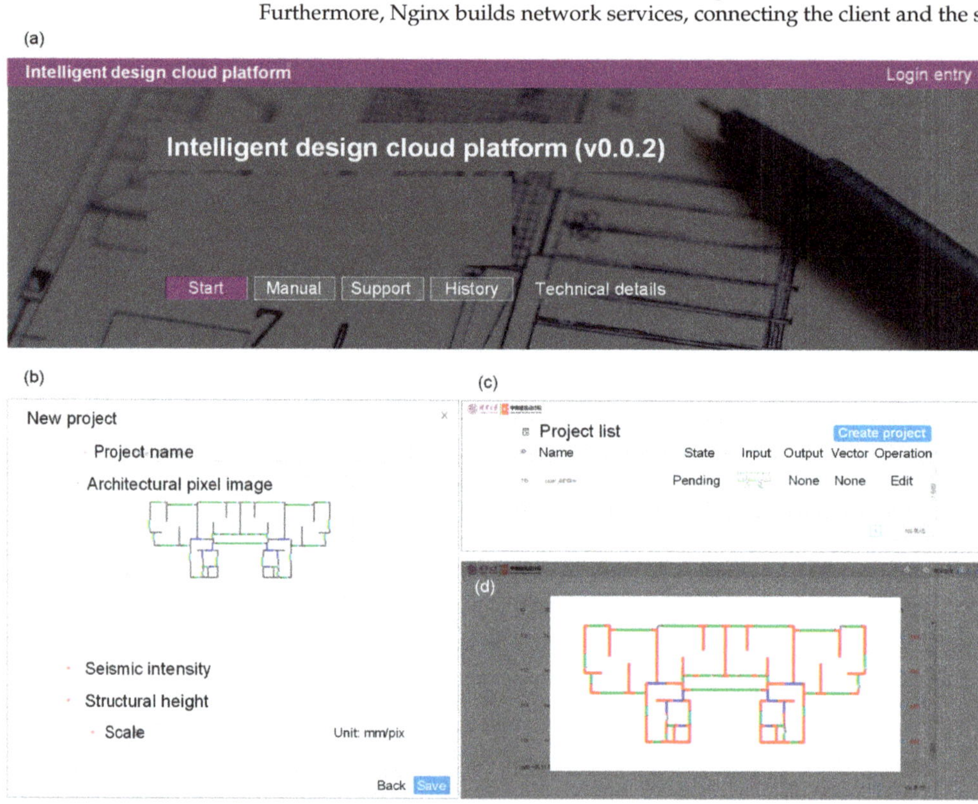

Figure 6. Cloud design platform: (**a**) homepage; (**b**) project creation; (**c**) project list; (**d**) design result. (The user interface of the platform is in Chinese, and the figure is translated into English for convenient reading).

6. Establishment of the Structural Analysis Model

6.1. Heterogeneous Data Transformation

Based on Lu et al. [14], this study proposes a heterogeneous data transformation method that considers architectural design information, as shown in Figure 7. This method involves the following steps:

Step 1: Extract shear wall pixels

First, the RGB pixel image is expressed in the HSV color. Second, the red pixels (i.e., shear walls) are stripped from the structural pixel image and binarized. Subsequently, the corrosion (cv2.erode()) and dilation (cv2.dilate()) functions in Python-OpenCV are used to remove the noise in the binary image. Finally, a binary pixel image of the shear walls is obtained.

Step 2: Extract shear wall axes

Based on the assumption that the shear walls can only be positioned at the location of the partition walls, the intersection points between the axis of the partition wall and the contour of the shear wall pixels in the binary image are searched pixel-by-pixel and used as the endpoints of the axis of the shear wall. All axes of the partition walls are traversed to complete the extraction of the shear wall axes.

Step 3: Assign frame and coupling beams

The axes of the partition walls, doors, and windows are collectively called the architectural axes. It is assumed that (1) beams are only positioned on architectural axes and (2) coaxial shear walls are connected by coupling beams. First, assign the beams over the architectural axes, excluding the positions where the shear walls are already positioned. Subsequently, check the topological relationship of the beams, delete cantilever beams, and confirm that both ends of the beams are connected to the shear walls or other beams.

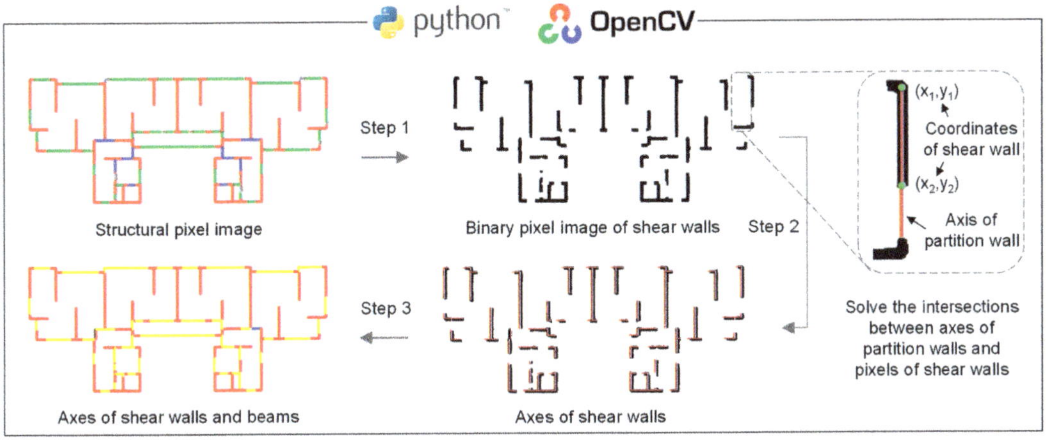

Figure 7. Heterogeneous data transformation.

Note that GAN-based methods focus on common residential buildings so that general design rules can be learned from existing design data. Therefore, the assumptions described in Step 3 are generally acceptable. If the assumptions are not applicable, the beam layout obtained with the proposed method might be unreasonable, resulting in inaccurate structural analysis results. However, in the schematic design phase of RC shear wall structures, the beam layout is less crucial than the shear wall layout. Additionally, inaccuracy in the structural analysis model is acceptable because it can be corrected in the subsequent detailed design phase by optimization algorithms or manual adjustments to fulfill special requirements.

Based on the above steps (Figure 7), the axis coordinates of the shear walls, frame beams, and coupling beams are derived. Furthermore, the shear wall thickness is estimated according to the empirical law proposed by Lu et al. [14]. The section heights of the frame and coupling beam can also be determined according to the empirical law presented by Qian et al. [4] as 1/12 and 1/8 of the beam span, respectively. The axis coordinates and section dimensions of the shear walls, frame beams, and coupling beams are used as the input data for the parametric modeling. Based on the initial structural layout and section size, subsequent adjustments and optimizations can be easily accomplished using a parametric model.

6.2. Parametric Modeling

Parametric modeling is a digital modeling method that builds structural models from input data according to predefined rules, thereby realizing real-time mapping between input data and structural models. By leveraging the human–computer interaction interface developed via visual programming, users can modify the structural design and corresponding model in real-time. Rhino and its Grasshopper plugin are the commonly used parametric modeling platforms in the industry, but they lack professional structural analysis capabilities. In this study, Swallow (ESD), a parametric modeling plugin for building structures based on Grasshopper, is used as a bridge between the Grasshopper and structural analysis software (e.g., ETABS) [27]. Using Swallow (ESD), structural properties can be defined in Grasshopper, and structural analysis models can be assembled. ETABS can be called in real-time through the ETABS API for structural analysis, and the analysis results of ETABS can be viewed.

Figure 8a demonstrates the parametric modeling procedure based on Swallow (ESD), and the final parametric model is shown in Figure 8b. (1) Firstly, the structural parameter interpretation module is built to read the axis coordinates and section sizes of structural components generated in Section 6.1; to convert them into structured data required for the modeling of shear walls, coupling beams, and frame beams; and to read the overall parameters of the structure specified by the user, including the story height and the number of stories. (2) Subsequently, the structural element modeling module is built to model shear walls and coupling beams using shell elements, model frame beams using beam elements, and model slabs using membrane elements; to define the properties of the material, section, and element; and to assemble all structural elements to complete the structural analysis model. (3) Next, a load definition module is built to distribute beam-end loads according to the structural layout and set seismic and wind loads according to the seismic and wind design requirements. (4) Finally, an ETABS calling module is constructed to complete the establishment and analysis of the ETABS model by calling the ETABS API.

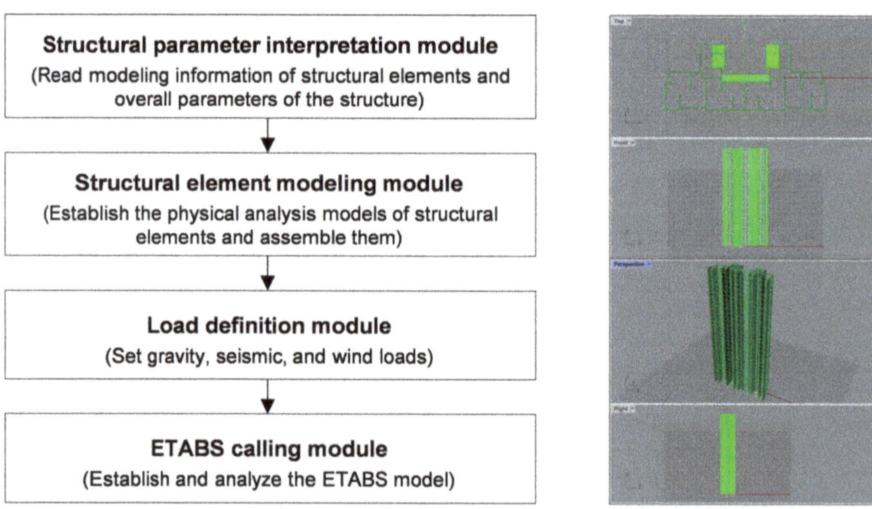

(a) Parametric modeling procedure (b) Parametric model in Rhino

Figure 8. Parametric modeling procedure based on Swallow (ESD).

7. Case Study

7.1. Evaluation Method of Shear Wall Layout

The schematic design of RC shear wall structures focuses on the shear wall layout. This study uses the IoU of the shear wall axes to evaluate the consistency of the model-generated and engineer-designed planar layouts, which is more reasonable than the IoU of pixel images [10,13]. Meanwhile, the shear wall layout significantly influences the vertical load-transfer mechanism of the floor system. A reasonable shear wall layout can effectively hold the slabs so that they are uniformly stressed [30]. Therefore, this study uses the supported floor ratio to evaluate the vertical load transferability of slabs. In addition, with the help of structural analysis models, the physical performance of the designed structure is also evaluated.

(1) Planar layout consistency

The heterogeneous data transformation method described in Section 6.1 is utilized to obtain the coordinates of the shear walls designed by the GAN. Subsequently, the shear wall planar layout is drawn as a set of rectangles, and the intersection and union areas of the GAN's and the engineer's designs are calculated. Furthermore, the difference between the total length of the shear walls designed by the GAN and the engineer is calculated as a correction coefficient. Figure 9a shows the calculation of the intersection and union areas, where green represents the intersection area, blue represents the exclusive part of the engineer's design, and red represents the exclusive part of the GAN's design. The union area is a combination of green, red, and blue colors. The modified planar layout consistency indicator $S_{\mathrm{IoU-M}}$ can be calculated using Equations (5) and (6).

$$S_{\mathrm{IoU-M}} = \eta_{\mathrm{DiffSwall}} \frac{A_{\mathrm{inter}}}{A_{\mathrm{union}}}, \tag{5}$$

$$\eta_{\mathrm{DiffSwall}} = 1 - \frac{|L_{\mathrm{GAN}} - L_{\mathrm{ENG}}|}{L_{\mathrm{ENG}}}, \tag{6}$$

where A_{inter} and A_{union} are the intersection and union areas of the shear walls designed by the GAN and the engineer, respectively; $\eta_{\mathrm{DiffSwall}}$ is the correction coefficient for the difference in total shear wall length; and L_{GAN} and L_{ENG} are the total lengths of the shear walls designed by the GAN and the engineer, respectively.

(2) Vertical load transferability

The vertical load transferability is assessed by the floor area supported by the shear walls, as shown in Figure 9b. First, the floor boundary (blue contour) is obtained; then, the floor area that each shear wall can support (red contour) is calculated, as shown in Figure 9c [30]. Furthermore, all the supported floor areas are subtracted from the floor area. Finally, the unsupported floor areas (green contours) are obtained, and the supported floor ratio under the vertical load (S_{FloorA}) is obtained based on the ratio of the green area to the blue area, as indicated in Equation (7).

$$S_{\mathrm{FloorA}} = 1 - \frac{A_{\mathrm{minus}}}{A_{\mathrm{floor}}}, \tag{7}$$

where A_{minus} is the floor area that is not supported by the shear walls, and A_{floor} is the total area of the floor.

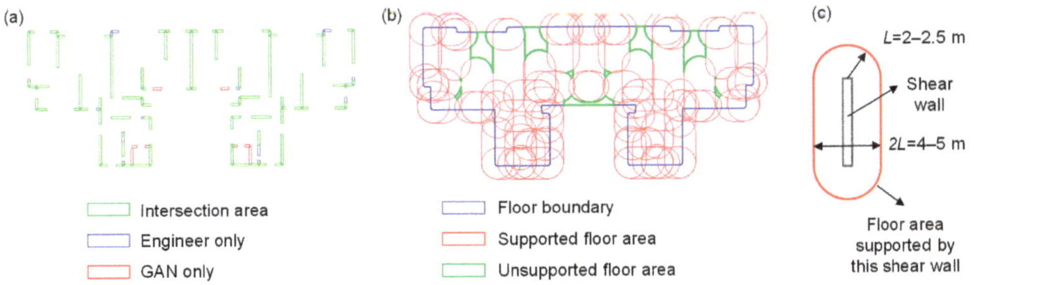

Figure 9. Evaluation method of shear wall layout: (**a**) planar layout consistency; (**b**) vertical load transferability; (**c**) floor area supported by a shear wall.

(3) Physical performance under horizontal seismic load

After the ETABS model is established, a refined structural analysis can be performed to evaluate the physical performance of the structure. For example, the inter-story drift under earthquakes is a critical indicator in evaluating the lateral resistance capacity of RC shear wall structures. Equation (8) calculates the consistency of the inter-story drifts of the structures designed by the GAN and the engineer.

$$S_{\text{IDR}} = 1 - \left(\left| 1 - \frac{\theta_{\text{GAN,X}}}{\theta_{\text{ENG,X}}} \right| + \left| 1 - \frac{\theta_{\text{GAN,Y}}}{\theta_{\text{ENG,Y}}} \right| \right) / 2, \tag{8}$$

where $\theta_{\text{GAN,X}}$ and $\theta_{\text{GAN,Y}}$ are the maximum inter-story drifts of the GAN's design in the X and Y directions, respectively, and $\theta_{\text{ENG,X}}$ and $\theta_{\text{ENG,Y}}$ are the maximum inter-story drifts of the engineer's design in the X and Y directions, respectively.

7.2. Basic Information of a Typical Case

This case study is based on a typical residential building in northern China. The building has a structural height of 96 m and 30 stories. It is lower than 100 m and is classified as a common high-rise building. Its floor has a bounding area of 41.2 m $\times$ 17.7 m (around 500 m^2). The seismic intensity is 8-degree, corresponding to a 0.20 g seismic design acceleration with an exceedance probability of 10% in 50 years. This seismic design acceleration a is medium-level (0.05 g $\leq a \leq$ 0.40 g). The characteristic site period T_g is 0.55 s, which is also medium-level (0.2 s $\leq T_g \leq$ 0.9 s). The architectural CAD drawing, corresponding pixel image, and structural design by the engineer are shown in Figure 10a–c, respectively.

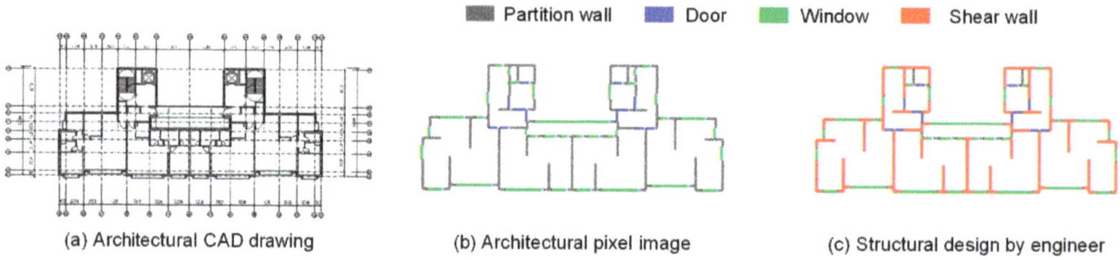

Figure 10. A typical residential building in northern China.

7.3. Detailed Analyses of a Typical Case

The building was designed using the proposed integrated design method. The structural pixel image downloaded from the cloud design platform (i.e., the output of the StructGAN-PHY model) is shown in Figure 11a, and the details in the black dashed box are shown in Figure 11b. Two heterogeneous data transformation methods, one proposed by Lu et al. [14] and another proposed in this study, were used to convert the structural pixel image into structured data. The results are compared in Figure 11c,d. Lu et al.'s method [14] results in the absence of several short shear walls and an undesirable offset of the shear wall axes, which is adverse for the subsequent modeling task. The proposed method prevents these problems and accurately extracts nearly all shear walls. In this case study, the StructGAN-PHY lays out a total of 73 shear walls. Lu et al.'s method [14] correctly extracted 44 shear walls with an accuracy of 60.3%. The proposed method correctly extracted 71 shear walls with an accuracy of 97.3%. The accuracy therefor increases significantly by 37.0%. It should be noted that the above results are obtained from the typical case study in Section 7.2.

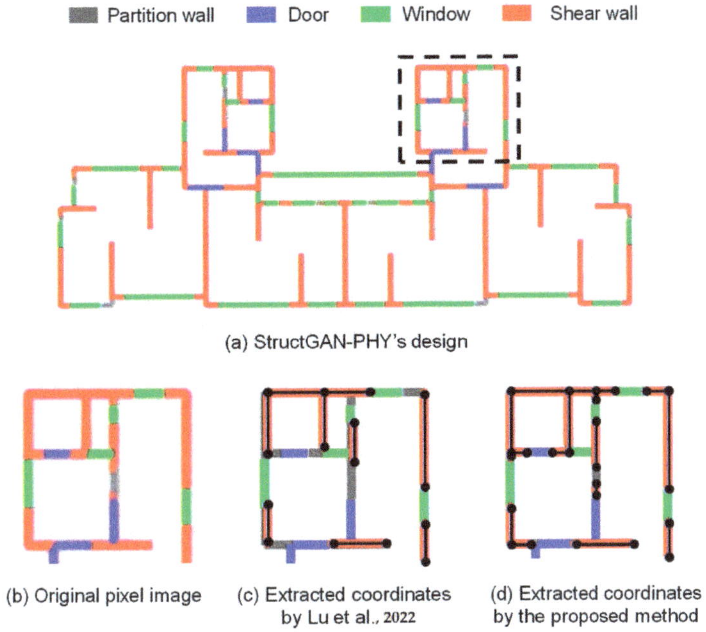

Figure 11. Comparison between different heterogeneous data transformation methods [14].

The design of the proposed integrated method is evaluated using the methods described in Section 7.1. The evaluation results are shown in Figure 12 and Table 1 (StructGAN-PHY). Regarding the planar layout consistency, generally, IoU > 0.5 implies that the consistency between the designs from GAN and the engineer is acceptable [10]. The $S_{\mathrm{IoU-M}}$ in this case is 0.9902, which shows that the design of the proposed method is very similar to that of the engineer. In terms of the vertical load transferability, S_{FloorA} is 0.9334, indicating that the designed shear walls can support vertical loads of the floor. In terms of physical performance, the S_{IDR} is 0.9602, and the inter-story drift is within the 1/1000 limit specified in the code [29]. It is noteworthy that with the help of the parametric model, the structural design can be manually adjusted by engineers and automatically optimized by algorithms [31,32] in the future.

Table 1. Comparison between designs of data-driven and physics-enhanced models.

Designer	$S_{\mathrm{IoU-M}}$	S_{FloorA}	S_{IDR}
StructGAN-PHY	0.9902	0.9334	0.9602
StructGAN	0.5855	0.8372	0.9380
Difference	69.1%	11.5%	2.4%

Furthermore, to illustrate the superiority of StructGAN-PHY adopted in this study, its evaluation results are compared with those of a data-driven GAN, i.e., StructGAN [10], as shown in Figure 12 and Table 1. The shear walls designed by StructGAN are insufficient in number and length, resulting in lower evaluation indicators. The evaluation indicators of StructGAN-PHY are improved by 69.1%, 11.5%, and 2.4%, respectively, compared with those of StructGAN.

Moreover, in terms of design efficiency, the times required for a competent engineer, StructGAN-PHY, and the proposed method to complete the schematic design of an RC shear wall structure are presented in Table 2. Compared with the results for engineers, the design efficiency is dramatically boosted by 225 times when the proposed method is used. Additionally, the preprocessing and modeling method proposed in this study boosts the efficiency by 2.5 times for the entire design phase compared to existing studies (i.e., StructGAN-PHY).

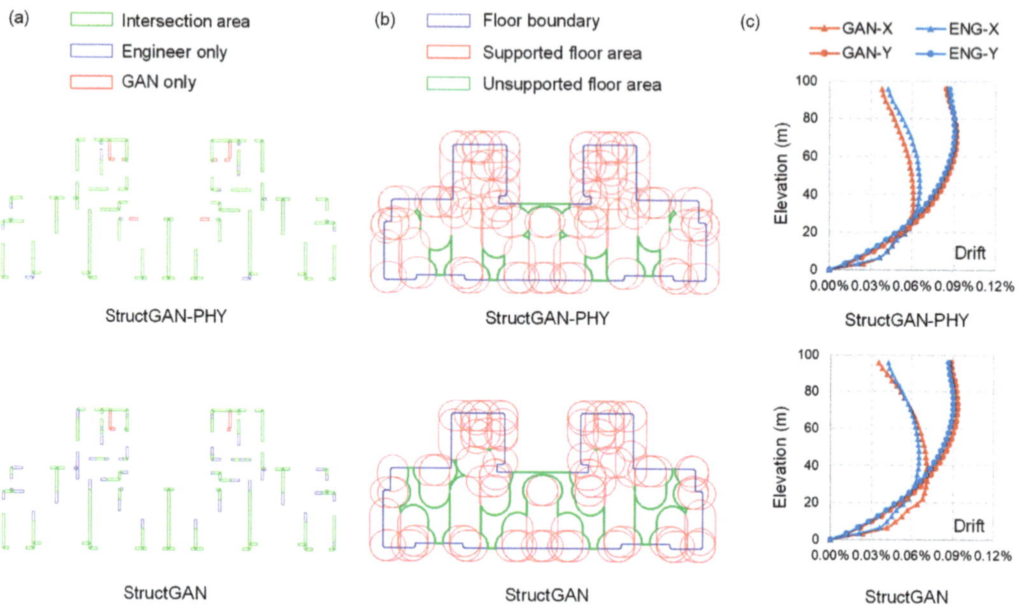

Figure 12. Evaluation results of the typical case: (**a**) planar layout consistency; (**b**) vertical load transferability; (**c**) physical performance.

Table 2. Comparison of design efficiency between different methods.

Designer	Preprocess	Design	Model	Total	Efficiency Enhanced
Engineer (manually)	0 min	20 h	10 h	30 h	/
StructGAN-PHY	15 min	1 min	4 min	20 min	90 times faster
Proposed method	5 min	1 min	2 min	8 min	225 times faster

Note that the design efficiencies of the proposed method and StructGAN-PHY were obtained under the conditions described in Section 3. The design efficiency of engineers is obtained by consulting several senior engineers from top architectural design institutes in China. It is also based on common RC shear wall structures with a floor area of around 500 m^2.

8. Conclusions

Despite showing potential in intelligent structural design, GAN-based methods are difficult to apply in the industry because of their heterogeneous data form with traditional CAD methods and high requirements in terms of the computer environment. This study proposes an integrated schematic design method based on GAN, enabling the entire schematic design phase of the RC shear wall structures to be intelligent and automated and providing a workable solution for the industrial application of GAN-based methods. First, a preprocessing method for architectural CAD drawings is proposed to connect GAN with upstream architectural design tasks. Second, a user-friendly cloud design platform is built to reduce the user's local computer environment requirements. Third, a heterogeneous data transformation method and a parametric modeling procedure are developed to establish the structural analysis model based on GAN's design, facilitating subsequent detailed design tasks. The following conclusions are drawn from the study:

(1) The cloud design platform and its pre- and post-processing methods have the advantage of being straightforward and efficient. For common RC shear wall structures with a floor area of around 500 m^2, as shown in the case study, the overall efficiency is 225 times higher than that of a competent engineer and 2.5 times higher than that of the existing intelligent design method.
(2) In a typical case, the heterogeneous data transformation method can convert the shear wall design from a pixel image to structured data with a high accuracy of 97.3% and enable the data transfer between GAN and parametric modeling.
(3) According to the case study, the shear wall layout obtained using the proposed method is close to the engineer's design, with a planar layout consistency $S_{\text{IoU-M}}$ of 0.9902. It can also support the vertical load of the floor system with a vertical load transferability S_{FloorA} of 0.9334. Additionally, the inter-story drift under design-based earthquakes can meet the requirements of the code.

Currently, the scope of this study is limited to the schematic design of RC shear wall structures. In the future, parametric modeling can be used to improve structural optimization algorithms. This will allow the proposed integrated design method to be used in the detailed design phase and take into account more design factors, such as structural stability. Additionally, the applicability of the proposed method to other material and structure types should be investigated further.

Author Contributions: Conceptualization, W.L. and X.L.; methodology, Y.F. and W.L.; software, S.Z., P.Y. and B.H.; investigation, Y.F., P.Z. and X.C.; writing—original draft preparation, Y.F.; writing—review and editing, W.L.; funding acquisition, X.L. and W.L. All authors have read and agreed to the published version of the manuscript.

Funding: This work is supported by the National Natural Science Foundation of China (52121005), the Science and Technology Program of the Ministry of Housing and Urban-Rural Development of the People's Republic of China (2022-K-073), the Tencent Foundation through the XPLORER PRIZE, and the Shuimu Tsinghua Scholar Program (2022SM005).

Institutional Review Board Statement: Not applicable.

Informed Consent Statement: Not applicable.

Data Availability Statement: The dataset used in this study is available at: https://github.com/wenjie-liao/StructGAN_v1 (accessed on 23 August 2022).

Acknowledgments: The authors would like to acknowledge Hui Cheng and Xingyu Wang (Central-South Architectural Design Institute), and Yitian Feng (Tsinghua University) for assisting in developing the cloud design platform.

Conflicts of Interest: The authors declare no conflict of interest.

References

1. Forcael, E.; Ferrari, I.; Opazo-Vega, A.; Pulido-Arcas, J.A. Construction 4.0: A literature review. *Sustainability* **2022**, *12*, 9755. [CrossRef]
2. Darko, A.; Chan, A.P.C.; Adabre, M.A.; Edwards, D.J.; Hosseini, M.R.; Ameyaw, E.E. Artificial intelligence in the AEC industry: Scientometric analysis and visualization of research activities. *Autom. Constr.* **2020**, *112*, 103081. [CrossRef]
3. Muñoz-La Rivera, F.; Mora-Serrano, J.; Valero, I.; Oñate, E. Methodological-technological framework for Construction 4.0. *Arch. Comput. Methods Eng.* **2021**, *28*, 689–711. [CrossRef]
4. Qian, J.; Zhao, Z.; Ji, X.; Ye, L. *Design of Tall Building Structures*; China Architecture & Building Press: Beijing, China, 2018. (In Chinese)
5. Aragaw, L.F.; Calvi, P.M. Comparing the performance of traditional shear-wall and rocking shear-wall structures designed using the direct-displacement based design approach. *Bull. Earthq. Eng.* **2020**, *18*, 1345–1369. [CrossRef]
6. Wang, L.; Shen, W.; Xie, H.; Neelamkavil, J.; Pardasani, A. Collaborative conceptual design—State of the art and future trends. *Comput. Aided Des.* **2002**, *34*, 981–996. [CrossRef]

7. Pizarro, P.N.; Hitschfeld, N.; Sipiran, I.; Saavedra, J.M. Automatic floor plan analysis and recognition. *Autom. Constr.* **2022**, *140*, 104348. [CrossRef]
8. Málaga-Chuquitaype, C. Machine learning in structural design: An opinionated review. *Front. Built Environ.* **2022**, *8*, 815717. [CrossRef]
9. Sun, H.; Burton, H.V.; Huang, H. Machine learning applications for building structural design and performance assessment: State-of-the-art review. *J. Build. Eng.* **2021**, *33*, 101816. [CrossRef]
10. Liao, W.J.; Lu, X.Z.; Huang, Y.L.; Zheng, Z.; Lin, Y.Q. Automated structural design of shear wall residential buildings using generative adversarial networks. *Autom. Constr.* **2021**, *132*, 103931. [CrossRef]
11. Pizarro, P.N.; Massone, L.M.; Rojas, F.R.; Ruiz, R.O. Use of convolutional networks in the conceptual structural design of shear wall buildings layout. *Eng. Struct.* **2021**, *239*, 112311. [CrossRef]
12. Liao, W.J.; Huang, Y.L.; Zheng, Z.; Lu, X.Z. Intelligent generative structural design method for shear-wall building based on "fused-text-image-to-image" generative adversarial networks. *Expert Syst. Appl.* **2022**, *210*, 118530. [CrossRef]
13. Zhao, P.J.; Liao, W.J.; Xue, H.J.; Lu, X.Z. Intelligent design method for beam and slab of shear wall structure based on deep learning. *J. Build. Eng.* **2022**, *57*, 104838. [CrossRef]
14. Lu, X.Z.; Liao, W.J.; Zhang, Y.; Huang, Y.L. Intelligent structural design of shear wall residence using physics-enhanced generative adversarial networks. *Earthq. Eng. Struct. Dyn.* **2022**, *51*, 1657–1676. [CrossRef]
15. Almasabha, G.; Alshboul, O.; Shehadeh, A.; Almuflih, A.S. Machine learning algorithm for shear strength prediction of short links for steel buildings. *Buildings* **2022**, *12*, 775. [CrossRef]
16. Zheng, H.; Moosavi, V.; Akbarzadeh, M. Machine learning assisted evaluations in structural design and construction. *Autom. Constr.* **2020**, *119*, 103346. [CrossRef]
17. Chang, K.H.; Cheng, C.Y. Learning to simulate and design for structural engineering. In Proceedings of the 37th International Conference on Machine Learning, Vienna, Austria, 12 July 2020; Available online: http://proceedings.mlr.press/v119/chang20a/chang20a.pdf (accessed on 23 August 2022).
18. Goodfellow, I.; Pouget-Abadie, J.; Mirza, M.; Xu, B.; Warde-Farley, D.; Ozair, S.; Courville, A.; Bengio, Y. Generative adversarial nets. In Proceedings of the Advances in Neural Information Processing Systems 27, Montréal, QB, Canada, 8 December 2014; Available online: https://proceedings.neurips.cc/paper/2014/file/5ca3e9b122f61f8f06494c97b1afccf3-Paper.pdf (accessed on 23 August 2022).
19. Sacks, R.; Barak, R. Impact of three-dimensional parametric modeling of buildings on productivity in structural engineering practice. *Autom. Constr.* **2008**, *17*, 439–449. [CrossRef]
20. Yu, R.; Gu, N.; Ostwald, M. Comparing designers' problem-solving behavior in a parametric design environment and a geometric modeling environment. *Buildings* **2013**, *3*, 621–638. [CrossRef]
21. Cavieres, A.; Gentry, R.; Al-Haddad, T. Knowledge-based parametric tools for concrete masonry walls: Conceptual design and preliminary structural analysis. *Autom. Constr.* **2011**, *20*, 716–728. [CrossRef]
22. Yuan, Z.; Sun, C.; Wang, Y. Design for manufacture and assembly-oriented parametric design of prefabricated buildings. *Autom. Constr.* **2018**, *88*, 13–22. [CrossRef]
23. Gan, V.J.L.; Wong, C.L.; Tse, K.T.; Cheng, J.C.P.; Lo, I.M.C.; Chan, C.M. Parametric modelling and evolutionary optimization for cost-optimal and low-carbon design of high-rise reinforced concrete buildings. *Adv. Eng. Inform.* **2019**, *42*, 100962. [CrossRef]
24. Khidmat, R.P.; Fukuda, H.; Kustiani. Design optimization of hyperboloid wooden house concerning structural, cost, and daylight performance. *Buildings* **2022**, *12*, 110. [CrossRef]
25. Pizarro, P.N.; Massone, L.M. Structural design of reinforced concrete buildings based on deep neural networks. *Eng. Struct.* **2021**, *241*, 112377. [CrossRef]
26. AutoCAD. NET Developer's Guide. Available online: http://docs.autodesk.com/ACD/2010/ENU/AutoCAD%20.NET%20Developer$'$s%20Guide/index.html (accessed on 13 July 2022).
27. Swallow (ESD). Introduction and Download of Swallow (ESD). Available online: https://www.food4rhino.com/en/app/swallowesd (accessed on 10 August 2022).
28. Wang, T.C.; Liu, M.Y.; Zhu, J.Y.; Tao, A.; Kautz, J.; Catanzaro, B. High-resolution image synthesis and semantic manipulation with conditional GANs. In Proceedings of the IEEE Conference on Computer Vision and Pattern Recognition, Salt Lake City, UT, USA, 18 June 2018. [CrossRef]
29. MOHURD. *Code for the Seismic Design of Buildings (GB50011-2010)*; China Architecture & Building Press: Beijing, China, 2010. (In Chinese)
30. Tafraout, S.; Bourahla, N.; Bourahla, Y.; Mebarki, A. Automatic structural design of RC wall-slab buildings using a genetic algorithm with application in BIM environment. *Autom. Constr.* **2019**, *106*, 102901. [CrossRef]
31. Luo, R.; Wang, Y.; Xiao, W.; Zhao, X. AlphaTruss: Monte Carlo tree search for optimal truss layout design. *Buildings* **2022**, *12*, 641. [CrossRef]
32. He, J.; Lin, S.; Li, Y.; Dong, X.; Chen, S. Genetic algorithm for optimal placement of steel plate shear walls for steel frames. *Buildings* **2022**, *12*, 835. [CrossRef]

buildings

MDPI

Article

Factors Influencing the Adoption of Blockchain in the Construction Industry: A Hybrid Approach Using PLS-SEM and fsQCA

Chunhao Li [1], Yuqian Zhang [1] and Yongshun Xu [2,*]

[1] School of Business and Management, Jilin University, Changchun 130012, China
[2] School of Civil Engineering and Architecture, Hainan University, Haikou 570228, China
* Correspondence: xys8023cm@163.com

Abstract: Blockchain is considered a breakthrough technology in the construction industry, with the potential to improve the trust environment and workflow of construction stakeholders. Although recent research offers hints regarding possible contributing elements to blockchain adoption in the construction industry, no specific study has addressed this topic. This knowledge gap hinders the adoption and promotion of blockchain in construction organizations. This study aimed to identify the determinants of blockchain adoption in the construction industry and verify the influence of the combination of various factors on adoption intention. Based on the technology–organization–environment framework, a conceptual model of blockchain adoption in the construction industry was constructed. Data were collected through the distribution of questionnaires, and 244 professionals in the construction field participated in this study. To evaluate the model hypotheses, we used a two-stage partial least squares structural equation modeling (PLS-SEM) and fuzzy-set qualitative comparative analysis (fsQCA) combination. The PLS-SEM revealed that factors such as compatibility, top management support, relative advantage, regulatory support, cost, competitive pressure, organizational readiness, and firm size significantly influence blockchain adoption. The fsQCA indicated that six causal conditions achieve high adoption intention. This is one of the first empirical studies on blockchain adoption in the construction industry, which can aid organizations, policymakers, and project participants in making informed decisions regarding the adoption of blockchain.

Keywords: blockchain; innovation adoption; construction industry; technology–organization–environment (TOE); PLS-SEM; fsQCA

Citation: Li, C.; Zhang, Y.; Xu, Y. Factors Influencing the Adoption of Blockchain in the Construction Industry: A Hybrid Approach Using PLS-SEM and fsQCA. *Buildings* **2022**, *12*, 1349. https://doi.org/10.3390/buildings12091349

Academic Editors: Hongling Guo, Jia-Rui Lin and Yantao Yu

Received: 25 July 2022
Accepted: 26 August 2022
Published: 1 September 2022

Publisher's Note: MDPI stays neutral with regard to jurisdictional claims in published maps and institutional affiliations.

Copyright: © 2022 by the authors. Licensee MDPI, Basel, Switzerland. This article is an open access article distributed under the terms and conditions of the Creative Commons Attribution (CC BY) license (https://creativecommons.org/licenses/by/4.0/).

1. Introduction

In recent years, as an emerging technology, blockchain has attracted the interest of practitioners and scholars from different industries [1–3]. Blockchain is essentially a decentralized database, a novel application paradigm that integrates computer technologies such as point-to-point transmission, consensus mechanisms, distributed data storage, and encryption algorithms [3,4]. Because blockchain has the advantages of multiparty maintenance, non-tampering, openness, transparency, auditability, and security, it has begun to subvert many traditional business processes [4]. As a breakthrough technology, blockchain provides valuable opportunities for companies and organizations. In particular, it is expected to address difficult problems in the construction industry, such as trust among stakeholders [5], delayed payment [6,7], poor bidding information channels, opaque transaction processes [8,9], unclear rights and responsibilities [10], and poor process traceability [11]. Shojaei et al. [12] evaluated the current implementation of a circular economy and highlighted blockchain as a potential technique in a built environment. Perera et al. [13] conducted a critical analysis of current information regarding blockchain technology and its applications, demonstrating that blockchain has significant promise in the construction industry. Hunhevicz and Hall [14] emphasized that blockchain can provide opportunities to integrate smart contracts and digital information into management, thereby enhancing collaboration among stakeholders in the construction industry.

However, although blockchain can provide many benefits to the construction industry and is recognized as a disruptive innovation technology that changes the industry [13], its adoption speed has not reached market expectations [11,15]. Construction companies are still hesitant and adopt wait-and-see attitudes regarding whether to adopt blockchain. Although some scholars have investigated the drivers and obstacles of blockchain adoption, few studies have explored the determinants of blockchain adoption in the construction industry [1,15]. In addition, although some scholars have conducted quantitative research, most have focused on supply chain management in non-construction fields [16,17]. In other words, organizations in the construction industry know little about adopting blockchain decisions. Therefore, a more in-depth research is required to identify the factors that impact blockchain adoption in the construction industry. More importantly, although the use of structural equation models and software technology to prioritize the predictive indicators of technology adoption is efficient and effective [18,19], no research has been conducted on the integration of these methods in the construction industry.

To fill the gap, this study aimed to achieve the following objectives: (1) use the partial least squares structural equation model (PLS-SEM) method to identify the determinants of blockchain adoption in the construction industry; (2) combine the fuzzy-set qualitative comparative analysis (fsQCA) approach and the PLS-SEM method to explore the synergistic effect among these determinants. To achieve the research goals, we extracted 11 factors from existing blockchain adoption studies. A theoretical model was established based on the technology–organization–environment (TOE) framework. Then, to test the theoretical model, an integration of the PLS-SEM and fsQCA methods was used. The methods are complementary because fsQCA offers an in-depth comprehension of the complicated, nonlinear, and synergistic effects, whereas PLS-SEM gives an explanation of the net effect of linear connections between variables. Moreover, we demonstrated that the intention for blockchain adoption in the construction industry can be encouraged by configuring many causal indicators rather than signal causal indicators. This research was the first attempt in the construction field to combine PLS-SEM and fsQCA technologies to identify the factors that affect blockchain adoption and provide richer insight into the effects of complex trade-offs. The results provide valuable insights for industry practitioners and decision-makers in related departments.

2. Literature Review

2.1. Blockchain Technology

Blockchain technology can be defined as an open, secure, and immutable distributed ledger [20]. It enables transactions without third-party involvement, eliminating the need for third-party trust. It is a decentralized network that runs on top of Internet protocols and records transactions in an immutable manner using cryptography and distributed consensus algorithms among a distributed set of users [21]. Suppliers and demanders can conduct peer-to-peer transactions using a blockchain. In a blockchain system, every transaction is recorded on a ledger and then placed into a block. Each block is connected to another block before and after it. When a block is linked to a chain, it becomes immutable. The blockchain is verified using automation and governance protocols, which cannot be changed or deleted by a single participant.

Depending on the type of access mechanism, blockchains can be broadly classified into permissionless and permissioned blockchains [22]. In the first type of blockchain, every transaction is public and users do not need permission to transact and reach a consensus. The users remain anonymous at all times and the public network encourages participation in the latex network through incentives. In the second type of blockchain, participants must be invited to join a network. Many private blockchains are permitted to control the types of users that can transact.

Owing to the characteristics of blockchain technology, its advantages are relatively significant. First, the blockchain is transparent [23]. Based on blockchain hashes, the transaction records of the participants can be checked in real time and cannot be forged. Second, blockchain reduces third-party dependencies in decentralized peer-to-peer network transactions [21]. Third, blockchain technology improves security. This establishes a consensus of trust in the entire network, making it difficult for hackers to penetrate the internal network. Simultaneously, the information recorded in the database is permanent and cannot be easily manipulated [24].

2.2. Blockchain in the Construction Industry

Blockchain has been studied by scholars in the construction industry since 2015. In recent years, most scholars have either conducted research on theoretical methods or conducted literature reviews, and few studies have explored blockchain adoption in the construction industry [25].

Several review papers have been published on this topic. Xu et al. [26] provided a comprehensive review of the application of blockchain technology in the construction field based on bibliometric and content analysis methods and discussed key research topics and future research directions. Perera et al. [13] analyzed the advantages and challenges of blockchain technology and concluded that it has compelling promise in the construction industry. Li et al. [11] identified the main research areas of blockchain in the built environment by presenting the latest blockchain technologies and conducting literature reviews and compiled an extensive list of the challenges and opportunities related to blockchain, in addition to forming a roadmap for implementing blockchain in the construction sector. Scott et al. [27] used an exploratory approach to examine 33 application categories of blockchain applications in construction, which were organized into seven thematic areas. Mahmudnia et al. [28] reviewed the characteristics of blockchain and explored its important role in solving interaction issues in payments, documentation, and interaction in the construction industry.

In addition, few studies on the potential advantages of blockchain have recently been conducted. For example, Qian and Papadonikolaki [5] suggested that blockchain can provide data tracking, transferring resources, and contracting in construction supply chain management. Meanwhile, some studies [7,8] focused on secure construction payments and indicated that the application of blockchain might create a transparent and efficient platform to guarantee secure payments for construction projects. According to Wang et al. [29], blockchain technology may improve traceability and make it easier for participants to share information during precast construction. Lee et al. [30] used a case study to demonstrate that integrating digital twins with blockchain can aid in ensuring traceability.

Despite extensive research and the rapid spread of blockchain technology in the construction industry, many challenges and barriers remain to its adoption. Sharma and Kumar [31] argued that in the early stages of adopting blockchain technology, inadequate knowledge and experience are key challenges that must be addressed. Xu et al. [32] indicated that barriers to blockchain adoption in the construction industry are prominent, centered on insufficient information technology infrastructure and legal and regulatory ambiguity. Yang et al. [9] indicated that the fragmentation and uncertainty of construction projects complicate the widespread adoption of blockchain technology. Tezel et al. [33] and Toufaily et al. [1] stated that construction companies lack the IT infrastructure and servers required for blockchain applications. Compared with the significant potential of blockchain technology, its application and research are still in the preliminary stage. Overall, the widespread use of blockchain technology has not yet occurred in the construction industry.

2.3. Adoption Model

Owing to the various determinants that may affect innovation adoption in a wide variety of domains, various theoretical models have been presented to investigate and comprehend the adoption of innovation in organizations [34]. As a generic theory used for innovation adoption, TOE theory has guided scholars in identifying and determining the drivers of innovative technologies [35]. From a business development perspective, TOE indicates that a company's decision to adopt new technology is based on technological characteristics and organizational and environmental considerations. The context of technology describes the technical characteristics that can influence the adoption of innovation, the organizational context relates to the organizational attributes that may hinder or foster adoption, and the context of the environment refers to external factors relative to the organization, which may present opportunities and challenges for innovation adoption [36,37]. In several studies on construction innovation adoption, the three contexts of the TOE framework have focused on identifying the factors affecting new technology adoption by construction companies [38].

In the subsequent analysis, the TOE framework served as an overarching theoretical underpinning for this research. It presents an extensive analysis of technology adoption as decisions to adopt technology in the organizational dimension depend on factors in the context of technology, environment, and organization. Specifically, because the TOE framework combines human and non-human factors into one framework, it has better advantages than other traditional models, such as the technology acceptance model, diffusion of innovation, and unified theory of acceptance model [39]. It provides a suitable foundation for considering and understanding the appropriate determinants for the adoption of innovation, and many of the results of innovation adoption studies support this [40]. Wong et al. [17] mentioned that the TOE framework can be used to better examine blockchain adoption in organizations.

3. Model Construction and Hypotheses Development

The construction industry is generally regarded as structurally fragmented, with low productivity and a lack of improvements. Blockchain exhibits the basic properties of traceability, transparency, and immutability. Therefore, it can facilitate a paradigm shift towards cooperation and trust in the construction industry. With researchers developing blockchain-based solutions, specific issues in the construction industry are being addressed, such as construction quality, supply chain management, and construction payments. Based on the above literature review and analysis, a conceptual model for this study is proposed. It considers both the factors of technology and organization and the influence of the external environment. The framework includes 12 different constructs, with the willingness to adopt blockchain serving as the dependent variable, and the 11 determinants that are considered as independent variables being determinants in the TOE context. Figure 1 shows the proposed model.

3.1. Context of Technology

3.1.1. Relative Advantage

The degree to which the adoption of innovation may provide an organization with greater benefits than the status quo is described as a relative advantage [37]. Relative advantage is considered a fundamental indicator of innovation adoption [41], as observed for supply chains [42], cloud computing services [43], and business intelligence systems [44]. The construction industry is considered to be in a state of continual reengineering [45], and the adoption of new technologies will promote industry flourishing. As a primary use of blockchain technology, smart contracts may efficiently resolve construction payment delays and contractual disputes [46]. Blockchain technology can also provide construction companies with trusted partnerships, information sharing during the design and construction phases, foster collaboration, enhance traceability and transparency, reduce transaction costs, and address late payment challenges, thereby improving operational and management efficiencies [13]. Consequently, we propose the following hypothesis:

H1: *Relative advantage positively influences the willingness of blockchain adoption.*

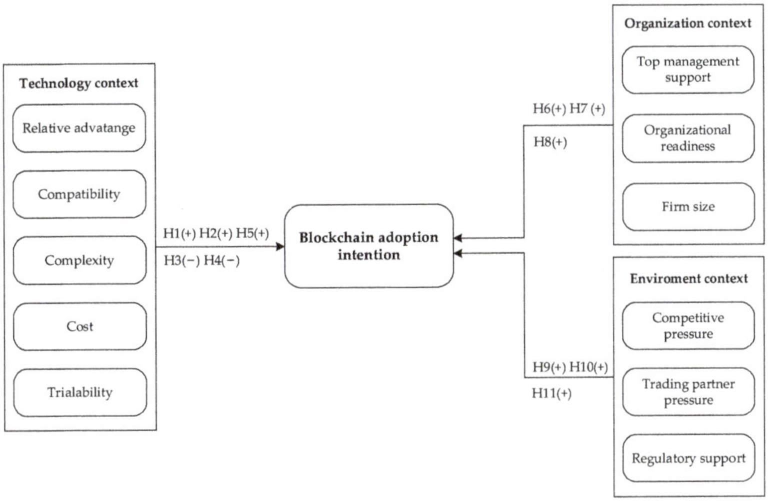

Figure 1. Research model of intention to adopt blockchain technology based on the TOE framework.

3.1.2. Compatibility

The extent to which an innovation system is considered compatible with the present system is referred to as compatibility [47]. Compatibility between the innovation and management requirements, as well as corporate culture and practices, is widely recognized as a critical factor in innovation adoption [37,48]. Fernando et al. [35] indicated that as the main motivation for technology adoption. Construction projects are highly fragmented and uncertain, and engineering changes during project implementation are common. Accordingly, construction companies are more likely to embrace and implement blockchain technology in various aspects of their operations if they believe that blockchain adoption is compatible with the current corporate culture and business practices. Thus, we formulate the following hypothesis:

H2: *Compatibility positively influences the willingness of blockchain adoption.*

3.1.3. Complexity

Complexity is the extent to which an innovation is difficult to comprehend and apply [49]. According to some studies, complexity is considered a key factor affecting innovation adoption [50]. Complexity is not positively associated with technology adoption as other elements of technology adoption, but rather negatively [51]. The blockchain's transaction mechanism is relatively complex and speed is a major problem to be considered; in addition, its implementation is challenged by its immature security properties [52]. Construction has long been considered a poorly performing and low-tech industry for innovation [53], and construction companies do not intend to adopt blockchain because of technical complexity. As construction companies move from traditional IT systems to blockchain-based ones, complex programming, integration challenges, and a lack of blockchain technology talent can hinder their adoption [54]. When the technology is complex, decision-makers consider whether to adopt it. Thus, construction companies have limited utility in participating in blockchain technology unless it can be readily incorporated into current building operating systems. Consequently, we propose the following hypothesis:

H3: *Complexity negatively influences the willingness of blockchain adoption.*

3.1.4. Cost

Various costs are associated with technology adoption, and costs influence the willingness to adopt technology. High costs discourage adoption [55]. The cost of obtaining and using blockchain technology is referred to as the cost of adopting blockchain [17]. Although digital technology has many benefits, its adoption in the construction industry remains quite low [56]. High costs are often a barrier for companies adopting new technologies [57]. Blockchain adoption requires the acquisition of the necessary hardware and software, which may be expensive for organizations. More importantly, several non-empirical studies on blockchain technology in the construction sector have indicated that cost is a significant factor that prevents construction companies from adopting blockchain technology [10]. Thus, the following hypothesis is proposed:

H4: *Cost negatively influences the willingness of blockchain adoption.*

3.1.5. Trialability

Trialability is the extent to which new technology can be attempted on a limited basis [58]. The likelihood of successful adoption increases when the organization has had the opportunity to test innovation before it is adopted [46]. Research has shown that trialability facilitates the successful adoption of innovations [47,59]. A high degree of trialability would make it less risky for companies to adopt the technology, which can increase the level of acceptance. Moreover, trialability is promising to enable companies to better understand the potential benefits and accurately determine the value of blockchain technology. Thus, we propose the following hypothesis:

H5: *Trialability positively influences the willingness of blockchain adoption.*

3.2. *Context of Organization*

3.2.1. Top Management Support

Top management support is the degree to which top management in an organization accepts and implements new technology [60]. The early adoption of blockchain inevitably encounters resistance, and top management support can motivate members of an organization by providing direction and satisfying the demand for resources and funding. Many studies related to the construction industry have highlighted the importance of adopting new technologies [61]. Top management support is essential for integrating emerging technologies into existing business processes to facilitate the learning and dissemination of innovative technologies [59]. Therefore, we propose the following hypothesis:

H6: *Top management support positively influences the willingness of blockchain adoption.*

3.2.2. Organizational Readiness

Organizational readiness is the capacity and intention of firms to adopt an innovation [62]. It denotes business management and investment readiness to invest in innovation technology, including cognitive readiness, resource readiness, and IT systems [59]. Pan and Pan [36] reported that organizational readiness positively influences the adoption of construction innovation. The awareness of change, financial resources, expertise, and technical capabilities of construction companies are the fundamental bases for ensuring the adoption and implementation of blockchain. Thus, we propose the following hypothesis:

H7: *Organizational readiness positively influences the willingness of blockchain adoption.*

3.2.3. Firm Size

Firm size is a critical condition in innovation adoption [55]. Many studies have indicated that firm size positively affects and controls the innovation process [35,40]. The adoption of blockchain technology involves a change from old to new systems and requires a large initial investment, the risks and costs of which may deter many small construction companies, whereas larger firms can often manage the costs of innovation and provide financial resources that occur in technology adoption. Meanwhile, larger firms have more skilled professionals to ensure that the implementation of innovation is smooth [63]. Accordingly, we propose the following hypothesis:

H8: *Firm size positively influences the willingness of blockchain adoption.*

3.3. Context of Environment

3.3.1. Competitive Pressure

Competitive pressure is the degree to which a company experiences pressure from competitors in the same field [17]. Intense competition among peers requires organizations to adopt innovation to improve quality, reduce costs, and increase effectiveness and efficiency [35]. As an emerging technology, blockchain can help early adopters to thrive in today's ultra-competitive market. The construction industry is competitive and fraught with challenges [64]. Competitive pressure is likely to increase construction companies' demand for blockchain technology, driving the aggressive adoption of blockchain. Thus, the following hypothesis is proposed:

H9: *Competitive pressure positively influences the willingness of blockchain adoption.*

3.3.2. Trading Partner Pressure

Trading partners influence the construction industry as project-based groups [65].

Pressure from partners has been proved to be a main factor in innovation adoption in various empirical investigations [40]. Badi et al. [46] indicated the beneficial role of partners in facilitating the adoption, implementation, and completion of projects. Wamba et al. [66] built a model to study the various factors influencing blockchain adoption in supply chain management, demonstrating that trading partner pressure significantly influenced blockchain adoption in India and the US. To facilitate collaboration between trading partners, construction companies would further decide whether to adopt blockchain technology, depending on whether blockchain is used by trading partners. Thus, the following hypothesis is proposed:

H10: *Trading partner pressure positively influences the willingness of blockchain adoption.*

3.3.3. Regulatory Support

Regulatory support is assistance offered by the government or its authority to encourage innovation adoption [37]. Regulatory policies and legislation, such as required rules or standards, have a crucial role in enabling blockchain implementation [57,61]. Interestingly, Gibbs and Kraemer [67] emphasized that regulatory support has a greater role in developing countries than in developed countries. Most studies suggest that social acceptance is a significant barrier to blockchain applications [68]. China is a developing country and the role of government regulations and guidance is critical for innovation adoption. Because of the novelty of blockchain technology, most construction companies have a wait-and-see attitude in the early stages of adoption, and blockchain adoption systems in construction would be further hindered by government regulation of what and how to regulate the process of adoption. Thus, we propose the following hypothesis:

H11: *Regulatory support positively influences the willingness of blockchain adoption.*

4. Research Design and Methodology

4.1. Measurement of Determinants

Owing to its accessibility and scientific nature, questionnaire research has been extensively adopted by researchers in the field of construction. To guarantee validity and reliability, we designed the questionnaire for this study by referring to well-established scales and reviewing the results of published literature in the context of blockchain adoption and the construction industry. Accordingly, three construction experts were invited to pretest the preliminary version of the questionnaire. The questionnaire was modified and utilized in the pilot research based on their feedback. Meanwhile, it was translated into Chinese by a language expert, considering that the study addressed blockchain adoption in the Chinese construction industry. Three other experts assessed the translated version to confirm that the content of the questionnaire was related to construction companies.

A questionnaire with 43 construct items was used to measure the variables within the TOE framework. To facilitate the judgment of respondents, we scored the questionnaire items on a five-point Likert scale, with values ranging from 1 (strongly disagree) to 5 (strongly agree). Respondents made judgments and decisions based on their experiences and were informed that there were no right or wrong answers and that they were used for academic research only. Technological factors included compatibility, relative advantages, cost, trialability, and complexity. Accordingly, the measurements of these five variables were adapted [17,36,46]. In addition, three organizational constructs—top management support, organizational readiness, and firm size—were adapted from [35,36,46]. Environmental factors included regulatory support, trading partner pressure, and competitive pressure. These three items were adapted from [17,46,63]. Appendix A presents a set of questions for each construct.

Moreover, this study used three variables, namely, years of experience, education, and job position. The years of experience and education could suggest a level of professionalism, which may be related to higher levels of self-efficacy [69], that has the potential to affect the dependent variable. Job position may influence the attitudes and behaviors of participants [70].

4.2. Sample and Data Collection

Because blockchain is an emerging technology in China's construction industry, this study adopted a snowball sampling technique to obtain more valid and extensive responses. The questionnaires were distributed in WeChat groups of relevant conferences and forums of engineering management majors, focusing on topics related to intelligent construction, digital transformation of construction companies, and blockchain, in which the participants were more aware and concerned about emerging technologies in the construction industry. We selected eligible experts, senior managers, directors, and chief executive officers in the conferences and forum as our key informants and they were encouraged to forward the questionnaire. To further increase the enthusiasm of the respondents, the research group promised to send the research conclusions to the respondents in the form of a report after the end of the study to enable the respondents to adopt and apply blockchain technology in the entire construction industry. Due to the pandemic of COVID-19, the survey lasted five months in total, and finally, 244 valid surveys were received after removing disqualified questionnaires, such as partial answers and nonsensical responses.

The demographic information of the participants is presented in Table 1. To test for non-response bias, we used a t-test to compare the early and late participants. The findings demonstrated that there were no significant differences between the two groups, indicating that nonresponse bias was not a problem in this study. Statistically, we used Harman's single-factor test to evaluate the common method bias (CMB) problem [71], which accounts for the vast majority of the model variance. Because only a signal factor accounted for 38.2%, which was less than 50%, the result revealed no substantial CMB. Additionally, each correlation coefficient was less than 0.90, which also indicated that there was no issue with CMB [17]. Meanwhile, variance inflation factors (VIFs) were used to check for multicollinearity. All VIFs in this study were lower than the threshold of 5, indicating that linear correlation was not a problem.

Table 1. Profiles of questionnaire participants.

Demographic	Categories	Frequency	Percentage (%)
Years of work experience	<5 years	11	4.5
	5–9 years	88	36.0
	10–15 years	89	36.5
	>15 years	56	23
Education	High school degree or below	3	1.2
	College degree	103	42.2
	Undergraduate degree	117	48.0
	Graduate degree	21	8.6
Job position	Senior manager	44	18.0
	Department manager	53	21.7
	Project manager	56	23.0
	Chief engineer	58	23.8
	Other	33	13.5
Employee number	Less than 100	56	23.0
	100–200	100	41.0
	More than 200	88	36.0

4.3. Analytical Approaches

A multiple-method approach was applied to validate the proposed model. First, data retrieved from the questionnaire were analyzed using PLS-SEM via SmartPLS software to test the model and hypotheses, and the Statistical Package for Social Sciences (SPSS) was used for descriptive analysis. SEM employs a confirmatory approach to analyze the phenomenon-based structure and could account for the measurement error, thereby providing valid conclusions on the structural patterns of multiple indicator variables than other analytical approaches such as linear regression [63]. PLS-SEM was chosen for this study for the following reasons: (1) PLS-SEM is a contemporary multivariate analytic approach that is capable of estimating theoretically proven causality models; (2) PLS-SEM is more favorable than covariance-based structural equation modeling techniques to determine the connection variance between dependent and independent variables [36]; (3) PLS-SEM is more suitable for research involving non-normally distributed data, such as the data of this study; and (4) PLS-SEM has been applied to solve construction management problems in recently published articles [72]. The data were then analyzed in two steps using SmartPLS. To ensure the goodness of the model, stage one examined the measurement model to determine its validity and reliability. Subsequently, the proposed hypotheses were tested using a bootstrapping procedure in phase two.

Second, fsQCA was conducted to obtain knowledge of the components that constitute adequate combinations for blockchain adoption. fsQCA leverages Boolean logic to uncover several paths that result in a common outcome [73]. It is an asymmetric approach different from traditional symmetric approaches, such as regression and structural equation modeling, which only permit the analysis of a single path of antecedent factors. Although synergies exist in the factors influencing blockchain adoption, using fsQCA can capture decision-making complexity in construction companies. The following phases were included in the modeling process when using the fsQCA software: Phase one involved calibrating the data from the survey into a fuzzy set (0 to 1) with three main points: full set membership, crossover point, and full non-membership. Phase two was the analysis of the necessary condition, which identified the determinants that may influence the achievement of the target outcome. Subsequently, a truth table algorithm was constructed to draw the study's suggested conclusion in the third phase.

5. Results

5.1. Results of PLS-SEM

5.1.1. Measurement Model

The quality of the research model was assessed in terms of convergent validity, reliability, and discriminant validity. These three aspects are discussed next, as suggested in [36] and [63]. To verify convergent validity, we applied the average variance extracted (AVE) and factor loadings. Table 2 shows that every AVE was above the 0.5 benchmark, and all factor loadings were above the 0.7 benchmark, which indicated satisfactory convergent validity. The constructs' reliability was examined by jointly analyzing Cronbach's alpha and composite construct reliability (CR). The tests both had acceptable values exceeding 0.7, as shown in Table 2, indicating that the reliability of the construct was validated. The cross-loadings and Fornell–Larcker criteria were used to estimate the discriminant validity. Figure 2 shows that the correlation coefficients were less than the square root of the AVE, which satisfied the requirements of the Fornell–Larcker criterion. Additionally, all cross-loadings were below each construct loading, which satisfied the cross-loading criterion in this study, and discriminant validity was further established. The results of validity and reliability are presented in Figure 3.

Table 2. Convergent validity and reliability results.

Constructs	Items	Loadings	Cronbach's α	CR	AVE
BI	BI1	0.926	0.889	0.931	0.819
	BI2	0.915			
	BI3	0.872			
RA	RA1	0.816	0.873	0.905	0.613
	RA2	0.833			
	RA3	0.748			
	RA4	0.747			
	RA5	0.786			
	RA6	0.762			
CB	CB1	0.897	0.897	0.929	0.765
	CB2	0.872			
	CB3	0.875			
	CB4	0.854			
CX	CX1	0.927	0.937	0.955	0.841
	CX2	0.920			
	CX3	0.921			
	CX4	0.902			
CT	CT1	0.916	0.942	0.959	0.853
	CT2	0.941			
	CT3	0.909			
	CT4	0.928			
TA	TA1	0.823	0.782	0.866	0.685
	TA2	0.908			
	TA3	0.743			
TMS	TMS1	0.902	0.912	0.938	0.790
	TMS2	0.880			
	TMS3	0.898			
	TMS4	0.875			
OR	OR1	0.862	0.863	0.907	0.709
	OR2	0.854			
	OR3	0.845			
	OR4	0.805			
FS	FS1	0.920	0.890	0.932	0.820
	FS2	0.915			
	FS3	0.881			

Table 2. *Cont.*

Constructs	Items	Loadings	Cronbach's α	CR	AVE
CP	CP1	0.741	0.799	0.869	0.624
	CP2	0.819			
	CP3	0.824			
	CP4	0.771			
TPP	TPP1	0.874	0.821	0.894	0.738
	TPP2	0.904			
	TPP3	0.797			
RS	RS1	0.828	0.848	0.898	0.687
	RS2	0.843			
	RS3	0.825			
	RS4	0.819			

Notes: BI: behavioral intention; RA: relative advantage; CB: compatibility; CX: complexity; CT: cost; TA: trialability; TMS: top management support; OR: organizational readiness; FS: firm size; CP: competitive pressure; TPP: trading partner pressure; RS: regulatory support.

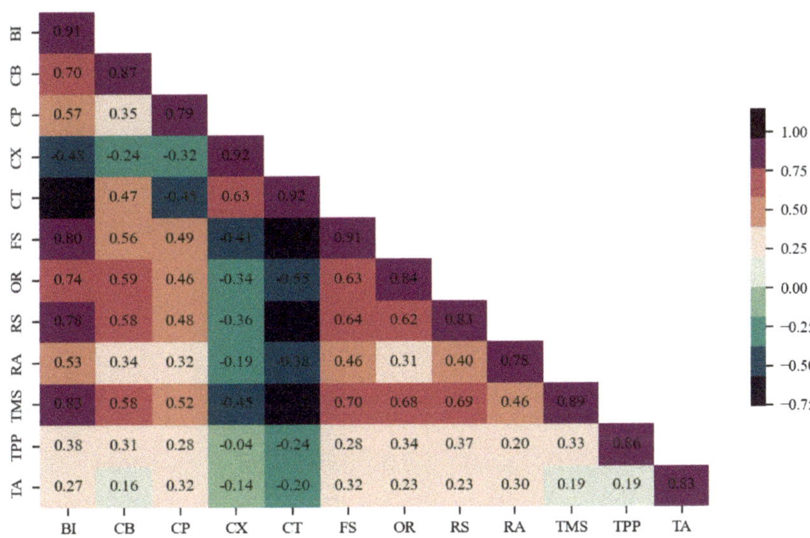

Figure 2. Fornell and Larcker criterion for discriminant validity results.

5.1.2. Structural Model

In this step, the proposed hypothesized relationships were examined using a bias-corrected bootstrap procedure with 5000 subsamples. As Table 3 and Figure 4 show, complexity, trialability, and trading partner pressure were non-significant factors at the 0.05 level. Among the remaining eight accepted relationships, compatibility, top management support, regulatory support, organizational readiness, relative advantage, firm size, and competitive pressure significantly positively influenced construction companies' willingness to adopt blockchain technology, whereas cost was negatively correlated with adoption. In addition, the R^2 value of 0.88 in Figure 4 indicates that the entire research model fits the survey data well. Moreover, we assessed the control variables' relevance by adding them separately to a model that includes all main variables. According to the results, none of the control variables exert a significant effect on blockchain adoption, thus all the control variables were excluded from the final model.

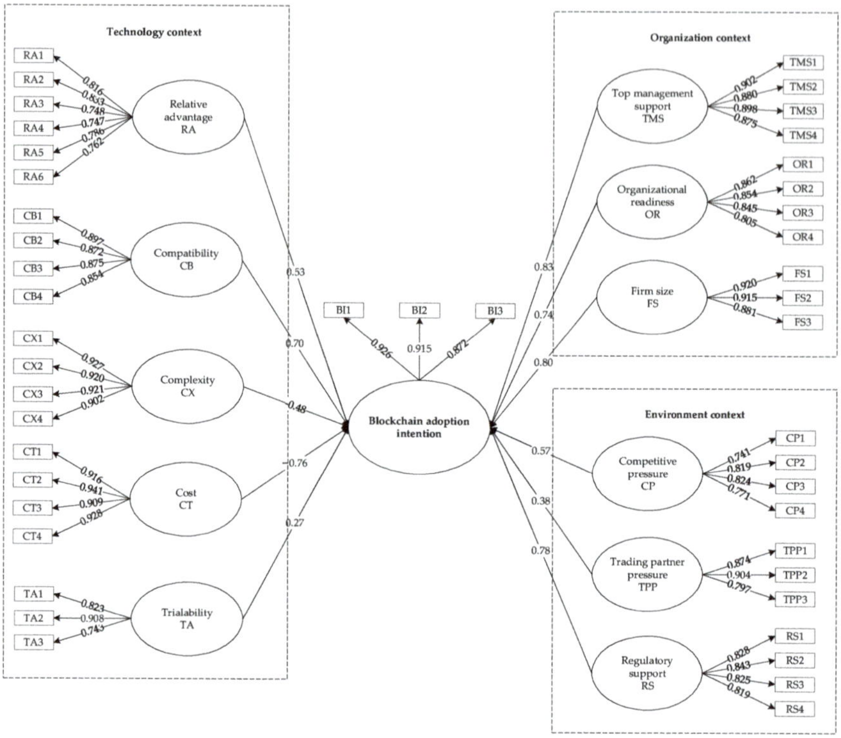

Figure 3. Measurement model.

Table 3. Path analysis results.

Hypotheses (Effects)	Hypothesis	Path Coefficient	t-Value	Conclusions
H1 (+)	RA → BI	0.102	3.960 ***	Supported
H2 (+)	CB → BI	0.166	4.314 ***	Supported
H3 (−)	CX → BI	−0.013	0.459	Not Supported
H4 (−)	CT → BI	−0.205	3.605 ***	Supported
H5 (+)	TA → BI	−0.013	0.579	Not Supported
H6 (+)	TMS → BI	0.207	4.208 ***	Supported
H7 (+)	OR → BI	0.112	2.489 **	Supported
H8 (+)	FS → BI	0.170	3.406 **	Supported
H9 (+)	CP → BI	0.064	2.367 **	Supported
H10 (+)	TPP → BI	0.030	1.175	Not Supported
H11 (+)	RS → BI	0.155	3.679 ***	Supported

Notes: * $p < 0.05$; ** $p < 0.01$; *** $p < 0.001$.

5.2. Results of fsQCA

The fsQCA, as a supplementary analysis, was further used to investigate the synergistic impact of numerous factors that may influence the willingness of construction companies to adopt blockchain. Some steps were necessary to perform an fsQCA analysis. The first step was data calibration. Ordinary data must be transformed into fuzzy sets with three meaningful thresholds: setting the original values from Likert scales to full membership, crossover anchors, and full non-membership. The 5th, 50th, and 95th percentiles were respectively used to show the level of membership among variables [74]. After calibration, the necessary conditions analysis (NCA) was performed to detect the conditions that might influence the achievement of the desired result. Table 4 lists the calibration and NCA results. Because all the consistency scores were below 0.90 as presented in Table 4, none of the conditions are necessary for high levels of blockchain adoption. In the next step, a truth table was constructed based on consistency and frequency. The number of cases threshold was set to five [75], and the lowest acceptable observation consistency was set at 0.9 [63].

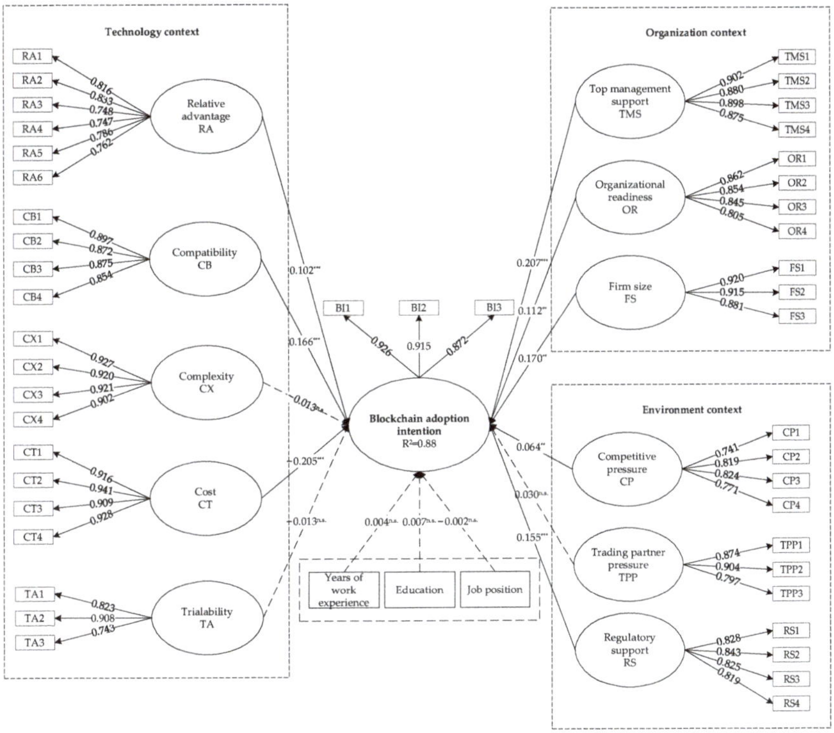

Figure 4. Structural path diagram for the hypothesized relationships. Notes: * $p < 0.05$; ** $p < 0.01$; *** $p < 0.001$; n.s. = not significant.

Table 4. Calibration thresholds of the measures and causal necessary conditions test.

Construct	Full Membership	Cross-Over	Full Non-Membership	Consistency	Coverage
RA	4.33	3.50	2.50	0.768	0.775
CB	4.50	3.25	2.00	0.868	0.836
CX	4.75	3.25	2.00	0.515	0.544
CT	4.46	3.00	2.00	0.380	0.459
TA	4.67	3.67	2.33	0.692	0.703
TMS	4.00	3.25	2.00	0.863	0.853
FS	4.33	3.33	2.00	0.867	0.899
OR	4.00	3.25	2.00	0.856	0.830
CP	4.33	3.00	2.00	0.760	0.807
TPP	4.25	3.50	2.04	0.765	0.717
RS	4.28	3.33	1.72	0.884	0.856

A parsimonious solution, a complex solution, and an intermediate solution were used through the fsQCA software. Considering the superiority over the other two solutions [75], intermediate solutions were selected to conduct the fsQCA analysis of high blockchain adoption intention. Two combinations of causal conditions resulting in a high willingness to adopt blockchain are listed in Table 5. According to the raw coverage, solution 1 accounts for 21.7%, and solution 2 accounts for 28.7% of cases associated with the outcome. The overall solution coverage in Table 5 indicates that the two solutions covered 34.6% of cases that had the willingness to adopt blockchain. The consistent values for the two solutions are above 0.8, which means the two solutions have achieved technology innovation with sufficient consistency [76]. Factors including compatibility, top management support, relative advantage, regulatory support, firm size, and organizational readiness are considered the main conditions for blockchain adoption because they appeared in both solutions, which indicated that these six factors together strengthen adoption willingness. Solution 1 suggested that high blockchain adoption can be attained through the six main conditions listed above: low levels of complexity and cost and low levels of trialability and trading partner pressure. Solution 2 revealed that the six main conditions listed above—low levels of complexity and cost, combined with trialability, trading partner pressure, and competitive pressure—can achieve high blockchain adoption.

Table 5. Configurations for high blockchain adoption intention.

Configuration	Solution	
	1	2
The context of technology		
RA	•	•
CB	•	•
CX	O	O
CT	O	O
TA	O	•
The context of organization		
TMS	•	•
OR	•	•
FS	•	•
The context of environment		
CP	O	•
TPP	O	•
RS	•	•
Consistency	0.999	0.999
Raw coverage	0.217	0.287
Unique coverage	0.059	0.130
Configuration	**Solution**	
Overall solution coverage	0.346	
Overall solution consistency	0.9995	

Notes: The black circles "•" indicate the presence of an element. The black circles "•" indicate the presence of an auxiliary condition. The circle "O" represents the absence of an element.

5.3. Comparing PLS-SEM and fsQCA Results

The PLS-SEM analysis indicated that compatibility, top management support, relative advantage, regulatory support, cost, firm size, organizational readiness, and competitive pressure can significantly influence the intention of blockchain adoption for construction companies in order of decreasing influence, whereas complexity, trialability, and trading partner pressure can inhibit the intention to adopt. The fsQCA results indicated that compatibility, top management support, relative advantage, regulatory support, firm size, and organizational readiness are considered core elements for adoption because the six variables mentioned above were included in both solutions. These results indicated that the fsQCA and PLS-SEM analyses agreed.

However, some differences were observed between the fsQCA and PLS-SEM analyses. FsQCA complemented the PLS-SEM analysis by revealing more than one complex configuration of antecedents to achieve a high adoption of blockchain. Corresponding to the concept of causal asymmetry, fsQCA indicated that factors such as trialability, competitive pressure, and trading partner pressure have opposite impacts on the willingness to adopt blockchain, based on how they are combined or interact with other attributes. For example, solution 1 indicated that, although the level of complexity, cost, trialability, competitive pressure, and trading partner pressure is low, the six core conditions can increase organizations' willingness to adopt blockchain. Similarly, solution 2 indicated that trialability, competitive pressure, and trading partner pressure as the auxiliary conditions can contribute to the high intention of blockchain adoption, as long as the level of complexity and cost is low and the levels of six core elements are high.

6. Discussion

This study identified significant factors of blockchain adoption across technological, organizational, and environmental dimensions, which can provide the foundation for promoting blockchain adoption in the construction industry. Consequently, these key results contribute to a deeper understanding of blockchain adoption in the construction industry. By combining PLS-SEM with fsQCA, we also gained a better understanding of the overall adoption process. According to an evaluation of the research model, compatibility, top management support, and relative advantage were observed to be the top three important determinants to influence the intention of blockchain adoption, whereas complexity, trialability, and trading partner pressure received no meaningful statistical support at a significant level. Next, all findings related to the hypotheses are discussed.

Within the technology dimension framework, relative advantage (H1) is positively correlated with construction companies' intentions to adopt blockchain. Previous studies on innovation adoption also support this conclusion [17,36]. For example, relative advantages such as enabling efficiency gains, cost reductions, instant tracking and tracing of assets, and automated contract enforcement have been demonstrated to be potential benefits in the construction industry. Compatibility (H2) has a considerable impact on the adoption of blockchain by construction companies. Several studies supported this finding [35,77]. Blockchain adoption would be facilitated if the existing business operating model of an organization is compatible with blockchain technology. The internal systems of the construction industry are complex, and if the blockchain application matches the existing information infrastructure, construction companies would be more active in implementing blockchain. Studies have shown that the effect of complexity (H3), which was considered in earlier studies, does not have a significantly negative influence on blockchain adoption [36,78]. The relationship between the complexity and intention to adopt blockchain technology was not supported by the data we collected. Although somewhat unusual, this insignificant relationship may be due to the following reasons. On one hand, most construction companies have no ability to develop blockchain technology on their own, and they purchase it directly from high-tech companies. To some extent, companies do not care much about the complexity of blockchain but more about its usefulness, highlighting the significance of the relative advantage. On the other hand, with the advent of the digital age, construction companies could acquire technology in multiple ways, and technical barriers no longer play a crucial role in a company's competitive advantage. Cost (H4) has a significantly negative impact on the intention to adopt blockchain, which is consistent with the findings of earlier research that identified high costs as a primary barrier to innovation adoption [17,36]. The construction industry is a collaborative stakeholder with a complex network of relationships. A significant advantage of blockchain technology is the elimination of third-party-related costs in the network. Additionally, based on interviews with professionals working in the construction industry, it is anticipated that blockchain technology will reduce the costs associated with data processing and management by 70% through the automation of compliance checks, payments, and project performance analyses [5]. Even with such cost savings, the adoption of blockchain will increase hardware and facility costs, and the costs of operation and maintenance will remain significant. Construction companies may make comprehensive decisions to weigh the costs of blockchain adoption against the cost savings. Many construction companies are reluctant to adopt blockchain technology, considering the large initial investment and uncertainty on whether the expected benefits could be achieved. Trialability (H5) was confirmed to have no effect on blockchain adoption for construction companies, which was consistent with the findings of [46] and [77]. The present relative immaturity of blockchain technology could explain this result. The construction industry is recognized for its lack of innovation, and anxiety about using emerging technology systems could discourage the adoption of blockchain technology. Although trialability was not a significant factor in this study, its significance may change if blockchain technology is applied more broadly in the construction industry.

Within the organizational context framework, this study revealed that top management support (H6) has a crucial role in the adoption of blockchain in construction organizations. It has been an essential component of the implementation of various technological advancements [36,47]. It had the second-highest path coefficient value of all the examined factors, confirming the significance of top management in innovation adoption in construction. The significant effect of organizational readiness (H7) on blockchain adoption corresponds with the findings of [35]. Blockchain adoption by construction companies that lack sufficient technical, financial, and trained human resources may be challenging. A company may not implement blockchain technology if it does not have the necessary resources and competencies. Corresponding with the findings of [40] and [35], firm size (H8) emerged as a critical factor affecting adoption. Combined with the characteristics of the construction industry, larger firms intend to adopt blockchain technology because their capabilities and sources are sufficient to utilize and implement blockchain technology.

Regarding the framework of the environmental context, competitive pressure (H9) was confirmed to have a significant positive impact on the blockchain adoption intentions of construction companies. Competitive pressure has been shown to be a crucial facilitator of technology adoption across a broad variety of businesses, as previous studies have shown [35,46]. Firms under intense competition are more likely to use and implement blockchain technology to increase their market share. A construction company may explore effective strategies to gain a long-term competitive edge. This shows that competition exists and that the ability to remain at the forefront of technical advancement influences decisions. Trading partner pressure (H10) was confirmed to have no significant effect on blockchain adoption, similar to the study conducted in [36]. This may be because blockchain is still a relatively new technology, with most companies involved being start-ups. Chinese construction companies currently have minimal adoption of blockchain, and it is difficult to assess the difficulty of using blockchain technology in new construction projects. It is impossible for trading partners to fully adopt blockchain within a short period. Consequently, companies are less sensitive to pressure from trading partners in the construction industry. Regulatory support (H11) has a significantly positive influence on blockchain adoption, similar to the conclusions of [63] and [78]. The significance of this link results from the fact that construction companies consider the adoption of blockchain technology to be a large investment, and regulatory support is essential for legitimizing adoption and implementation. The reason for this significant relationship is that construction companies consider blockchain adoption a significant investment and regulatory support as necessary to ensure smooth implementation across the board.

6.1. Theoretical Implications

As blockchain research is still in its early phases in terms of empirical testing, theoretical processes, and methodological diversity, this paper provides timely important theoretical and methodological contributions. Although there are many previous qualitative studies on blockchain in construction [11,27], there is a lack of quantitative studies in this investigation. Based on the theoretical perspective of the TOE framework and empirical evidence from Chinese construction companies, the results of this study provide researchers, practitioners, and policymakers with relevant guidance for the construction industry by investigating the relationship between various determinants and the willingness to adopt blockchain. The results of the study revealed that factors such as compatibility, top management support, relative advantage, regulatory support, cost, competitive pressure, organizational readiness, and firm size significantly influence the intention of blockchain adoption whereas complexity, trialability, and trading partner pressure have no effect. These findings are in line with those reported by the vast majority of earlier studies on the process of technology innovation diffusion [36,46]. The inconsistency of these findings with the results of previous studies about the application of blockchain to other fields reflected the characteristics of the construction industry [17,35]. More importantly, a comprehensive analysis of the fsQCA and PLS-SEM results deepens our understanding of the adoption process. Furthermore, this research can provide organizations or businesses with a clearer picture of the factors influencing blockchain adoption, which can enhance the transformational capabilities of construction companies and provide insights into the impact of emerging technologies on the construction industry.

6.2. Practical Implications

With the rapid development of smart construction, the construction industry has experienced unprecedented disruptive innovation in recent years. Blockchain is recognized as an emerging technology that promises to solve pain points in the construction industry. It promises to have a significant impact on operations, trust management among stakeholders, and business processes. This study evaluated various factors that influence the intent to adopt blockchain, and our findings have many practical implications. Overall, these conclusions can help practitioners make better blockchain adoption decisions. This study evaluated the causes and situations that drive the migration of the construction industry to blockchain.

In the context of technology, construction companies should actively recognize the benefits of blockchain as the first step in its adoption. Our findings suggest that the relative advantages that blockchain brings to construction companies can incentivize them to adopt blockchain. The positive correlation between the comparative advantage and adoption intentions can help decision-makers recognize the value of blockchain in construction businesses; moreover, compatibility is a significant predictor of blockchain adoption. If blockchain technology is compatible with a firm's business philosophy and operating system, it contributes to the smooth operation of the firm. Compatibility: Our findings suggest that compatibility has a greater impact on the willingness to adopt blockchain than top management. For strategic deployment, executives first assess a company's compatibility before adopting blockchain technology. Additionally, cost is a key concern for construction companies, and can influence their willingness to adopt blockchain technology. Although the cost is negatively correlated with the willingness to adopt blockchain, construction companies still comprehensively decide whether to adopt blockchain technology in a dialectical and systematic way.

In the organizational dimension, the attitude of the top management is critical to the willingness of construction companies to adopt blockchain. Our findings suggest that, for senior managers to conclude that adopting blockchain will bring relative advantages, they should focus on improving the technical capabilities of R and D personnel to meet the company's needs for blockchain technology. Blockchain adoption is not a simple technology implementation process and involves various aspects of organizational readiness to create a foundation for adoption and application. Construction companies should focus on improving their technological capabilities to successfully implement blockchain adoption. More scientists with relevant knowledge must be recruited to facilitate the implementation of blockchain. Large-scale firms are more willing to change their adoption of new technologies than smaller ones. Large-scale companies can use blockchain for business expansion to solve their business challenges. Even with the uncertainties and risks associated with blockchain technology adoption, large construction companies are actively adopting big data analytics to gain a competitive advantage and open up new business opportunities.

In an environmental context, competitive pressure compels the construction industry to adopt blockchain technology to enhance the strength of companies. Learning and applying blockchain technology is gaining popularity as companies seek to gain a competitive advantage over their competitors. In an increasingly competitive market, our findings suggest that companies should adopt blockchain. Regulatory support is indispensable, and the improvement of relevant laws and regulations will provide effective protection for the construction industry to adopt blockchain. Governments play a pivotal role in encouraging the development of new technologies, and the adoption of blockchain technology requires the support of government entities.

7. Conclusions and Limitations

Based on the TOE framework, this paper attempted to fill a knowledge gap by identifying the determinants of blockchain adoption and presenting an empirical foundation for future blockchain adoption in the construction sector. The survey data for this study were obtained from Chinese construction companies, and 11 components in three different contexts were examined using a hybrid approach of PLS-SEM and fsQCA. The PLS-SEM findings show that factors such as compatibility, top management support, relative advantage, regulatory support, cost, firm size, organizational readiness, and competitive pressure significantly influence adoption. In addition, three factors (complexity, trialability, and trading partner pressure), because the relevant hypotheses did not receive support from the evidence, were confirmed to have no statistically significant influence. Further research on these three factors is required to obtain a clearer understanding. From the fsQCA results, a combination of compatibility, relative advantage, top management support, regulatory support, organizational readiness, and firm size achieves the highest level of blockchain adoption intention. These findings are intended to assist researchers, developers, and decision-makers in better comprehending the key blockchain adoption factors in the construction industry and addressing negative adoption factors more effectively.

This study had some limitations. First, the data were cross-sectional rather than longitudinal and the sample size and diversity were limited. However, the adoption of blockchain is a dynamic process and blockchain technology is constantly evolving. Thus, more studies should be conducted to extend the generalization of the findings in the current study. For example, involving more stakeholders in the progress of adopting emerging technologies; using different research methods, such as using multiple case studies or a system dynamics approach to replicate this study and verify the findings obtained from it. Second, except for those identified in this study, the research cannot be exhaustive because of many other technical, organizational, and environmental factors that may influence blockchain adoption; more precisely, some of the factors cannot be used to differentiate the construction industry from others. In the future, we will consider factors that are more specific to the construction industry, such as security and privacy issues. Furthermore, this study only considered the impact of factors on adoption decisions, and linkages may exist between factors, for example, whether the factors in the environmental dimension moderate the factors in the dimensions of technology and organization to impact adoption decisions. Additionally, the sample used in this study was from China. Because of the cultural differences between China and other countries, the study model should be further examined and contrasted using samples from other nations to offer more credible support for the hypotheses.

Author Contributions: Conceptualization, C.L. and Y.X.; methodology, Y.X. and Y.Z.; software, Y.Z.; validation, Y.Z. and Y.X.; formal analysis, Y.Z.; investigation, Y.Z. and Y.X.; resources, Y.Z.; data curation, Y.Z.; writing—original draft preparation, Y.Z.; writing—review and editing, Y.X.; visualization, Y.Z.; supervision, C.L. All authors have read and agreed to the published version of the manuscript.

Funding: This research was supported by the Humanities and Social Science Foundation of Ministry of Education of China under Grant No. 20YJA630028.

Institutional Review Board Statement: Not applicable.

Informed Consent Statement: Not applicable.

Data Availability Statement: The data presented in this research can be requested from the corresponding author.

Conflicts of Interest: The authors declare no conflict of interest.

Appendix A

Table A1. Survey items.

Construct	Measurement Items	Adapted From
Relative advantage	Adopting blockchain can enable my company to accomplish project tasks more efficiently and effectively Adopting blockchain can enhance the traceability of my company's projects Adopting blockchain can increase the transparency of my company's projects Blockchain can increase trust among stakeholders in construction Adopting blockchain can improve deferred payment issues	[17]
Compatibility	Blockchain can provide privacy protection and security of my company Blockchain is compatible with the business operating model in my company Blockchain is compatible with the management requirements of the company Blockchain fits with the existing values of my company Blockchain is compatible with my company's existing infrastructure	[36,46]
Complexity	Blockchain would be too complex for my company to use Learning how to use blockchain in my company is not easy It will take considerable time and effort for my company to learn how to use blockchain My company believes that blockchain adoption requires many skills	[17,36]
Cost	Adopting blockchain in my company will increase the cost of facility and hardware Adopting blockchain in my company will increase the cost of operations and maintenance The cost of adopting blockchain will be expensive for my company The cost of adopting blockchain is unknown and difficult to comprehend	[17,36]
Trialability	My company intends to try out some blockchain technology in a small scope before fully adopting and implementing it A trial period before blockchain adoption will reduce risks The ability to experiment with blockchain adoption is critical in deciding whether to adopt it	[46]
Top management support	Top management in my company will be responsive and attentive to blockchain adoption Top management in my company could take the risks associated with blockchain adoption My top management will provide the necessary human resources, finances and materials for blockchain adoption My top management will look at blockchain as strategically important	[35,46]
Organizational readiness	My company has resources necessary to use blockchain My company has possessed the necessary expertise and skills to adopt blockchain The technology staff in the company have the sufficient experience and skills to conduct the adoption of blockchain My company's existing technologies support blockchain adoption	[36,46]
Firm size	My company's capital is higher than others in the construction industry My company's revenue is higher than others in the construction industry My company has more competent staff than others in the construction industry	[35,36]
Competitive pressure	The adoption of blockchain will offer my company a stronger competitive advantage My company believes it is important to adopt blockchain to be competitive My company is forced to adopt blockchain due to competitive pressure My company believes that competitors have recently started exploring blockchain technology	[17,63]

Table A1. *Cont.*

Construct	Measurement Items	Adapted From
Regulatory support	The government or competent agencies provide financial assistance for blockchain development The government or relevant authorities provide technical guidance for adopting blockchain technology Blockchain technology can be implemented with the current set of laws and regulations Government encourages the adoption of blockchain in procurement and projects	[17]
Trading partner pressure	My company's major trading partners recommend blockchain adoption My company's major trading partners encourage blockchain adoption My company's major trading partners request blockchain adoption	[46,63]
Behavioral intention	My company intends to adopt blockchain technology actively in the future My company intends to digitally transform management My company is willing to utilize blockchain technology in various projects	[17,46]

References

1. Toufaily, E.; Zalan, T.; Ben Dhaou, S. A framework of blockchain technology adoption: An investigation of challenges and expected value. *Inf. Manag.* **2021**, *58*, 103444. [CrossRef]
2. Ali, O.; Ally, M.; Clutterbuck; Dwivedi, Y. The state of play of blockchain technology in the financial services sector: A systematic literature review. *Int. J. Inf. Manag.* **2020**, *54*, 102199. [CrossRef]
3. Lu, Y. The blockchain: State-of-the-art and research challenges. *J. Ind. Inf. Integr.* **2019**, *15*, 80–90. [CrossRef]
4. Casino, F.; Dasaklis, T.K.; Patsakis, C. A systematic literature review of blockchain-based applications: Current status, classification and open issues. *Telemat. Inform.* **2019**, *36*, 55–81. [CrossRef]
5. Qian, X.A.; Papadonikolaki, E. Shifting trust in construction supply chains through blockchain technology. *Eng. Constr. Arch. Manag.* **2020**, *28*, 584–602. [CrossRef]
6. Das, M.; Luo, H.; Cheng, J.C. Securing interim payments in construction projects through a blockchain-based framework. *Autom. Constr.* **2020**, *118*, 103284. [CrossRef]
7. Ahmadisheykhsarmast, S.; Sonmez, R. A smart contract system for security of payment of construction contracts. *Autom. Constr.* **2020**, *120*, 103401. [CrossRef]
8. Chong, H.-Y.; Diamantopoulos, A. Integrating advanced technologies to uphold security of payment: Data flow diagram. *Autom. Constr.* **2020**, *114*, 103158. [CrossRef]
9. Yang, R.; Wakefield, R.; Lyu, S.; Jayasuriya, S.; Han, F.; Yi, X.; Yang, X.; Amarasinghe, G.; Chen, S. Public and private blockchain in construction business process and information integration. *Autom. Constr.* **2020**, *118*, 103276. [CrossRef]
10. Sheng, D.; Ding, L.; Zhong, B.; Love, P.E.; Luo, H.; Chen, J. Construction quality information management with blockchains. *Autom. Constr.* **2020**, *120*, 103373. [CrossRef]
11. Li, J.; Greenwood, D.; Kassem, M. Blockchain in the built environment and construction industry: A systematic review, conceptual models and practical use cases. *Autom. Constr.* **2019**, *102*, 288–307. [CrossRef]
12. Shojaei, A.; Ketabi, R.; Razkenari, M.; Hakim, H.; Wang, J. Enabling a circular economy in the built environment sector through blockchain technology. *J. Clean. Prod.* **2021**, *294*, 126352. [CrossRef]
13. Perera, S.; Nanayakkara, S.; Rodrigo, M.; Senaratne, S.; Weinand, R. Blockchain technology: Is it hype or real in the construction industry? *J. Ind. Inf. Integr.* **2020**, *17*, 100125. [CrossRef]
14. Hunhevicz, J.J.; Hall, D.M. Do you need a blockchain in construction? Use case categories and decision framework for DLT design options. *Adv. Eng. Inform.* **2020**, *45*, 101094. [CrossRef]
15. McNamara, A.J.; Sepasgozar, S.M. Intelligent contract adoption in the construction industry: Concept development. *Autom. Constr.* **2021**, *122*, 103452. [CrossRef]
16. Kouhizadeh, M.; Saberi, S.; Sarkis, J. Blockchain technology and the sustainable supply chain: Theoretically exploring adoption barriers. *Int. J. Prod. Econ.* **2021**, *231*, 107831. [CrossRef]
17. Wong, L.-W.; Leong, L.-Y.; Hew, J.-J.; Tan, G.W.-H.; Ooi, K.-B. Time to seize the digital evolution: Adoption of blockchain in operations and supply chain management among Malaysian SMEs. *Int. J. Inf. Manag.* **2020**, *52*, 101997. [CrossRef]
18. Yadegaridehkordi, E.; Nilashi, M.; Shuib, L.; Nasir, M.H.N.B.M.; Asadi, S.; Samad, S.; Awang, N.F. The impact of big data on firm performance in hotel industry. *Electron. Commer. Res. Appl.* **2020**, *40*, 100921. [CrossRef]
19. Leong, L.-Y.; Hew, T.-S.; Ooi, K.-B.; Wei, J. Predicting mobile wallet resistance: A two-staged structural equation modeling-artificial neural network approach. *Int. J. Inf. Manag.* **2020**, *51*, 102047. [CrossRef]
20. Choi, T.-M. Blockchain-technology-supported platforms for diamond authentication and certification in luxury supply chains. *Transp. Res. Part E Logist. Transp. Rev.* **2019**, *128*, 17–29. [CrossRef]
21. Crosby, M.; Pattanayak, P.; Verma, S.; Kalyanaraman, V. Blockchain technology: Beyond bitcoin. *Appl. Innov. Rev.* **2016**, *2*, 71. Available online: http://blossom.informatik.uni-rostock.de/id/eprint/58 (accessed on 21 August 2022).

22. Fernández-Caramés, T.M.; Fraga-Lamas, P. Towards the Internet of Smart Clothing: A Review on IoT Wearables and Garments for Creating Intelligent Connected E-Textiles. *Electronics* **2018**, *7*, 405. [CrossRef]
23. Kamath, R. Food Traceability on Blockchain: Walmart's Pork and Mango Pilots with IBM. *J. Br. Blockchain Assoc.* **2018**, *1*, 3712. [CrossRef]
24. Kouhizadeh, M.; Zhu, Q.; Sarkis, J. Blockchain and the circular economy: Potential tensions and critical reflections from practice. *Prod. Plan. Control* **2020**, *31*, 950–966. [CrossRef]
25. Elghaish, F.; Hosseini, M.R.; Matarneh, S.; Talebi, S.; Wu, S.; Martek, I.; Poshdar, M.; Ghodrati, N. Blockchain and the 'Internet of Things' for the construction industry: Research trends and opportunities. *Autom. Constr.* **2021**, *132*, 103942. [CrossRef]
26. Xu, Y.; Chong, H.-Y.; Chi, M. Blockchain in the AECO industry: Current status, key topics, and future research agenda. *Autom. Constr.* **2022**, *134*, 104101. [CrossRef]
27. Scott, D.J.; Broyd, T.; Ma, L. Exploratory literature review of blockchain in the construction industry. *Autom. Constr.* **2021**, *132*, 103914. [CrossRef]
28. Mahmudnia, D.; Arashpour, M.; Yang, R. Blockchain in construction management: Applications, advantages and limitations. *Autom. Constr.* **2022**, *140*, 104379. [CrossRef]
29. Wang, Z.; Wang, T.; Hu, H.; Gong, J.; Ren, X.; Xiao, Q. Blockchain-based framework for improving supply chain traceability and information sharing in precast construction. *Autom. Constr.* **2020**, *111*, 103063. [CrossRef]
30. Lee, D.; Lee, S.H.; Masoud, N.; Krishnan, M.S.; Li, V.C. Integrated digital twin and blockchain framework to support accountable information sharing in construction projects. *Autom. Constr.* **2021**, *127*, 103688. [CrossRef]
31. Sharma, M.G.; Kumar, S. The Implication of Blockchain as a Disruptive Technology for Construction Industry. *IIM Kozhikode Soc. Manag. Rev.* **2020**, *9*, 177–188. [CrossRef]
32. Xu, Y.; Chong, H.-Y.; Chi, M. Modelling the blockchain adoption barriers in the AEC industry. *Eng. Constr. Arch. Manag.* 2021; *ahead of print*. [CrossRef]
33. Tezel, A.; Papadonikolaki, E.; Yitmen, I.; Hilletofth, P. Preparing construction supply chains for blockchain technology: An investigation of its potential and future directions. *Front. Eng. Manag.* **2020**, *7*, 547–563. [CrossRef]
34. Hameed, M.A.; Counsell, S.; Swift, S. A conceptual model for the process of IT innovation adoption in organizations. *J. Eng. Technol. Manag.* **2012**, *29*, 358–390. [CrossRef]
35. Fernando, Y.; Rozuar, N.H.M.; Mergeresa, F. The blockchain-enabled technology and carbon performance: Insights from early adopters. *Technol. Soc.* **2021**, *64*, 101507. [CrossRef]
36. Pan, M.; Pan, W. Understanding the determinants of construction robot adoption: Perspective of building contractors. *J. Constr. Eng. Manag.* **2020**, *146*, 04020040. [CrossRef]
37. Oliveira, T.; Thomas, M.; Espadanal, M. Assessing the determinants of cloud computing adoption: An analysis of the manufacturing and services sectors. *Inf. Manag.* **2014**, *51*, 497–510. [CrossRef]
38. Taylor, J.E. Antecedents of Successful Three-Dimensional Computer-Aided Design Implementation in Design and Construction Networks. *J. Constr. Eng. Manag.* **2007**, *133*, 993–1002. [CrossRef]
39. Awa, H.O.; Uko, J.P.; Ukoha, O. An Empirical Study of Some Critical Adoption Factors of ERP Software. *Int. J. Hum. Comput. Interact.* **2017**, *33*, 609–622. [CrossRef]
40. Wang, Y.-M.; Wang, Y.-S.; Yang, Y.-F. Understanding the determinants of RFID adoption in the manufacturing industry. *Technol. Forecast. Soc. Chang.* **2010**, *77*, 803–815. [CrossRef]
41. Kapoor, K.K.; Dwivedi, Y.K.; Williams, M.D. Rogers' Innovation Adoption Attributes: A Systematic Review and Synthesis of Existing Research. *Inf. Syst. Manag.* **2014**, *31*, 74–91. [CrossRef]
42. Kshetri, N. 1 Blockchain's roles in meeting key supply chain management objectives. *Int. J. Inf. Manag.* **2018**, *39*, 80–89. [CrossRef]
43. Gutierrez, A.; Boukrami, E.; Lumsden, R. Technological, organisational and environmental factors influencing managers' decision to adopt cloud computing in the UK. *J. Enterp. Inf. Manag.* **2015**, *28*, 788–807. [CrossRef]
44. Puklavec, B.; Oliveira, T.; Popovič, A. Understanding the determinants of business intelligence system adoption stages: An empirical study of SMEs. *Ind. Manag. Data Syst.* **2018**, *118*, 236–261. [CrossRef]
45. Begić, H.; Galić, M. A Systematic Review of Construction 4.0 in the Context of the BIM 4.0 Premise. *Buildings* **2021**, *11*, 337. [CrossRef]
46. Badi, S.; Ochieng, E.; Nasaj, M.; Papadaki, M. Technological, organisational and environmental determinants of smart contracts adoption: UK construction sector viewpoint. *Constr. Manag. Econ.* **2021**, *39*, 36–54. [CrossRef]
47. Maroufkhani, P.; Tseng, M.-L.; Iranmanesh, M.; Ismail, W.K.W.; Khalid, H. Big data analytics adoption: Determinants and performances among small to medium-sized enterprises. *Int. J. Inf. Manag.* **2020**, *54*, 102190. [CrossRef]
48. Thiesse, F.; Staake, T.; Schmitt, P.; Fleisch, E. The rise of the "next-generation bar code": An international RFID adoption study. *Supply Chain Manag. Int. J.* **2011**, *16*, 328–345. [CrossRef]
49. Corrocher, N. The diffusion of Internet telephony among consumers and firms: Current issues and future prospects. *Technol. Forecast. Soc. Chang.* **2003**, *70*, 525–544. [CrossRef]
50. Asiaei, A.; Rahim, N.Z.A. A multifaceted framework for adoption of cloud computing in Malaysian SMEs. *J. Sci. Technol. Policy Manag.* **2019**, *10*, 708–750. [CrossRef]
51. Alshamaila, Y.; Papagiannidis, S.; Li, F. Cloud computing adoption by SMEs in the north east of England: A multi-perspective framework. *J. Enterp. Inf. Manag.* **2013**, *26*, 250–275. [CrossRef]

52. Saberi, S.; Kouhizadeh, M.; Sarkis, J.; Shen, L. Blockchain technology and its relationships to sustainable supply chain management. *Int. J. Prod. Res.* **2019**, *57*, 2117–2135. [CrossRef]
53. Martínez-Román, J.A.; Tamayo, J.A.; Gamero, J. Innovativeness and its influence on growth and market extension in construction firms in the Andalusian region. *J. Eng. Technol. Manag.* **2017**, *43*, 19–33. [CrossRef]
54. Sahebi, I.G.; Masoomi, B.; Ghorbani, S. Expert oriented approach for analyzing the blockchain adoption barriers in humanitarian supply chain. *Technol. Soc.* **2020**, *63*, 101427. [CrossRef]
55. Lin, H.-F. Understanding the determinants of electronic supply chain management system adoption: Using the technology–organization–environment framework. *Technol. Forecast. Soc. Chang.* **2014**, *86*, 80–92. [CrossRef]
56. Elrefaey, O.; Ahmed, S.; Ahmad, I.; El-Sayegh, S. Impacts of COVID-19 on the Use of Digital Technology in Construction Projects in the UAE. *Buildings* **2022**, *12*, 489. [CrossRef]
57. Shi, P.; Yan, B. Factors affecting RFID adoption in the agricultural product distribution industry: Empirical evidence from China. *SpringerPlus* **2016**, *5*, 2029. [CrossRef]
58. Laurell, C.; Sandström, C.; Berthold, A.; Larsson, D. Exploring barriers to adoption of Virtual Reality through Social Media Analytics and Machine Learning—An assessment of technology, network, price and trialability. *J. Bus. Res.* **2019**, *100*, 469–474. [CrossRef]
59. Ramdani, B.; Chevers, D.; Williams, D.A. SMEs' adoption of enterprise applications: A technology-organisation-environment model. *J. Small Bus. Enterp. Dev.* **2013**, *20*, 735–753. [CrossRef]
60. Ifinedo, P. Impacts of business vision, top management support, and external expertise on ERP success. *Bus. Process Manag. J.* **2008**, *14*, 551–568. [CrossRef]
61. Ozorhon, B.; Oral, K. Drivers of Innovation in Construction Projects. *J. Constr. Eng. Manag.* **2017**, *143*, 04016118. [CrossRef]
62. Gangwar, H. Understanding the determinants of big data adoption in India: An analysis of the manufacturing and services sectors. *Inf. Resour. Manag. J. IRMJ* **2018**, *31*, 1–22. [CrossRef]
63. Sun, S.; Hall, D.J.; Cegielski, C.G. Organizational intention to adopt big data in the B2B context: An integrated view. *Ind. Mark. Manag.* **2020**, *86*, 109–121. [CrossRef]
64. Pan, M.; Linner, T.; Pan, W.; Cheng, H.; Bock, T. A framework of indicators for assessing construction automation and robotics in the sustainability context. *J. Clean. Prod.* **2018**, *182*, 82–95. [CrossRef]
65. Sepasgozar, S.M.E.; Davis, S.R.; Li, H.; Luo, X. Modeling the Implementation Process for New Construction Technologies: Thematic Analysis Based on Australian and U.S. Practices. *J. Manag. Eng.* **2018**, *34*, 05018005. [CrossRef]
66. Wamba, S.F.; Queiroz, M.M.; Trinchera, L. Dynamics between blockchain adoption determinants and supply chain performance: An empirical investigation. *Int. J. Prod. Econ.* **2020**, *229*, 107791. [CrossRef]
67. Gibbs, J.L.; Kraemer, K.L. A Cross-Country Investigation of the Determinants of Scope of E-commerce Use: An Institutional Approach. *Electron. Mark.* **2004**, *14*, 124–137. [CrossRef]
68. Cheng, M.; Liu, G.; Xu, Y.; Chi, M. When Blockchain Meets the AEC Industry: Present Status, Benefits, Challenges, and Future Research Opportunities. *Buildings* **2021**, *11*, 340. [CrossRef]
69. Farashah, A.D.; Thomas, J.; Blomquist, T. Exploring the value of project management certification in selection and recruiting. *Int. J. Proj. Manag.* **2019**, *37*, 14–26. [CrossRef]
70. Chi, M.; Chong, H.-Y.; Xu, Y. The effects of shared vision on value co-creation in megaprojects: A multigroup analysis between clients and main contractors. *Int. J. Proj. Manag.* **2022**, *40*, 218–234. [CrossRef]
71. Podsakoff, P.M.; MacKenzie, S.B.; Lee, J.Y.; Podsakoff, N.P. Common method biases in behavioral research: A critical review of the literature and recommended remedies. *J. Appl. Psychol.* **2003**, *885*, 879–903. [CrossRef]
72. Zeng, N.; Liu, Y.; Gong, P.; Hertogh, M.; König, M. Do right PLS and do PLS right: A critical review of the application of PLS-SEM in construction management research. *Front. Eng. Manag.* **2021**, *8*, 356–369. [CrossRef]
73. Woodside, A.G. Moving beyond multiple regression analysis to algorithms: Calling for adoption of a paradigm shift from symmetric to asymmetric thinking in data analysis and crafting theory. *J. Bus. Res.* **2013**, *66*, 463–472. [CrossRef]
74. Chuah, S.H.-W.; Tseng, M.-L.; Wu, K.-J.; Cheng, C.-F. Factors influencing the adoption of sharing economy in B2B context in China: Findings from PLS-SEM and fsQCA. *Resour. Conserv. Recycl.* **2021**, *175*, 105892. [CrossRef]
75. Fiss, P.C. Building Better Causal Theories: A Fuzzy Set Approach to Typologies in Organization Research. *Acad. Manag. J.* **2011**, *54*, 393–420. [CrossRef]
76. Ganter, A.; Hecker, A. Configurational paths to organizational innovation: Qualitative comparative analyses of antecedents and contingencies. *J. Bus. Res.* **2014**, *67*, 1285–1292. [CrossRef]
77. Besklubova, S.; Skibniewski, M.J.; Zhang, X. Factors Affecting 3D Printing Technology Adaptation in Construction. *J. Constr. Eng. Manag.* **2021**, *147*, 04021026. [CrossRef]
78. Lai, Y.; Sun, H.; Ren, J. Understanding the determinants of big data analytics (BDA) adoption in logistics and supply chain management: An empirical investigation. *Int. J. Logist. Manag.* **2018**, *29*, 676–703. [CrossRef]

buildings

MDPI

Article

Expanding Domain Knowledge Elements for Metro Construction Safety Risk Management Using a Co-Occurrence-Based Pathfinding Approach

Na Xu [1], Bo Zhang [2,*], Tiantian Gu [1], Jie Li [3] and Li Wang [1]

1 School of Mechanics and Civil Engineering, China University of Mining and Technology, Xuzhou 221000, China
2 School of Computer Science and Technology, China University of Mining and Technology, Xuzhou 221000, China
3 School of Civil Engineering, Neijiang Normal Universities, Neijiang 641100, China
* Correspondence: 4090@cumt.edu.cn

Abstract: Knowledge is a contribution factor leading to more effective and efficient construction safety management. Metro construction practitioners always find it difficult to determine what specialized knowledge is needed in order to lead to better safety risk management. Currently, domain knowledge elements are generally determined by experts, which is coarse-grained and uncomprehensive. Therefore, this paper aims to provide a structure of domain knowledge elements, using an automatic approach to expand domain knowledge elements (DKEs) from a big dataset of unstructured text documents. First, the co-word co-occurrence network (CCN) was used to find the connected knowledge elements, and then the association rule mining (ARM) was compiled to prune the weakly related subnetworks, leaving the strong associated elements. Finally, a list of DKEs in the metro construction safety risk management was obtained. The result shows that the obtained DKEs are more comprehensive and valuable compared to previous studies. The proposed approach provides an automatic way to expand DKEs from a small amount of known knowledge, minimizing the expert bias. This study also contributes to building a fine-grained knowledge structure for metro construction safety risk management. The structure can be used to guide safety training and help knowledge-based safety risk management.

Keywords: metro construction project; safety risk management; knowledge expansion; co-occurrence analysis; association rule mining

Citation: Xu, N.; Zhang, B.; Gu, T.; Li, J.; Wang, L. Expanding Domain Knowledge Elements for Metro Construction Safety Risk Management Using a Co-Occurrence-Based Pathfinding Approach. *Buildings* **2022**, *12*, 1510. https://doi.org/10.3390/buildings12101510

Academic Editors: Hongling Guo, Jia-Rui Lin and Yantao Yu

Received: 3 August 2022
Accepted: 20 September 2022
Published: 22 September 2022

Publisher's Note: MDPI stays neutral with regard to jurisdictional claims in published maps and institutional affiliations.

Copyright: © 2022 by the authors. Licensee MDPI, Basel, Switzerland. This article is an open access article distributed under the terms and conditions of the Creative Commons Attribution (CC BY) license (https://creativecommons.org/licenses/by/4.0/).

1. Introduction

Metro construction has presented a powerful momentum for rapid economic development worldwide [1]. Due to various uncertainty factors, especially complex underground geological conditions, metro construction is inherently complicated and high-risk. Many safety accidents and near-miss events are related to ineffective risk management [2,3], resulting in serious social impact, many casualties, and huge economic losses [4]. It is therefore significant to improve the safety risk management in metro construction to avoid and reduce safety accidents and near-miss events.

The architecture, engineering, and construction (AEC) industry is becoming increasingly knowledge-driven and information-intensive [5], especially for metro construction safety risk management, owing to its characteristics of complexity. Research and practice (such as by health and safety executives in the UK) have indicated that up to 80% of accidents are attributed to the actions or omissions of people [6,7]. One of the main causes behind unsafe behaviors is a lack of domain knowledge [8,9], i.e., specific subject-matter knowledge [10]. Hence, increasing domain knowledge leads to the promotion of the level of safety management and a decrease in the rate of accidents and injuries.

Owing to the one-off nature of metro construction projects and organizations, knowledge learning is required every day to meet the demands of addressing new issues. The term domain knowledge elements (DKEs) (e.g., soil mixing wall and shield cut) refers to knowledge that has complete logic and cannot be divided [11]. DKEs can be used as the domain-particular concepts (i.e., knowledge structure) in a domain knowledge graph (DKG), which underpins the knowledge base of safety risk management [12]. However, study on DKEs has received little attention in the literature and practice. Metro construction practitioners always find it difficult to determine what specialized knowledge is needed to lead to better performance and fewer errors.

Since domain knowledge is immense and new DKEs are generated every now and then, the huge mass of information has resulted in a form of information overload for safety risk managers. It is easy to list some DKEs, but it is hard to list all DKEs and update the list in time. Information in text format remains a greatly underutilized form of knowledge in the metro construction safety risk management domain. Hundreds of research works are uploaded to public websites every day, containing rich but unstructured domain knowledge elements. Consequently, this situation calls for an automatic approach for expanding immense unknown DKEs from a small set of existing DKEs.

The main contributions of this work are summarized as follows:

1. Practically, we built a fine-grained knowledge structure for metro construction safety risk management. The structure can be used to guide safety training and help the construction of domain knowledge graphs, etc.
2. Theoretically, we propose an automatic approach to expand domain knowledge elements from massive documents, minimizing the expert bias.

In the following sections, a literature review is first given on related research. Then, a pathfinding approach of knowledge expansion is proposed, followed by a step-by-step experiment and its results. Lastly, conclusions are drawn, informing the reader of opportunities for future research.

2. Literature Review

2.1. Knowledge-Based Safety Risk Management in Metro Construction

Knowledge management research in the AEC industry has significantly blossomed in the last two decades [13]. Since ineffective risk management in metro construction projects is partly due to a lack of knowledge [14], knowledge-based safety risk management is becoming an important method for risk prevention and mitigation. Substantial progress has been made recently. Studies can be separated into three major categories.

The first category centers around using effective knowledge management to increase organization performance regarding safety risk management, such as using knowledge sharing to improve the safety climate [15], exploring knowledge transfer factors to benefit cooperation networks [16], and using a knowledge dynamics-integrated map to clarify the fluidity of knowledge through the risk management process [17].

The second category focuses on knowledge-based intelligent systems to implement safety risk management processes, including automatic risk identification, supervision, and warning. For instance, Ding et al. developed a safety risk identification system for metro construction from construction drawings [18]. Zhong et al. extracted safety risk factors from construction specifications and developed an ontology-based system to match the potential hazards implied in photography images [19]. Current research mainly extracts specified knowledge units related to risk factors and their attributes, e.g., in Ref. [19] construction equipment and its quality, materials, and bearing were identified and extracted as knowledge units.

The third category explores domain knowledge elements. As the core component of knowledge-based systems, a knowledge base is a warehouse of domain-specific knowledge [20]. Four types of knowledge elements were mentioned that lead to successful projects in the AEC industry: Technical fundamentals, materials of construction, construction-applied resources, and field construction operations [21]. Also, key phrases (i.e., domain-specific compounds of words) were extracted from unstructured text documents with relations based on association frequencies of co-occurring word pairs [22]. Another interesting study put forward a building information modeling (BIM) body of knowledge (BOK) to present common knowledge, skills, and abilities using the Delphi method [23,24].

It is acknowledged that knowledge-based safety risk management is an important and effective method to assist metro construction safety. Knowledge-based intelligent systems have been developed to address safety risk issues. Yet the establishment of DKEs was mainly based on empirical data collected from experts. It is expected that this study helps facilitate the automatic construction and expansion of DKEs.

2.2. Automatic Methods for Safety Knowledge Discovery

Benefiting from the big data and artificial intelligence technologies, many automatic methods have been developed to deal with knowledge discovery. The current study mainly focuses on the two categories: (i) data-driven safety risk identification and analysis, and (ii) knowledge extraction from publication works.

Data-driven safety risk analysis is prone to integrate data mining technologies and risk assessment models. Na et al. improved the term frequency (TF) model with information entropy values to extract safety risk factors from construction accident reports [25]. Zhipeng et al. utilized Cramer's V and Phi coefficients to uncover statistical correlations between risk factors [26]. Alshboul et al. [27] combined machine learning (ML) techniques and multiple linear regression (MLR) to predict liquidated damages for construction projects. Wen-hui et al. proposed a comprehensive risk assessment framework incorporating credal networks (CNs) and an improved evaluation based on the distance from average solution (EDAS) method [28]. Additionally, many hybrid models have been developed to improve the accuracy of risk evaluation. Alshboul et al. complied genetic algorithms to optimize the associated decision variables for earthmoving equipment [29]. Zhang et al. optimized t-squares support vector machines (LSSVM) using quantum-behaved particle swarm optimization (QPSO) to perform early risk warning in subway station construction [30]. Shehadeh et al. developed a Gaussian mixture model to estimate the construction companies' capabilities in performing construction and maintenance activities during the pandemic [31]. Li et al. provided a second-order structural model using structural equation modeling (SEM) to determine the safety level in metro subway projects [32]. However, domain knowledge elements are far broader than safety risk factors. Knowledge units required for safety risk management should also be explored, such as the description of metro structures and construction equipment. Hence, the amount of processed data in this study is much larger.

Information extraction (IE) methods are mainly used to extract knowledge from scientific publications [33,34]. Two prevailing tasks of IE are named entity recognition (NER) and relation extraction (RE). The NER approach focuses on finding and classifying relevant knowledge units at a semantic level [35], such as names, organizations, and locations, whereas RE extracts the relationships between entities [36]. NER tasks require highly accurate and domain-specific part-of-speech (POS) tagging results [37], which is laborious and time-consuming. Thus, the linguistic characteristics of DKEs are more complicated. As for RE, the relationship types are fixed and limited within a range of semantic predefined rules, such as IsA, SubClassOf, and AtLocation [38]. For example, Yoo and Jeong utilized ConceptNet to extract relationships, including RalatedTo, IsA, part of, and HasA, between existing words and neologisms from news sites and social media, in order to add new neologisms to existing knowledge [39]. These studies utilized verbs in sentences to extract specified results based on a semantic labeling system, using text mining and natural language processing methods. If two entities in one sentence are related to the main verb, and the main verb is included in the predefined verb lexicon, those entities and relationships are annotated and extracted [40]. However, only very few sentences in the literature, such as definitions, adopt the narratives with specified verbs. Most of the related knowledge elements appear in one document rather than in one sentence.

Thus, a novel method needs to be brought forward to address the two issues of knowledge extraction in the metro construction domain: (i) the obtained DKEs should cover as many knowledge units as possible, and (ii) the process should use as little manpower as possible. To achieve such demands, this study aimed to provide an automatic approach for expanding immense unknown DKEs from a small set of seed words.

3. Methodology

3.1. Co-Word Co-Occurrence Analysis

Among the various NER and RE techniques, many studies have been based on co-word co-occurrence analysis [41,42]. In the AEC industry, CCNs are widely used to extract and visualize the potential relationships of topics and keywords from large-scale literature works in order to find research trends and gaps [43,44]. The advantage of CCNs is that they visualize the knowledge element network. However, they are considered to typically have limitations in terms of the quality of keywords and the selection of strong linkage. Regarding keywords, the results from indexing are more akin to the conceptualizations of indexers than to those of the scientists whose work is being studied. As for the linkage strength, the frequency of co-occurrences is counted to evaluate the strength of linkages in co-occurrence relationships. For example, in Ref. [43], the number of articles in which two topics tended to co-occur was calculated to evaluate the interlinkage strengths among all topics. Moreover, social network analysis (SNA) was put forward to calculate the linkage of nodes. The density and centrality of high-frequency words were counted to measure the co-occurrence strength [45,46]. However, neither of the above methods is capable of dealing with voluminous data.

To overcome the limitations of CCNs and to enhance the expanding performance, the proposed co-occurrence-based pathfinding approach made the following improvements:

(1) To enlarge the scope of the dataset, a web crawler was utilized to search the literatures on the World Wide Web across related domain platforms. Moreover, the entire abstract was retrieved and used to build a corpus to mitigate the bias of improper keywords. Moreover, because the collected abstracts were unstructured text, a domain lexicon closely related to domain-specific documents was constructed to improve the performance of text segmentation.

(2) A CCN was generated by a huge binary matrix. Both co-occurrence frequency and the centrality of the node have limitations in dealing with big data. To accomplish this, ARM, as a typical data mining method, was integrated to evaluate the strength of network linkage and to prune the redundant subnetworks.

3.2. Association Rule Mining

Association rule mining (ARM) is a data-mining method that is widely used in commercial settings (e.g., the purchase tendencies of customers) to find interesting associations that often occur in large datasets [47]. In the construction safety risk management field, Ayhan et al. identified the correlations between the attributes and accident types in occupational accidents [48]. Guo et al. analyzed the relationships among unsafe behaviors, worker types, and construction phases [49]. Zhou et al. investigated the associations between safety risk monitoring types and the coupling of risks [50]. In conclusion, ARM is normally used to analyze the coupling safety factors or the co-occurrences of causes and accidents. This paper aimed to use ARM to evaluate the association strength and highlight the strongly linked DKEs in a CCN.

3.3. Architecture of the Co-Occurrence-Based Pathfinding Approach

The idea of co-occurrence-based pathfinding is to assume that DKEs frequently co-occur in one document. The method is based on the integration of a CCN and ARM for finding frequently co-occurring pairwise domain-specific terms in a big dataset of text documents. Figure 1 presents the architecture of the co-occurrence-based pathfinding approach.

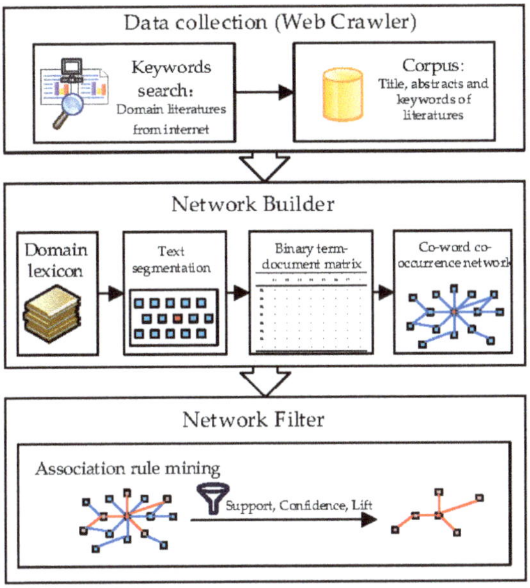

Figure 1. Architecture of the co-occurrence-based pathfinding approach.

(1) Web crawler. Select some known DKEs as seed words, and then use the web crawler to collect domain literature, taking the selected seed words as search words. The title, abstract, and keywords of each literature work are collected and stored in the corpus.

(2) Network builder. A general lexicon rarely contains proper or highly technical terms. Thus, a domain lexicon is built to improve the performance of text segmentation. Then, text segmentation is performed and word pairs and a binary term–document matrix (B-TDM) are generated. A CCN is then built based on the B-TDM by counting frequencies of pairwise term co-occurrence.

(3) Network filter. The association rule mining method serves as a filter to remove weakly related subnets in the CCN. Support, confidence, and lift are the three important indicators used for filtering rules by setting thresholds [51]. The threshold values are context-specific and user-defined [52]. The newly discovered DKEs are compared and added to the existing domain knowledge elements by using string matching.

3.4. Integration of a CCN and ARM

DKEs usually co-occur in one domain document simultaneously because they describe one subject-specific topic. For example, if the term "geological conditions" appears, then related terms such as "geotechnical structure," "adverse geology," and "collapse accident" may be found in the same document. Therefore, we assumed that word pairs generated by co-occurrence are prone to representing the same specific subject (i.e., domain knowledge). As DKEs and their co-occurrence relationships form a CCN [53], the expansion of DKEs can be achieved by finding the propagated paths of co-occurrence relationships.

Figure 2 displays a CCN of DKEs. The CCN comprises nodes, representing DKEs, and links, representing the co-occurrence relationships between said DKEs. As an example, suppose we want to find unknown DKEs related to the known element A. We begin by exploring the paths from the node labeled A, which leads us to the new nodes B, C, D, E, and F according to the strength of the link. We then expand from node F, leading to node G and subsequently node F. This finding process is repeated until an expansion path no longer appears and the network cannot be expanded any more—at which point it is considered that all domain knowledge elements have been found. Finally, we can find the following five paths: A→B, A→C, A→D, A→E, and A→F→G→H. Therefore, we can expand the domain knowledge elements from A to B, C, D, E, F, G, and H.

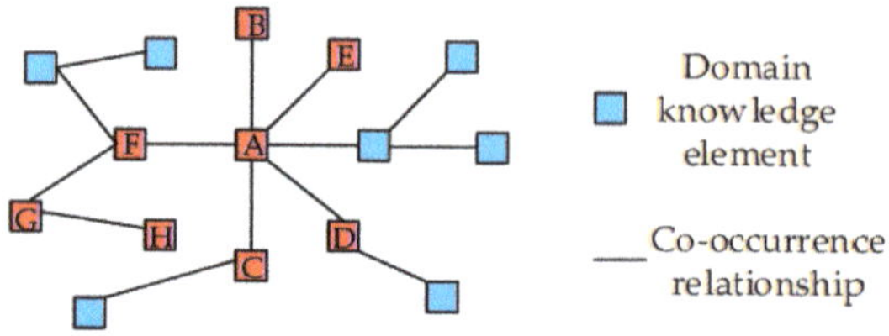

Figure 2. Co-word co-occurrence network of domain knowledge elements.

The basic concept of the association rule is to generate rules based on items with frequent occurrence [48]. Association rule mining is utilized here to filter out the weak linage of the CCN based on the support, confidence, and lift value. D is the total number of documents. X and Y represent the two different knowledge elements. An association rule $X \rightarrow Y$ is defined as: if X occurs, then Y occurs. X is the antecedent node of the CCN, and Y is the node. Support, confidence, and lift are the common indicators employed to evaluate the strength of an association rule [54].

The support shows the number of documents supporting the association rule of co-occurrence (X and Y occur in one document together) over the total number of documents (D). $P(X \cup Y)$ means the possibility of the co-occurrence of both X and Y. As for confidence, it is defined as the frequency with which X and Y occur together over the frequency with which X occurs in isolation. The value of support is an indication of how frequently the word appears in the corpus, while confidence shows how often the co-occurrence is found to be true [55]. Moreover, the value of lift reflects the influence of the occurrence of X on the occurrence of Y. The indicators are defined as in Equations (1)–(3) [51,56].

$$Support_{(X \rightarrow Y)} = D_{(X \cup Y)}/D = P(X \cup Y) \tag{1}$$

$$Confidence_{(X \rightarrow Y)} = D_{(X \cup Y)}/D_X = P(Y|X) \tag{2}$$

$$Lift_{(X \rightarrow Y)} = \left(D_{(X \cup Y)}/D_X\right)/\left(D_y/D\right) = P(Y|X)/P(Y) \tag{3}$$

The Apriori algorithm is adopted to mine the pairwise item sets $X \rightarrow Y$. The path $X \rightarrow Y$ will remain as a frequent item set when it meets the minimum thresholds of the support, confidence, and lift. Otherwise, the path will be eliminated from the CCN.

Choosing a threshold value is one of the most difficult aspects of applying ARM. Currently, thresholds are determined intuitively by users, according to the dataset's characteristics and the user's desires. Normally, the threshold values of support and confidence are usually set around 1%~4% and 10%~20% in safety risk factor discovery [56], while the threshold value of lift is often set around 1~1.5. In the experiment, statistical analysis was utilized to help generate the candidate threshold groups. Then, the domain experts decided the best threshold group according to the mined outputs. It is expected that the designed process can minimize the user's uncertainty.

4. Experiment and Results

4.1. Data Collection

Compared to engineering documents in the field, the academic literature is large in terms of quantity of works—more formally, it is easier to obtain and contains more new emerging knowledge. Thus, academic literature work was selected to build the corpus.

(1) Selection of the seed words

Using "metro construction" as the topic word to search the literature in the database of China National Knowledge Infrastructure (CNKI) during the years 2008–2018, we selected the top 100 indexed literature works and extracted the keywords. The, duplicated words were deleted, synonyms were normalized, and general words were deleted (e.g., control, management, and simulation analysis). Then, a list of seed words (No. = 188) was determined.

(2) Corpus building

With the CNKI, Vepu, and Wanfang databases (three of the main Chinese academic databases) taken as the data sources, web crawling was conducted separately using the seed words as search words to match the keywords in the academic literature published during 2008–2018. Finally, 68,817 literature works were collected, including journal articles, dissertation papers, newspaper reports, and conference proceedings. The title, abstract and keywords of the literature works were collected and transformed to text-type documents to construct a corpus. Moreover, the crawler supports the "Feed Adapter" function for data ingestion, so it continuously integrates data from external sources.

4.2. Network Building

According to the architecture (Figure 1) in the Methodology section, four steps were conducted to build a CCN:

(1) Domain lexicon

A domain lexicon was built according to Refs. [57,58]. Not only were subject-specific terms listed in the lexicon, but so were the synonym terms and stopwords. The domain lexicon benefits the generation of subject-specific tokens.

(2) Text segmentation

The corpus was divided into linguistically meaningful units (tokens) in this step. JiebaR, a Chinese tokenization toolkit, was used to implement the segmentation. The created domain lexicon was deployed in the program to improve the performance of text segmentation.

(3) Binary term–document matrix (B-TDM)

B-TDM is a numeric two-dimensional matrix representing the occurrence of a term appearing in a document [59]. JiebaR was used to calculate the occurrence count of each token for each document. The row refers to the sequence number of the document, while the column shows the tokens obtained after text segmentation. The number "1" represents the token occurring in the document, while the number "0" represents the token not occurring. It should be noted that the B-TDM of this case is a large sparse matrix because the distribution of terms is scattered in the massive dataset. Table 1 shows the B-TDM of the case. "Tunnel engineering" and "construction technology" both appear in documents No. 1, No. 2, and No. 5. Therefore, "construction technology" is considered a new candidate DKE related to "tunnel engineering".

Table 1. Binary term–document matrix.

	Tunnel Engineering	Construction Technology	Shallow Burying	Mining Method	Shield Method	...	Foundation Support
1	1	1	0	1	0		1
2	1	1	0	0	0		0
3	1	0	0	0	1		1
4	1	0	1	0	0		0
5	1	1	1	0	0		0
...				...			
68,817	0	0	1	0	1		1

(4) Co-word co-occurrence network

A CCN was created based on the data of B-TDM by counting the frequencies of pairwise term co-occurrence. Because of the large number of redundant co-relations within the tokens, the network needs to be largely pruned to highlight the most important co-related item sets.

4.3. Network Filtering

The Arules toolkit was used to perform association rule mining. The B-TDM was taken as the input data, and the Apriori algorithm was chosen to generate the pairwise item sets according to Equations (1)–(3). To set the thresholds of frequent item sets, 10 of the 188 seed words were randomly selected as the training data. Figure 3 displays the support value of 10 seed words by different colors. The mining results reflect that the support value shows a long tail distribution, i.e., only a small number of item sets have a high frequency of co-occurrence. The curve decreases sharply at the beginning, then gradually decreases to a low level after entering the inflection area, and finally forms a nearly horizontal straight line.

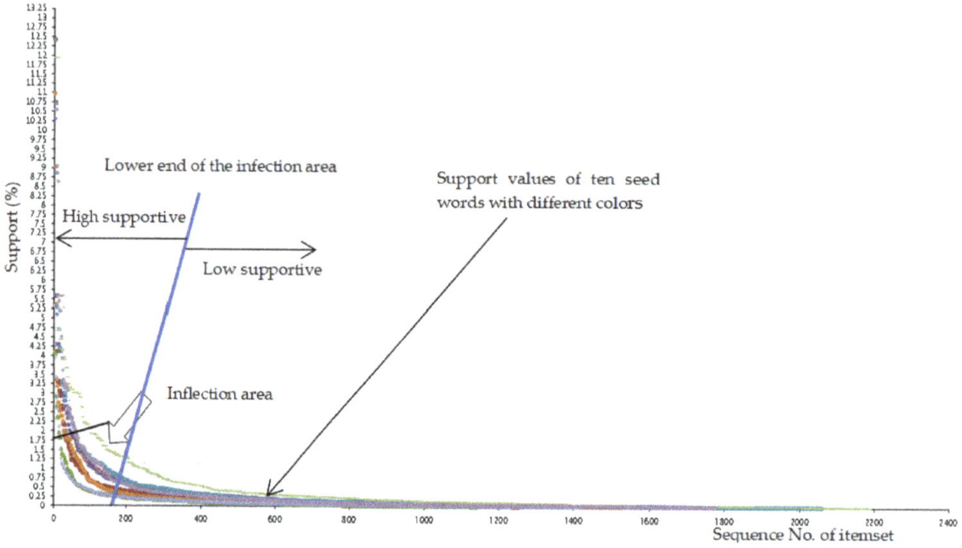

Figure 3. Distribution of support values.

According to the dataset's distribution in Figure 3, we attempted to use the inflection area to set around the high/low frequent items. In order not to lose the valuable item sets, the lower end of the inflection area was selected as the boundary to define high and low supportive item sets.

Different seed words have different inflection areas. Figure 4 displays all of the support values at the lower end of the inflection area. To simplify the mining rule, the benchmark of support value was defined as the average of them, which was 0.5%. Thus, the potential mining rule set was defined as the combination of support $\geq \{0.5\%, 1\%\}$ and confidence $\geq \{0, 10\%\}$ and lift $\geq \{1, 1.5\}$. This means that there are eight candidate mining rules. Eight experiments were conducted on the above 10 words, as shown in Figure 3. Expert knowledge was required to evaluate the strong association item sets under different mining rules and to choose the best of them. Finally, the threshold value of the three indicators in this study was set as support $\geq$ 0.5% and lift $\geq$ 1. No limitation for confidence (confidence $=$ 0) means that as long as Y occurs, the co-occurrence is valid. Compared to previous studies, the relatively lower threshold value of support and confidence led to more co-related terms. Otherwise, some item sets would have been missed because the B-DTM in this case was very sparse. The value of lift (lift $=$ 1) generated those terms whose occurrence increased with the occurrence of the seed words.

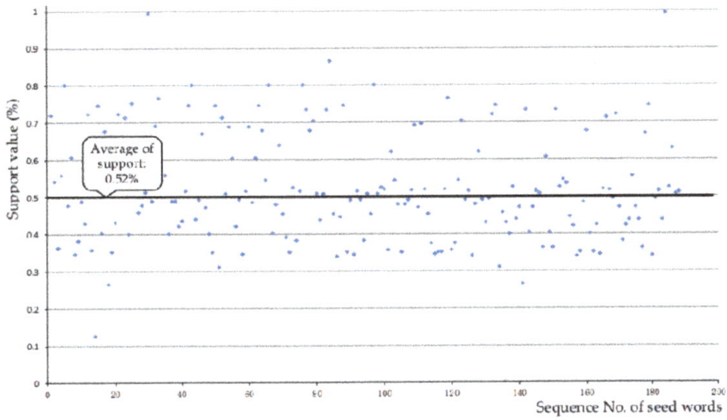

Figure 4. Support values of seed words at the lower end of the inflection area.

Therefore, the threshold group was finally set as support $\geq$ 0.5% and confidence $\geq$ 0 and lift $\geq$ 1. The minimum support value is lower than most cases (normally above 1%). This is because of the sparse matrix of B-DTM in this case. It is acknowledged that datasets with a low level of density required a smaller minimum support value when compared to datasets with high density [60].

4.4. Results

(1) Knowledge structure of metro construction safety risk management

Applying the association rule to the whole corpus, 2914 strong item sets were obtained. Then, the duplicate terms that already existed in previous rounds of mining were merged by using string matching. Finally, a list of 1583 DKEs was obtained.

To benefit the sharing and reuse of knowledge, the obtained DKEs were grouped into 11 themes and five categories by experts, according to the descriptions and meanings of the knowledge elements. Correspondingly, the knowledge structure of metro construction safety risk management was established (Figure 5). Limited by the length of this paper, the number of DKEs is displayed in parentheses instead of in detail. The knowledge structure can be used in many practical scenarios, such as safety training, ontology establishment in knowledge-based systems, and concept construction in domain knowledge graphs.

(2) The pathfinding process of the proposed approach

To verify the validity of the proposed approach, the process of one of the seed word was taken as an example. Figure 6 displays the co-occurrence-based pathfinding process that the seed word "tunnel engineering" experienced. For the one-round pathfinding, 17 strong associated items related to the word "tunnel engineering" were retrieved. The itemset of bearing capacity and tunnel engineering had the highest support value of 10.11%, indicating that the knowledge of "bearing capacity" is highly related to tunnel engineering in the metro safety risk management domain. Moreover, the knowledge of "construction management" had the highest lift value of 4.32, signifying that it is more likely to appear with "tunnel engineering" than to appear alone. Then, the new collected terms were used as root terms to match other pairwise item sets to retrieve the new DKEs, performing the second, third, and fourth rounds of searching until there were no matching item sets. One item may lead to several new items as long as the itemset meets the predefined mining rules. Finally, 39 new terms were found through the seed word "tunnel engineering".

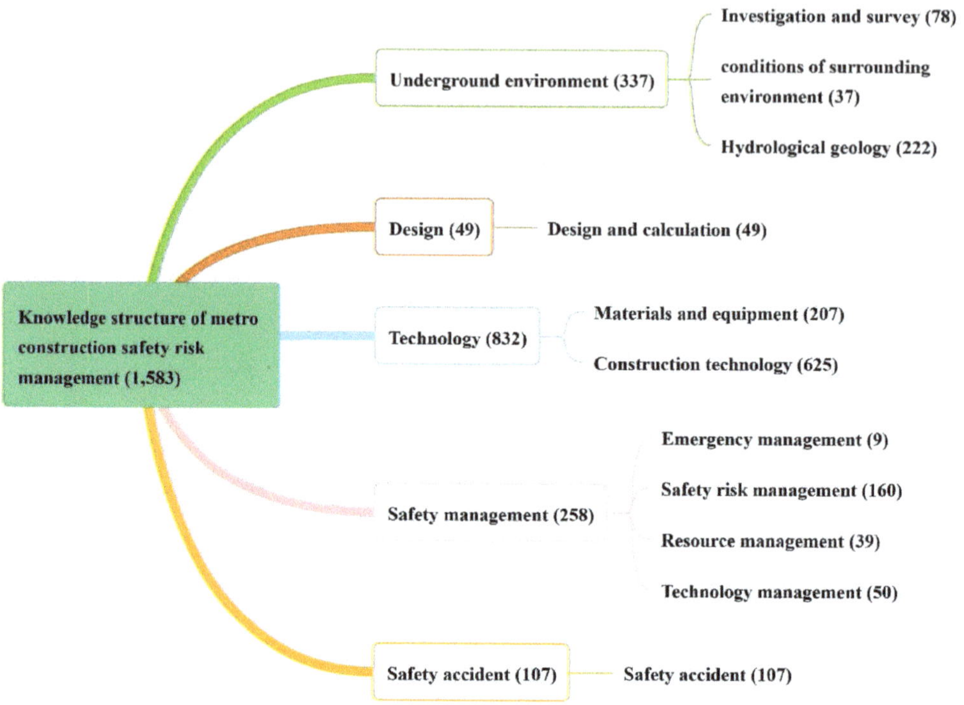

Figure 5. Knowledge structure of metro construction safety risk management.

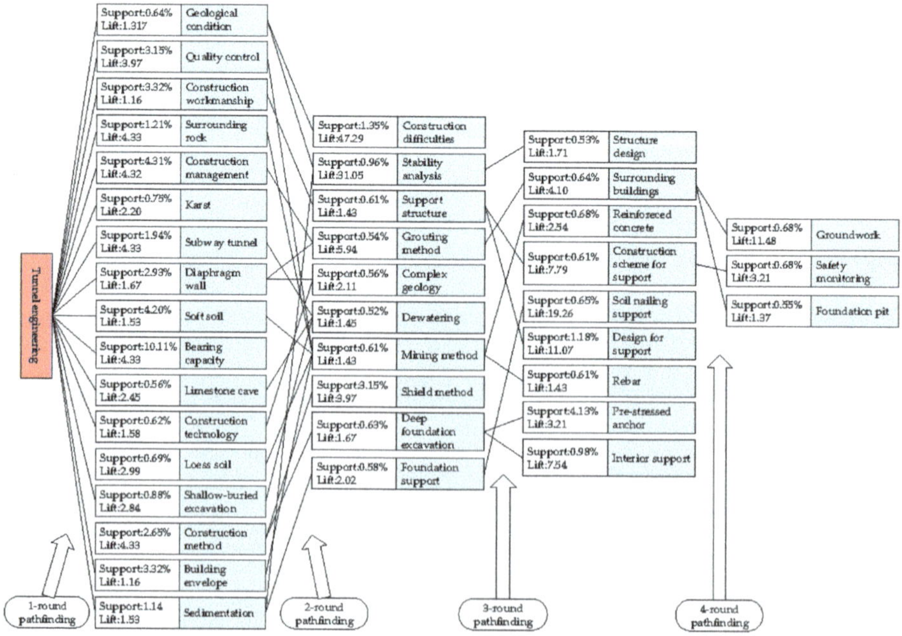

Figure 6. Co-occurrence-based pathfinding process for "tunnel engineering".

The result shows that the retrieved knowledge elements are comprehensive and valuable. Additionally, the number of newly found DKEs is rich without being overly redundant, demonstrating that the threshold of the ARM is reasonable. In brief, the proposed pathfinding approach performed well in the experiment.

5. Discussion

Knowledge elements are beneficial for the adjustment of the personal knowledge structure and for the acceleration of the process of knowledge sharing and reuse, and even knowledge innovation [61]. It is very important for each organization to reinforce the domain knowledge of technical workers and consolidate the accumulation of knowledge in metro construction. Moreover, extensive knowledge of practitioners can help all participants communicate more easily in order to effectively minimize the sources of risk in metro construction projects [62]. However, knowledge structures were manually established by experts in research and practice. In this study, a systematical and fine-grained knowledge structure is provided for practitioners.

Knowledge on technology takes the lead regarding the number of DKEs, followed by knowledge on the surrounding environment, knowledge on safety management, knowledge on safety accidents, and knowledge on design. Comparisons with previous studies are as follows.

(1) Knowledge on technology

Many previous studies have stated the importance of technical knowledge for construction safety management using questionnaires or expert interviews [63,64]. As Liang et al. stated, technical knowledge and skills constitute the third critical item affecting workers' safety competency, next only to physical conditions and safety awareness [65]. A few studies have tried to increase knowledge on safety behavior and safety conditions to reduce safety risks in metro construction. For example, Guo et al. built a behavioral risk knowledge base for a metro construction project in Wuhan city in order to classify and identify unsafe behavior [66]. Zhou and Ding established a safety barrier warning system for underground construction sites using Internet of Things (IoT) technologies [67].

From the perspective of domain knowledge elements, we confirmed that knowledge on technology, including 625 DKEs about construction technology and 207 DKEs about materials and equipment, occupies the largest number of knowledge elements in the metro construction safety risk management domain. This is probably because technical knowledge can help practitioners discover safety risk factors in the operation context and lead to safe behavior. Similarly, knowledge on materials and equipment, such as tunnel boring machines (TBMs) and shield blades, can help practitioners know better the materials and equipment on construction sites, discover unsafe conditions, and take proactive measures.

(2) Knowledge on the surrounding environment

Many severe safety accidents happen due to the surrounding environment. For example, a tunnel shaft of Xi'an metro collapsed because of excessive excavation and poor geological conditions [67]. Li et al. (2018) determined that underground pipelines are the most frequent reason for metro construction safety accidents based on an analysis of 156 accident reports [59].

We confirmed that knowledge on the underground environment plays a significant role in metro construction safety. This part of knowledge consists of 222 hydrological geology DKEs, 78 investigation survey DKEs, and 37 surrounding environment DKEs. Knowledge on hydrological geology, such as karst, soft soil, and geologic hazards, occupies the second largest number of knowledge themes, indicating that the unique nature of underground hydrological geology is of great uncertainty and the related knowledge plays a very import role for metro construction safety risk management. Knowledge on investigation surveys (advanced surveys, special surveys, and additional surveys, for example) and knowledge on the conditions of the surrounding environment (such as surface settlement and deformation observation) refer to the construction procedures and operation rules that practitioners should obey. Underground risks are hard to predict and prevent. However, knowledge on the underground environment can help practitioners understand how to perform investigation work and follow the work procedure and standards.

(3) Knowledge on safety management

Safety management is usually considered an indirect reason leading to a safety accident [68]. To decrease such risks, practitioners need to learn and understand how to perform safety risk management, especially the accurate identification of potential safety risks and safety management decision making during the construction process [21].

Knowledge on safety management focuses on knowledge elements about management, including 160 safety risk management DKEs (e.g., risk identification and risk loss), 50 technology management DKEs (e.g., safety inspection and safety supervision), 39 resource management DKEs (e.g., safety training), and 9 emergency management DKEs (e.g., emergency responsibility). This part of knowledge focuses on the management theory, method, and procedure, such as safety inspection and safety risk identification and analysis during metro construction. This part of knowledge may help practitioners build an identify for the organization's safety culture and perform good teamwork.

(4) Knowledge on safety accidents

Learning from past accidents is considered an effective way to prevent the occurrence of similar accidents and to promote construction safety [69,70]. Knowledge on safety accidents, such as accident type, near-miss accidents, and accident causation theory, is needed to identify hazards in the workplace and to take actions to prevent the occurrence of accidents. It is noted that knowledge on proper actions (e.g., knowledge on rescue and recovery) after the occurrence of accidents is also important to prevent further damage.

(5) Knowledge on design

Although design is considered one of the most important potential risks for construction safety [71], knowledge on design seems less important for workplace practitioners compared to other types of knowledge. This may be due to design and construction phrases being separate in most metro construction projects [62]. Little knowledge on design is needed for project managers and workers on construction sites, except for some essential concepts, such as bearing capacity and stability analysis. However, designers need to improve their knowledge of hazards because many construction accidents are connected to the design.

6. Conclusions

The current study developed a hybrid model to expand domain knowledge elements (DKEs) from a big dataset of text documents for metro construction safety risk management. First, the CCN was used to build the pathfinding network of candidate DKEs, and then the ARM was compiled to prune the weak related subnets, leaving the valuable ones. A case study was conducted using the Chinese academic literature as the corpus. The result verifies that the proposed approach is applicable to automatically expand domain knowledge elements from a big dataset of text documents. The advantage of the proposed approach is that it minimizes the expert bias.

Moreover, a list of knowledge elements was obtained. Knowledge on construction technology, hydrological geology, and construction resources constitutes the top three largest groups of knowledge elements. They play the most important role in metro construction safety risk management from the perspective of required knowledge. The obtained DKEs compose a fine-grained knowledge structure for practitioners. The knowledge structure can be used in various fields, such as safety education and training, construction of domain knowledge graphs, knowledge-based intelligent systems, and domain lexicon supplementation.

This approach can be extended to other projects to help engineers build a domain knowledge structure. There are three major limitations in this study that could be addressed in future research. First, the quantitative comparison between the method proposed in this paper and other methods is lacking, because the domain knowledge is too immense to check systematically. Second, the threshold determination of ARM still requires experts' recognition. A more general and automated approach is in need. Third, among the useful item sets extracted from a database, frequent item sets are usually thought to unfold "regularities" in the data [72]. In some situations, however, it may be interesting to search for "rare" item sets. These correspond to new emerging words, which might evolve into very important trends in the future of the domain. Moreover, exploring the semantic relationships of knowledge elements based on natural language processing and deep learning may advance the relation establishment for a knowledge graph.

Author Contributions: Conceptualization, N.X. and J.L.; methodology, B.Z.; software, L.W.; validation, T.G. and L.W.; formal analysis, N.X.; investigation, B.Z.; resources, B.Z.; data curation, B.Z.; writing—original draft preparation, B.Z.; writing—review and editing, N.X.; visualization, T.G.; supervision, N.X.; project administration, N.X.; funding acquisition, N.X. All authors have read and agreed to the published version of the manuscript.

Funding: This research was supported by the National Natural Science Foundation of China (grant number 71901206) and the Social Science Fund of Jiangsu Province (22GLB023).

Institutional Review Board Statement: Not applicable.

Informed Consent Statement: Not applicable.

Data Availability Statement: Some or all of the data, models, and codes that support the findings of this study are available from the corresponding author upon reasonable request.

Conflicts of Interest: The authors declare no conflict of interest.

References

1. Zhang, L.; Skibniewski, M.J.; Wu, X.; Chen, Y.; Deng, Q. A probabilistic approach for safety risk analysis in metro construction. *Saf. Sci.* **2014**, *63*, 8–17. [CrossRef]
2. Sousa, R.L.; Einstein, H.H. Risk analysis during tunnel construction using Bayesian networks: Porto Metro case study. *Tunn. Undergr. Space Technol.* **2012**, *27*, 86–100. [CrossRef]
3. Wang, Z.Z.; Chen, C. Fuzzy comprehensive Bayesian network-based safety risk assessment for metro construction projects. *Tunn. Undergr. Space Technol.* **2017**, *70*, 330–342. [CrossRef]
4. Qian, Q.; Lin, P. Safety risk management of underground engineering in China: Progress, challenges and strategies. *J. Rock Mech. Geotech. Eng.* **2016**, *8*, 423–442. [CrossRef]
5. Nepal, M.P.; Staub-French, S. Supporting knowledge-intensive construction management tasks in BIM. *J. Inf. Technol. Constr. ITcon* **2016**, *21*, 13–38.
6. Bellamy, L.J. Exploring the relationship between major hazard, fatal and non-fatal accidents through outcomes and causes. *Saf. Sci.* **2015**, *71*, 93–103. [CrossRef]
7. Health and Safety Executive. *Reducing Error and Influencing Behaviour*; Health and Safety Executive: London, UK, 2009.
8. Ahmed, S. Causes of accident at construction sites in Bangladesh. *Organ. Technol. Manag. Constr.* **2019**, *11*, 1933–1951. [CrossRef]
9. Wong, L.; Yuhong Wang, P.E.; Law, T.; Lo, C.T. Association of root causes in fatal fall-from-height construction accidents in Hong Kong. *J. Constr. Eng. Manag.* **2016**, *142*, 04016018. [CrossRef]
10. Durlach, P.J.; Lesgold, A.M. *Adaptive Technologies for Training and Education*; Cambridge University Press: Cambridge, UK, 2012.
11. Gao, G.; Wang, Y.; Li, Y. Review of the research of domestic knowledge element. *Inf. Sci.* **2016**, *34*, 161–165. [CrossRef]
12. Xu, N.; Ma, L.; Wang, L.; Deng, Y.; Ni, G. Extracting domain knowledge elements of construction safety management: Rule-based approach using Chinese natural language processing. *J. Manag. Eng.* **2021**, *37*, 04021001. [CrossRef]
13. Yu, D.; Yang, J. Knowledge management research in the construction industry: A review. *J. Knowl. Econ.* **2016**, *9*, 782–803. [CrossRef]
14. Serpella, A.F.; Ferrada, X.; Howard, R.; Rubio, L. Risk management in construction projects: A knowledge-based approach. *Proc. Soc. Behav. Sci.* **2014**, *119*, 653–662. [CrossRef]
15. Ni, G.; Cui, Q.; Sang, L.; Wang, W.; Xia, D. Knowledge-sharing culture, project-team interaction, and knowledge-sharing performance among project members. *J. Manag. Eng.* **2018**, *34*, 04017065. [CrossRef]
16. Sun, J.; Ren, X.; Anumba, C.J. Analysis of knowledge-transfer mechanisms in construction project cooperation networks. *J. Manag. Eng.* **2019**, *35*, 04018061. [CrossRef]
17. Dong, C.; Wang, F.; Li, H.; Ding, L.; Luo, H. Knowledge dynamics-integrated map as a blueprint for system development: Applications to safety risk management in Wuhan metro project. *Autom. Constr.* **2018**, *93*, 112–122. [CrossRef]
18. Ding, L.Y.; Yu, H.L.; Li, H.; Zhou, C.; Wu, X.G.; Yu, M.H. Safety risk identification system for metro construction on the basis of construction drawings. *Autom. Constr.* **2012**, *27*, 120–137. [CrossRef]

19. Zhong, B.; Li, H.; Luo, H.; Zhou, J.; Fang, W.; Xing, X. Ontology-based semantic modeling of knowledge in construction: Classification and identification of hazards implied in images. *J. Constr. Eng. Manag.* **2020**, *146*, 04020013. [CrossRef]
20. Ahmed, A.; Al-Masri, N.; Abu Sultan, Y.S.; Akkila, A.N.; Almasri, A.; Mahmoud, A.Y.; Zaqout, I.S.; Abu-Naser, S.S. Knowledge-based systems survey. *Int. J. Acad. Eng. Res. (IJAER)* **2019**, *3*, 1–22.
21. Tatum, C.B. Core elements of construction engineering knowledge for project and career success. *J. Constr. Eng. Manag.* **2011**, *137*, 745. [CrossRef]
22. Nedeljković, Đ.; Kovačević, M. Building a construction project key-phrase network from unstructured text documents. *J. Comput. Civ. Eng.* **2017**, *31*, 04017058. [CrossRef]
23. Wu, W.; Mayo, G.; McCuen, T.L.; Issa RR, A.; Smith, D.K. Building information modeling body of knowledge I: Background, framework, and initial development. *J. Constr. Eng. Manag.* **2018**, *144*, 04018065. [CrossRef]
24. Wu, W.; Mayo, G.; McCuen, T.L.; Issa RR, A.; Smith, D.K. Building information modeling body of knowledge II: Consensus building and use cases. *J. Constr. Eng. Manag.* **2018**, *144*, 04018066. [CrossRef]
25. Na, X.; Ling, M.; Qing, L.; Li, W.; Yongliang, D. An improved text mining approach to extract safety risk factors from construction accident reports. *Saf. Sci.* **2021**, *138*, 105216. [CrossRef]
26. Zhipeng, Z.; Yang Miang, G.; Qianqian, S.; Haonan, Q.; Song, L. Data-driven determination of collapse accident patterns for the mitigation of safety risks at metro construction sites. *Tunn. Undergr. Space Technol.* **2022**, *127*, 104616. [CrossRef]
27. Alshboul, O.; Mohammad, A.A.; Rabia Emhamed Al, M.; Ghassan, A.; Ali Saeed, A.; Ali, S. Forecasting liquidated damages via machine learning-based modified regression models for highway construction projects. *Sustainability* **2022**, *14*, 5835. [CrossRef]
28. Wen-Hui, H.; Xiao-Kang, W.; Hong-Yu, Z.; Jian-Qiang, W.; Lin, L. Safety risk assessment of metro construction under epistemic uncertainty: An integrated framework using credal networks and the EDAS method. *Appl. Soft Comput.* **2021**, *108*, 107436. [CrossRef]
29. Alshboul, O.; Shehadeh, A.; Tatari, O.; Almasabha, G.; Saleh, E. Multiobjective and multivariable optimization for earthmoving equipment. *J. Facil. Manag.* 2022; *ahead-of-print*. [CrossRef]
30. Zhang, L.; Wang, J.; Wu, H.; Wu, M.; Guo, J.; Wang, S. Early warning of the construction safety risk of a subway station based on the LSSVM optimized by QPSO. *Appl. Sci.* **2022**, *12*, 5712. [CrossRef]
31. Shehadeh, A.; Alshboul, O.; Hamedat, O. A Gaussian mixture model evaluation of construction companies' business acceptance capabilities in performing construction and maintenance activities during COVID-19 pandemic. *Int. J. Manag. Sci. Eng. Manag.* **2022**, *17*, 112–122. [CrossRef]
32. Li, X.; Liao, F.; Wang, C.; Alashwal, A. Managing Safety Hazards in Metro Subway Projects under Complex Environmental Conditions. *J. Risk Uncertain. Eng. Syst. Part A Civ. Eng.* **2022**, *2022*, 04021079. [CrossRef]
33. Ahn, H.; Song, M.; Heo, G.E. Inferring undiscovered public knowledge by using text mining analysis and main path analysis: The case of the gene-protein 'brings_about' chains of pancreatic cancer. *J. Korean BIBLIA Soc. Libr. Inf. Sci.* **2015**, *26*, 217–231. [CrossRef]
34. Jean Lieber, A.N.; Szathmary, L.; Toussaint, Y. First elements on knowledge discovery guided by domain knowledge. In Proceedings of the International Conference on Concept Lattices and Their Applications, Tunis, Tunisia, 30 October–1 November 2006. [CrossRef]
35. Moon, S.; Lee, G.; Chi, S.; Oh, H. Automatic review of construction specifications using natural language processing. In Proceedings of the ASCE International Conference on Computing in Civil Engineering 2019, Atlanta, GA, USA, 17–19 June 2019; pp. 401–407. [CrossRef]
36. Zhang, J.; El-Gohary, N.M. Semantic NLP-based information extraction from construction regulatory documents for automated compliance checking. *J. Comput. Civ. Eng.* **2016**, *30*, 04015014. [CrossRef]
37. Xue, X.; Zhang, J. Building codes part-of-speech tagging performance improvement by error-driven transformational rules. *J. Comput. Civ. Eng.* **2020**, *34*, 04020035. [CrossRef]
38. Shi, F.; Chen, L.; Han, J.; Childs, P. A data-driven text mining and semantic network analysis for design information retrieval. *J. Mech. Des.* **2017**, *139*, 111402. [CrossRef]
39. Yoo, S.; Jeong, O. Automating the expansion of a knowledge graph. *Expert Syst. Appl.* **2020**, *141*, 112965. [CrossRef]
40. Hwan, K.Y.; Min, S. A context-based ABC model for literature-based discovery. *PLoS ONE* **2019**, *14*, e0215313. [CrossRef]
41. Matsuo, Y.; Ishizuka, M. Keyword extraction from a single document using word co-occurrence statistical information. *Int. J. Artif. Intell. Tools* **2004**, *13*, 157–169. [CrossRef]
42. Sedighi, M. Application of word co-occurrence analysis method in mapping of the scientific fields (case study: The field of Informetrics). *Libr. Rev.* **2016**, *65*, 52–64. [CrossRef]
43. Ezzeldin, M.; El-Dakhakhni, W. Metaresearching structural engineering using text mining: Trend identifications and knowledge gap discoveries. *J. Struct. Eng.* **2020**, *146*, 04020061. [CrossRef]
44. Zhao, L. Mapping knowledge domains of international knowledge integration outputs, 2007–2012. In Proceedings of the ICCREM 2014: Smart Construction and Management in the Context of New Technology, Kunming, China, 27–28 September 2014; pp. 886–896. [CrossRef]
45. Song, W.; Man, Q. Comparative Research of integration application with BIM based on mapping knowledge domains at home and abroad. In Proceedings of the International Conference on Construction and Real Estate Management 2019, Banff, AL, Canada, 21–24 May 2019; pp. 125–134. [CrossRef]

46. Wu, H.; Xue, X.; Shen, G.Q.; Luo, Y. Mapping the knowledge structure in megaproject management research using complex network analysis. In Proceedings of the International Conference on Construction and Real Estate Management 2017 (ICCREM 2017), Guangzhou, China, 10–12 November 2017; pp. 82–88. [CrossRef]
47. Altay, E.V.; Alatas, B. Performance analysis of multi-objective artificial intelligence optimization algorithms in numerical association rule mining. *J. Ambient. Intell. Humaniz. Comput.* **2019**, *11*, 3449–3469. [CrossRef]
48. Ayhan, B.U.; Doğan, N.B.; Tokdemir, O.B. An association rule mining model for the assessment of the correlations between the attributes of severe accidents. *J. Civ. Eng. Manag.* **2020**, *26*, 315–330. [CrossRef]
49. Guo, S.; Zhang, P.; Ding, L. Time-statistical laws of workers' unsafe behavior in the construction industry: A case study. *Phys. A Stat. Mech. Its Appl.* **2019**, *515*, 419–429. [CrossRef]
50. Zhou, Y.; Li, C.; Ding, L.; Sekula, P.; Love PE, D.; Zhou, C. Combining association rules mining with complex networks to monitor coupled risks. *Reliab. Eng. Syst. Saf.* **2019**, *186*, 194–208. [CrossRef]
51. Hosseini, M.R.; Martek, I.; Papadonikolaki, E.; Sheikhkhoshkar, M.; Banihashemi, S.; Arashpour, M. Viability of the BIM manager enduring as a distinct role: Association rule mining of job advertisements. *J. Constr. Eng. Manag.* **2018**, *144*, 04018085. [CrossRef]
52. Gollapudi, S. *Practical Machine Learning*; Packt Publishing Ltd.: Birmingham, UK, 2016.
53. Wang, Q.; Zhou, W.; Zeng, C.; Li, T.; Shwartz, L.; Grabarnik, G.Y. Constructing the knowledge base for cognitive IT service management. In Proceedings of the IEEE 14th International Conference on Services Computing, Honolulu, HI, USA, 25–30 June 2017; pp. 410–417. [CrossRef]
54. Lipeng, F.; Xueqing, W.; Heng, Z.; Mengnan, L. Interactions among safety risks in metro deep foundation pit projects: An association rule mining-based modeling framework. *Reliab. Eng. Syst. Saf.* **2022**, *221*, 108381. [CrossRef]
55. Meng, H.; Hong, Y.; Ma, Y.; Li, Z.; Lu, J.; Siddiqui, N.A. Association rule-based traffic accident impact factors analysis on low-grade highways. In Proceedings of the CICTP, Nanjing, China, 6–8 July 2019; pp. 3549–3559. [CrossRef]
56. Yao, Z.; Deng, W.; Wu, D. Association rule analysis of contributory factors to severe traffic accidents. In Proceedings of the CICTP 2018: Intelligence, Connectivity, and Mobility, Beijing, China, 5–8 July 2018; pp. 1875–1884. [CrossRef]
57. Esmaeili, B.; Hallowell, M.R.; Rajagopalan, B. Attribute-based safety risk assessment. I: Analysis at the fundamental level. *J. Constr. Eng. Manag.* **2015**, *141*, 04015021. [CrossRef]
58. Na, X.; Wenshun, W.; Jianping, W.; Jie, L.; Ruopeng, H. Using text mining to extract safety accident causes and to assess importance on urban rail transit construction project. *Sci. Technol. Prog. Policy* **2018**, *35*, 134–138. [CrossRef]
59. Li, J.; Wang, J.; Xu, N.; Hu, Y.; Cui, C. Importance degree research of safety risk management processes of urban rail transit based on text mining method. *Information* **2018**, *9*, 26. [CrossRef]
60. Hikmawati, E.; Maulidevi, N.U.; Surendro, K. Minimum threshold determination method based on dataset characteristics in association rule mining. *J. Big Data* **2021**, *8*, 146. [CrossRef]
61. Zheng, M.; Qin, C.; Ma, X. Research on knowledge unit representation and extraction for unstructured abstracts of Chinese scientific and technical literature: Ontology theory based on knowledge unit. *Inf. Theory Appl.* **2019**, *43*, 157–163. [CrossRef]
62. Yu, Q.Z.; Ding, L.Y.; Zhou, C.; Luo, H.B. Analysis of factors influencing safety management for metro construction in China. *Accid. Anal. Prev.* **2014**, *68*, 131–138. [CrossRef]
63. Jeelani, I.; Albert, A.; Azevedo, R.; Jaselskis, E.J. Development and testing of a personalized hazard-recognition training intervention. *J. Constr. Eng. Manag.* **2017**, *143*, 04016120. [CrossRef]
64. Loosemore, M.; Malouf, N. Safety training and positive safety attitude formation in the Australian construction industry. *Saf. Sci.* **2019**, *113*, 233–243. [CrossRef]
65. Liang, K.; Fung, I.W.H.; Xiong, C.; Luo, H. Understanding the factors and the corresponding interactions that influence construction worker safety performance from a competency-model-based perspective: Evidence from scaffolders in China. *Int. J. Environ. Res. Public Health* **2019**, *16*, 1885. [CrossRef] [PubMed]
66. Guo, S.Y.; Ding, L.Y.; Luo, H.B.; Jiang, X.Y. A Big-Data-based platform of workers' behavior: Observations from the field. *Accid. Anal. Prev.* **2016**, *93*, 299–309. [CrossRef]
67. Zhou, C.; Ding, L.Y. Safety barrier warning system for underground construction sites using Internet-of-Things technologies. *Autom. Constr.* **2017**, *83*, 372–389. [CrossRef]
68. Xu, N. Occurrence tendency and cause analysis of safety accidents in rail transit projects. *J. Huaqiao Univ. Nat. Ed.* **2016**, *37*, 6. [CrossRef]
69. Xie, Y.; Lee, Y.-C.; Shariatfar, M.; Zhang, Z.; Rashidi, A.; Lee, H.W. Historical accident and injury database-driven audio-based autonomous construction safety surveillance. In Proceedings of the Computing in Civil Engineering 2019: Data, Sensing, and Analytics, Atlanta, GA, USA, 17–19 June 2019; pp. 105–113. [CrossRef]
70. Zhang, F.; Fleyeh, H.; Wang, X.; Lu, M. Construction site accident analysis using text mining and natural language processing techniques. *Autom. Constr.* **2019**, *99*, 238–248. [CrossRef]
71. Etemadinia, H.; Tavakolan, M. Using a hybrid system dynamics and interpretive structural modeling for risk analysis of design phase of the construction projects. *Int. J. Constr. Manag.* **2018**, *21*, 1–20. [CrossRef]
72. Pereira, E.; Ahn, S.; Han, S.; Abourizk, S. Identification and association of high-priority safety management system factors and accident precursors for proactive safety assessment and control. *J. Manag. Eng.* **2018**, *34*, 04017041. [CrossRef]

buildings

MDPI

Article

Maturity Assessment of Intelligent Construction Management

Chao Lin [1], Zhen-Zhong Hu [1], Cheng Yang [2], Yi-Chuan Deng [3], Wei Zheng [2] and Jia-Rui Lin [4,*]

1 Tsinghua Shenzhen International Graduate School, Tsinghua University, Shenzhen 518055, China
2 Zhejiang Supervision on Highway and Water Transportation Construction Engineering Co., Ltd., Hangzhou 310000, China
3 School of Civil Engineering and Transportation, South China University of Technology, Guangzhou 510641, China
4 Department of Civil Engineering, Tsinghua University, Beijing 100084, China
* Correspondence: lin611@tsinghua.edu.cn

Abstract: In the new era of Construction 4.0, the application of a large number of intelligent information technologies (ITs) and advanced managerial approaches have brought about the rapid development of intelligent construction management (ICM). However, it is still unclear how to assess the maturity of ICM. In this study, a maturity assessment system for ICM was formulated through literature reviews, questionnaires, expert discussions and a case study. A maturity scoring table containing five assessment dimensions and twenty assessment indicators was developed, and corresponding maturity levels and a radar chart of dimensions were designed. A case study of the assessments of two construction enterprises was conducted to validate that the proposed assessment system could be used by construction enterprises to quantitatively assess their ICM maturities and obtain both overall and specific assessment results. This study also proposed practical improvement methods to improve ICM maturities for construction enterprises with different maturity levels. Furthermore, the study also discussed the development direction of ICM at present and in the short-term future, which should be paid more attention to by the construction industry.

Keywords: intelligent construction management (ICM); construction industry; maturity assessment system; improvement plan

Citation: Lin, C.; Hu, Z.-Z.; Yang, C.; Deng, Y.-C.; Zheng, W.; Lin, J.-R. Maturity Assessment of Intelligent Construction Management. *Buildings* **2022**, *12*, 1742. https://doi.org/10.3390/buildings12101742

Academic Editor: Bo Xia

Received: 30 September 2022
Accepted: 18 October 2022
Published: 19 October 2022

Publisher's Note: MDPI stays neutral with regard to jurisdictional claims in published maps and institutional affiliations.

Copyright: © 2022 by the authors. Licensee MDPI, Basel, Switzerland. This article is an open access article distributed under the terms and conditions of the Creative Commons Attribution (CC BY) license (https://creativecommons.org/licenses/by/4.0/).

1. Introduction

The construction industry is a traditional pillar industry in many countries, and its contribution to economic growth and long-term national development is widely acknowledged [1]. In China, for example, the construction industry contributed about 25.6% to the national gross domestic product (GDP) in 2021 [2]. However, the construction industry involves a large number of participants and covers multiple professions, so improper and bad management of any aspect often causes huge losses. The fatal injury rate for the construction industry is higher than the average for all other industries due to its labor-intensive characteristics and poor safety management during production processes [3]. Careless quality management will cause a hidden danger to the later operation of structures. Many construction projects worldwide were completed with significant time and cost overruns due to bad schedule management [4]. Furthermore, conflicts, disputes and arbitration between construction parties caused by poor construction management greatly lower the construction productivity on site. On the contrary, proper construction management can reduce potential risks when implementing investment and construction projects and make necessary conditions for the timely and high-quality delivery of projects within the planned budget. Construction management is a professional service that provides a project's owner(s) with effective management of the project's schedule, cost, quality, safety, scope and function [5], and it plays an increasingly important role in various construction projects.

With the rise of Industry 4.0 [6], the rapid development of information technologies (ITs) has greatly promoted and improved the construction industry. As a result, the terms Construction 4.0 and intelligent construction came into being. Construction 4.0 is a concept that was proposed in reference to Industry 4.0. The definition of Construction 4.0 is dynamically evolving. For example, Sawhney et al. [7] defined Construction 4.0 as a framework that is a confluence and convergence of three broad themes: industrial production, cyber-physical systems and digital and computing technologies. Wu et al. [8] regarded Construction 4.0 as the integration of information and automation technologies in construction projects. There are many advanced technologies involved in Construction 4.0, and Forcael et al. [9] concluded that four essential technologies are needed to understand Construction 4.0 at the present time: 3D printing, big data, virtual reality (VR) and the Internet of Things (IoT). Except for advanced technologies, Construction 4.0 also brought advanced managerial approaches; García de Soto et al. [10] indicated that Construction 4.0 pushes construction organizations and roles to be transformed in many aspects. The evolution from digitalization to intelligence is the mainstream of the development of Industry 4.0 [11]. As a derivative of Industry 4.0, the development direction of Construction 4.0 is the same, so intelligent construction is the ultimate goal of Construction 4.0. The comprehensive development of intelligent construction requires intelligence in every segment, among which intelligent construction management (ICM) plays an essential and inevitable part; it is the foundation of Construction 4.0 and intelligent construction. ICM is the intelligent pattern of construction management; it is a comprehensive evolution of traditional construction management in management concepts, working mode and supporting measures which are achieved by the introduction of intelligent ITs and congenial managerial approaches.

Maturity is the competency, capability and level of sophistication of a selected domain based on a comprehensive set of criteria [12]. The ICM maturity of a certain construction enterprise is its ability to conduct ICM, and it should be considered comprehensively from the technological perspective and from the managerial perspective. Therefore, the maturity assessment of ICM is the comprehensive consideration of the development condition of IT and the application condition of managerial approaches. The purpose of maturity assessment is to identify a gap that can then be closed by subsequent improvement actions [13]. Many construction enterprises have been developing ICM, and the fierce competition among them requires more efficient improvement plans for their ICM maturities. Only when the ICM maturity is accurately assessed can an enterprise select IT and managerial approaches it needs to improve rather than extensively and aimlessly involving all kinds of intelligent ITs and managerial approaches, leading to a waste of human, material and financial resources. Therefore, the maturity assessment of ICM is of great significance for construction enterprises to find out shortcomings and formulate future improvement plans thereafter.

However, there is still a lack of effective systems, methods or even indicators to systematically assess the maturity of ICM, which has encumbered the development of the construction industry. Existing studies and explorations towards ICM just focused on the innovation or application of one or several types of IT. Due to the differences between construction enterprises or projects, as well as the diversity and complexity of advanced ITs, the application depth and breadth of relevant IT are different, and their values and benefits remain uncertain. At the same time, the introduction of advanced IT often leads to a change in managerial approach, including organizational form and workflow. The mismatch between the managerial approach and IT may also greatly limit the efficiency and value of ICM. Therefore, it is difficult to effectively assess the ICM maturities between different construction enterprises and discover their potential problems at the same time. In contrast, available maturity assessment models are increasingly being applied in other informational, digital or intelligent fields as approaches for continuous process improvements [14], such as the building information modeling (BIM) capability maturity model [15] and the digital maturity model [16]. These maturity models enable relevant organizations to clearly understand their development maturity and to make appropriate developing plans later.

In view of the above problems, this study formulated a maturity assessment system for ICM. An intelligent maturity scoring table was established for the quantitative maturity assessment of ICM. The scoring table consisted of five assessment dimensions and twenty assessment indicators. To present the assessment results in both overall and specific aspects, the levels of ICM maturity were set, and the radar chart of assessment dimensions was designed. Finally, a case study of the assessments of two construction enterprises was conducted to validate the usage of the proposed assessment system and intelligent maturity improvement strategies were discussed. The assessment system can be used for leaders in construction enterprises to assess their ICM maturities and obtain vivid assessment results as well as improvement plans. For every construction enterprise, the scoring table transformed its ICM maturity into a score. The corresponding maturity level plotted the position of its ICM maturity in the whole industry. The radar chart of dimensions visualized its strengths and weaknesses in dimensions. Finally, the improvement strategies guided it to improve its ICM maturity according to the assessment results.

The rest of this paper is organized as follows. Section 2 reviews and summarizes previous studies related to ICM and mature assessment systems in other fields. Section 3 introduces the methodology of this research to formulate the maturity assessment system. Section 4 presents the rationality and effectiveness of the proposed maturity assessment system through expert discussion. Section 5 enumerates the components of this maturity assessment system. Section 6 recounts a case study to validate the usage of the proposed assessment systems and discusses the methods to improve ICM maturity. Finally, Section 7 summarizes the main contributions, limitations and future improvements of this research.

2. Literature Review

In this section, studies concerning ICM are reviewed, and so are investigations about assessment systems, including methods and models in other informational, digital or intelligent fields to show mature examples.

2.1. ICM

Wu et al. [8] emphasized that Construction 4.0 heavily relies on data to build and maintain the interaction between the physical and virtual worlds. Because intelligent construction is the ultimate goal of Construction 4.0, data is also essential for ICM [17]. Intelligent IT for data collection, transmission, aggregation, analysis and sharing can contribute to ICM [18], so can advanced managerial approaches supporting the data-oriented work mode, such as corresponding working post setting and personnel training, online personnel management and workflow interaction, etc. Therefore, the essence of ICM can be concluded as the review and feedback of various types of relevant construction information and data, which includes the collection, transmission and statistics of them, with the support of visualization, intelligent analysis and other technical means in this process.

A number of researchers have investigated the attributes and development direction of ICM from the perspectives of intelligent IT and advanced managerial approaches, respectively. Aiming at intelligent IT, Sawhney et al. [7] illustrated representative IT that is used in Construction 4.0: BIM, cloud-based project management, augmented reality (AR), VR, artificial intelligence (AI), cybersecurity, big data and analytics, blockchain, laser scanner, IoT, etc. These ITs can also be applied to ICM. Aiming at managerial approaches, Woo et al. [19] reviewed different construction management methods by analyzing the efficiency of various methods currently applied to public construction projects. They concluded that the direct supervision method is the most efficient construction management method because of lower cost and less time. García de Soto et al. [10] analyzed the transformation of construction organizations and roles in Construction 4.0. Existing roles evolved, and new roles were created; for example, more employees with digital skills were needed. Many kinds of traditional construction work were automated with the application of robotic systems. Furthermore, current fragmented projects evolved into project-based integrations and eventually into a platform-based integration.

There are other studies that focused on the application of just one certain IT or managerial approach towards ICM. In this study, we reviewed nearly all the existing IT or managerial approaches from the literature. Furthermore, we also discussed with experts in the construction industry to supplement novel IT or managerial approaches which have not been published yet. All ITs or managerial approaches researched are presented in Table 1. Their effects on ICM and sources are also listed.

Table 1. IT or managerial approaches supporting ICM.

Effect on ICM	IT or Managerial Approach	Source
Management platform	Platforms with terminals for a personal computer (PC), mobile and website Use a firewall and virus scan against intrusion AI voice assistant Application of 5G technology	[20,21] [22] [23] [24]
Personnel management	Intelligent attendance system Human resource training and assessment Manage personnel information and user permissions in the platform Monitoring of personnel health and performance Warning of overdue personnel age and qualification Incorporation of COVID-19 guidelines into site health policies	[25] [26,27] [28] [29,30] Expert Discussion [31]
Visualization	Establish BIM or digital twin (DT) in the platform Construction simulation in a multidimensional BIM environment Construction information sharing in the platform Application of VR, AR and mixed reality (MR) Information carrier and displayed on the site	[20] [32] [33] [34,35] [36]
Workflow	Submit and receive information through the platform Fill and modify documents in the platform Task management through the platform High-performance communication facilities on site	[37] [38] [36] [39]
Production	Machinery operation and work tracking and monitoring Materials management using emerging technologies	[40,41] [42]
Environmental impact	Waste and pollutant monitoring on site Site workplace environmental situation monitoring	[43,44] [45]
Quality control	Automated data acquisition technologies on the site Application of robots Mark locations of quality problems in the models Declared quality problems tracking Vision-based inspection and real-time quality assessment Application of personal mobile devices	[46] [47] [36] [48] [49,50] [51]
Schedule and contract	Real-time schedule, contract and payment tracking and monitoring Warning of overdue schedule and contract	[52,53] [54]
Time and cost	Optimization of time and cost using a learning curve	[55]
Information management	Record of engineering data and personnel operation Information decentralization, forgery and alteration prevention Intelligent search engine Data integration and simplification Application of information extraction (IE)	[56] [57] [58] [59] [60]
Work safety	Real-time video surveillance on the site Worker safety device makes warning in proximity to certain areas Warning of unsafe behavior by real-time smart video surveillance Equipment collision prevention Warning of real-time fire, smoke, etc., on the site Warning of abnormal value in data collected	[61,62] [63] [61,62] [64] [65] [66]
Construction coordination	Dispatch list by intelligent work breakdown structure (WBS) calculation Time-space conflicts management	Expert Discussion [67]
Risk prevention	Preventive measures with the use of the prediction model	[68]

While all existing studies individually render positive influences on ICM, their research directions, in general, are too scattered to establish sufficient cooperation and connection with each other. Specifically, they suffer from several shortcomings—they

- Neglect the combined effects of IT and managerial approach;
- Do not summarize all ITs and managerial approaches available for ICM;
- Lack methods to assess the application maturities of IT and managerial approaches in ICM;
- Lack practical and appropriate plans to improve the ICM maturity of construction enterprises.

In this study, existing intelligent IT and advanced managerial approaches available for ICM were reviewed, and the combined effect of IT and managerial approaches were considered. Therefore, the assessment system could be established by extracting assessment objects from these contents, and then maturity improvement plans were provided.

2.2. Assessment Systems in Other Fields

This study reviewed some representative maturity assessment systems in other informational, digital or intelligent fields, as listed in Table 2. Berghaus and Back [69] indicated that a maturity assessment model should consist of dimensions and criteria that describe the areas of action and maturity stages that indicate the evolution path toward maturity. Though these assessment systems have different assessment targets using different assessment methods, they all set assessment dimensions and criteria, as well as corresponding development guides, to improve maturity. Furthermore, [15,70] set maturity levels to present the overall assessment results and [70] designed a radar chart of dimensions to visualize the strengths and weaknesses in each dimension. The advantage and priority of each assessment system can offer important references to the assessment methods needed in this research: (1) Scoring on dimensions is a quantitative assessment method that has been proven popular and easy to use. (2) Dimension(s) to assess the management capacity should be set, including organizational framework, personnel management, workflow, etc. (3) Assessment results should be displayed clearly from both overall and specific perspectives. For example, maturity levels and radar charts of dimensions can be applied for the overall and specific perspective.

Table 2. Maturity assessment systems in other informational, digital or intelligent fields.

Research	Assessment Target	Assessment Method	Dimension	Advantage and Priority
[71,72]	BIM adoption across markets	Score on dimensions	5	Comprehensive consideration of policies, management and technologies
[15]	BIM capability maturity	Score on dimensions	11	Needed dimensions can be selected from the given 11 dimensions
[16]	Digital maturity for companies	Single choice questions	2	Rapid assessment process
[73]	Digital readiness maturity for manufacturing	Yes or no questions	5	High objectivity
[74]	Project complexity	Analyze from dimensions	5	Detailed assessment results
[75]	Digital maturity of construction projects	Score on dimensions	4	Introduction of the frequency of assessment objects
[70]	Digital maturity on construction site	Score on dimensions	11	Comprehensive assessment objects

3. Methodology

In this study, assessment indicators were set with criteria from different dimensions. The steps to develop the assessment system are listed below in Figure 1.

The first step was the determination of assessment indicators. In the whole assessment system, the maturity scoring table was the most important part, whose basic elements were assessment indicators. There were a large number of assessment indicators extracted, so it was necessary to set up assessment dimensions and reasonably classify all indicators to facilitate the viewing and use of them and also help construction enterprises to assess their own ICM maturity from the perspective of each dimension.

The second step was the calculation of the weights of assessment indicators. This study used questionnaires designed in correspondence with the precedence chart method

(PCM) [76,77] to consult experts on each indicator's importance to the maturity assessment of ICM, and then the weight of each assessment indicator and dimension could be calculated with the results of the questionnaires. Scores of the assessment indicators and dimensions in the maturity scoring table could later be calculated by converting their weights.

The third step was the design of the maturity scoring table. The indicators could not be directly used, so the assessment criteria were set to instruct assessors when scoring. Arranging each indicator according to its dimension and then adding the corresponding score and assessment criterion made a complete maturity scoring table. Necessary instructions for each content should also be written to guide assessors to use it correctly.

The last step was the analysis of the assessment results. The presentation of assessment results should take into account overall and specific aspects. This research used maturity levels to plot the position of the ICM maturity in the whole industry and a radar chart [78] of dimensions to visualize the strengths and weaknesses in each dimension. Therefore, this step included the setting of appropriate maturity levels and the corresponding relationship between score intervals and maturity levels, as well as the design of the radar chart.

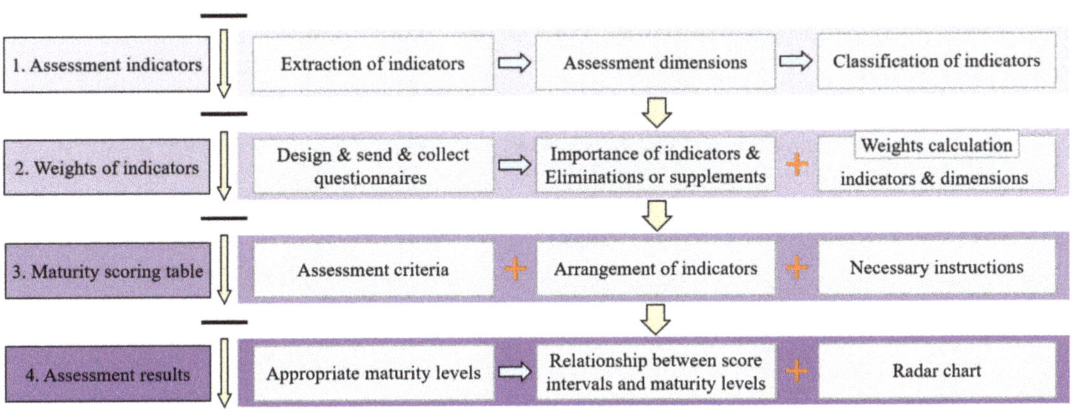

Figure 1. The construction process of the assessment system.

3.1. Determination of the Assessment Indicators

A complete assessment indicator included the assessment object, weight and criterion. The scope of construction management was confined, but intelligent ITs and advanced managerial approaches supporting ICM over the scope at present were unlimited and uncountable, not to mention new ones are being developed. The responsibility of indicators was to screen objects and contents that could best reflect the development of ICM. Therefore, assessment indicators in this study did not include ITs and managerial approaches available for the scope but abstract attributes that reflect the developing situation of these ITs and managerial approaches instead. In a word, it is not detailed ITs, and managerial approaches that enterprises use but attributes enterprises satisfy that develop ICM.

3.1.1. Extraction of Assessment Indicators

The extraction of assessment indicators needed to comprehensively contain factors from the following three aspects.

(1) Construction management scope to determine the assessment scope of indicators. According to the regulations and requirements of the construction industry, aspects and fields that construction management should be responsible for are clarified, which should be covered by assessment indicators.

(2) ITs and managerial approaches supported ICM to abstract attributes as assessment objects. Comprehension of the application status of relevant IT and managerial

approaches could fully tap the application potential and highest maturity of each one, that is, the scale of attributes.

(3) Existing assessment methods and systems refer to successful experiences. As mentioned, there were already advanced maturity assessment methods and systems in other fields, as shown in Table 2. Among them were successful experiences in indicator extraction, assessment dimension setting and assessment methods.

The above factors were extracted from both literature and expert discussions. Referring to the literature provided a comprehensive grasp of the relevant contents, and discussing with experts supplemented details omitted in the literature. The latest management technologies are not published yet, and these matters require attention for practical application. These factors should be considered together during collection. First, discover ITs and managerial approaches that could be applied to fields according to the management scope, as shown in Table 1. Then search the application for relevant ITs and managerial approaches to discern their abstract attributes, which are regarded as preliminary assessment objects. Finally, with reference to the framework of other maturity assessment systems, establish indicators by adjusting and reorganizing preliminary assessment objects according to service objects of all IT and managerial approaches. Ensure that the complete independence of indicators was obtained and there was no overlap between them. The indicator extraction process is shown in Figure 2. The assessment indicators were extracted from the literature and expert discussions through these three procedures. Each indicator does not have direct sources of literature or expert discussion because of these fused procedures.

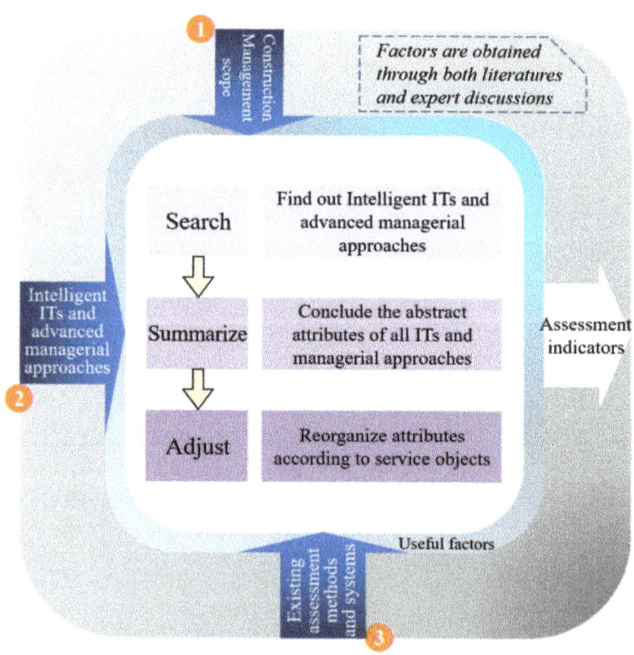

Figure 2. The indicator extraction process.

3.1.2. Setting of Assessment Dimension and Indicator Classification

The literature review indicated that a dimension to assess the management capacity should be set. Among all indicators, there were some that described the personnel organization and management or workflow of construction management, so first, we set a dimension for them. The remaining indicators described ITs and managerial approaches related to construction management itself. Blanco et al. [79] illustrated many specific and clear activities to different technologies used during different construction phases, including the phases of design, preconstruction, construction and operations. These specific activities classified ITs and managerial approaches for construction management properly, but they were too scattered. According to the essence of ICM, dimensions for the remaining indicators could be set by composing these specific activities (definitions of these specific activities can be seen in [79]). Considering the service objects and application fields of the remaining indicators, set four dimensions for them. Each dimension and its components are shown in Table 3. Five assessment dimensions were set following strict internal logic to ensure that there was no overlap between each other. The meaning and description of each assessment dimension are shown in Table 4.

Table 3. Assessment dimensions and their components for IT and managerial approaches.

Assessment Dimension	Activities in [79]
Information collection and monitoring	Materials management, equipment management
Information transmission and aggregation	Field productivity, performance dashboard
Decision-making supported by visualization	Digital design, design management, contract management, document management
Intelligent analysis and deduction	Estimating, construction relationship management, market intelligence, quality control, safety

Table 4. Meaning and description of each assessment dimension.

Assessment Dimension	Meaning and Description
I. Organizational framework and working process	A more suitable organizational framework, more powerful personnel management and more efficient work mode are required by ICM.
II. Information collection and monitoring	Collection and monitoring of various types of construction information and data through collection and measurement equipment arranged on the construction site.
III. Information transmission and aggregation	Transmission and aggregation of information and data collected on-site within the time limit, proper storage and archiving to prevent loss and tampering.
IV. Decision-making supported by visualization	Visualization and modeling of engineering information and simulation of construction operation to support decision-making.
V. Intelligent analysis and deduction	Analyze engineering information with the application of intelligent technologies to provide calculation, detection, prediction, optimization, etc.

In order to make each indicator more consistent with the meaning and description of the corresponding dimension when classifying, indicators were appropriately adjusted after determining the dimensions so that each indicator was clearly and uniquely classified into a certain dimension, and the number of indicators contained in each dimension was as close as possible. After the determination of dimensions and the adjustment of indicators, available indicators can be classified into each dimension.

3.2. Calculation of the Weights of Assessment Indicators

The questionnaire in this study mainly investigated respondents from three aspects: (1) Basic information of respondents, including professional field and title, working post and year, to show the objectivity of the questionnaire. (2) Eliminations or supplements for existing indicators and judgment of the suitability of the setting of the assessment dimensions and the classification of each assessment indicator to ensure the rationality of the assessment dimensions and indicators. (3) The consultation of respondents about the importance of each indicator to assess the maturity of the ICM. The questionnaire is designed following the PCM. A seven-point Likert scale [80] was used for respondents to choose from very low to very high on the importance of each assessment indicator, where each choice corresponded to a point from one to seven, as shown in Figure 3.

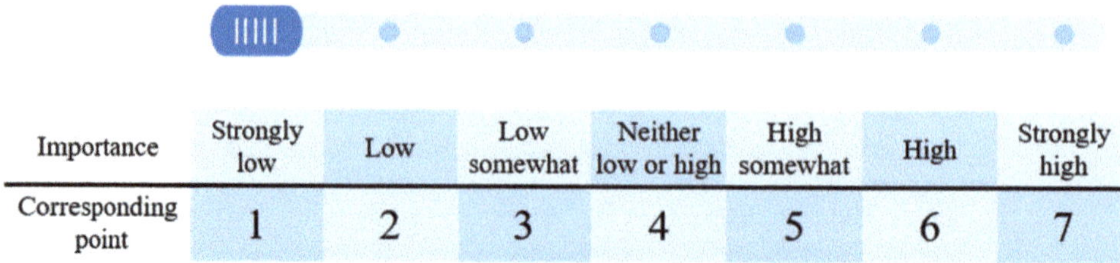

Figure 3. The seven-point Likert scale.

The PCM used a comparison matrix to calculate the weights of target objects, as shown in Table 5. The left columns of the table refer to the comparing objects, while the upper rows are the compared objects. In this study, the respondents' opinions on the importance of each assessment indicator were converted into a point from one to seven by the seven-point Likert scale. If there are n indicators for comparison and the average points of all respondents were calculated, it is easy to know that the indicator with the higher point possesses higher importance. Choose I_1 and I_4 as an example for a pairwise comparison: if I_1 is more important, then $a_{14} = 1$ and $a_{41} = 0$; if I_4 is more important, then $a_{14} = 0$ and $a_{41} = 1$; if I_1 and I_4 are equally important, then $a_{14} = a_{41} = 0.5$. Finally, the weight of each indicator can be calculated:

$$w_i = \frac{s_i}{\sum_{i=1}^{n} s_i}$$

In this function, w_i is the weight of the indicator i, s_i is the sum of all elements in the row i.

Table 5. Comparison matrix of PCM.

Comparison Indicator	I_1	I_2	I_3	I_4	...	I_n	Sum
I_1	$a_{11} = 0.5$	a_{12}	a_{13}	a_{14}		a_{1n}	$s_1 = \sum_{i=1}^{n} a_{1i}$
I_2		0.5					
I_3			0.5				
I_4	a_{41}			0.5			
...					0.5		
I_n						0.5	

3.3. Design of the Scoring Table

The assessment indicator itself was the summary of ITs and managerial approaches that construction enterprises used, and it does not contain the description of the highest intelligent maturity of each IT or managerial approach. Setting assessment criteria for each indicator was essential for assessors to make more accurate judgments when scoring each assessment indicator. After that, the assessment indicators with scores and assessment criteria could be reasonably arranged according to their dimensions. Finally, the complete scoring table was finished when the necessary instructions for each content were written for correct use.

3.4. Analysis of Assessment Results

The scores obtained by the assessors on the assessment of construction enterprises using the scoring table represent their ICM maturities. The score intuitively reflected the overall ICM maturity of each construction enterprise; however, as each enterprise commonly only knows its own score, it cannot plot its position in the whole industry without comparison with other enterprises. Besides, its strengths and weaknesses between different assessment dimensions remain unclarified.

In this study, scores were converted into corresponding maturity levels as the overall presentation of assessment results. The division of maturity levels must ensure that enterprises at the same level possess ICM maturities at roughly the same standard. How many maturity levels should be set? Whether the score intervals between levels should be consistent? How to allocate them if they are inconsistent? These questions can be answered only when preset maturity levels and the corresponding relationship between score intervals and maturity levels are further verified and corrected. Verification and correction of mentioned contents were also realized through expert discussion, so for convenience, they are discussed together in the verification section. Also, in order to clarify the strengths and weaknesses of enterprises in assessment dimensions, the radar chart of dimensions was designed as the specific presentation of assessment results.

4. Verification of the Assessment System

Until now, determined contents were mainly obtained by theoretical analysis, so crucial attributes of this assessment system have not been verified through an application. Wernicke et al. [70] developed a framework for assessing the digital maturity of construction site operations. To examine and verify the framework, they conducted a case study on one construction site. The digital maturity of that site was firstly calculated by the proposed framework. Then they discussed, with the assessor, the detailed status of the digitalization of that site as well as the strengths and weaknesses. The consistency between the assessment results and the discussion results verified the proposed framework. Furthermore, the usability and benefits of this framework were also discussed. In this research, we also used expert discussions to verify our proposed assessment system. However, rather than conducting just one discussion with one assessment target, we conducted two rounds of expert discussions with experts from several construction enterprises. The first round was conducted to verify the scoring table and preset the maturity levels. The second round was conducted to verify those preset maturity levels. The overall verification process is shown in Figure 4. More details were discussed in our expert discussions.

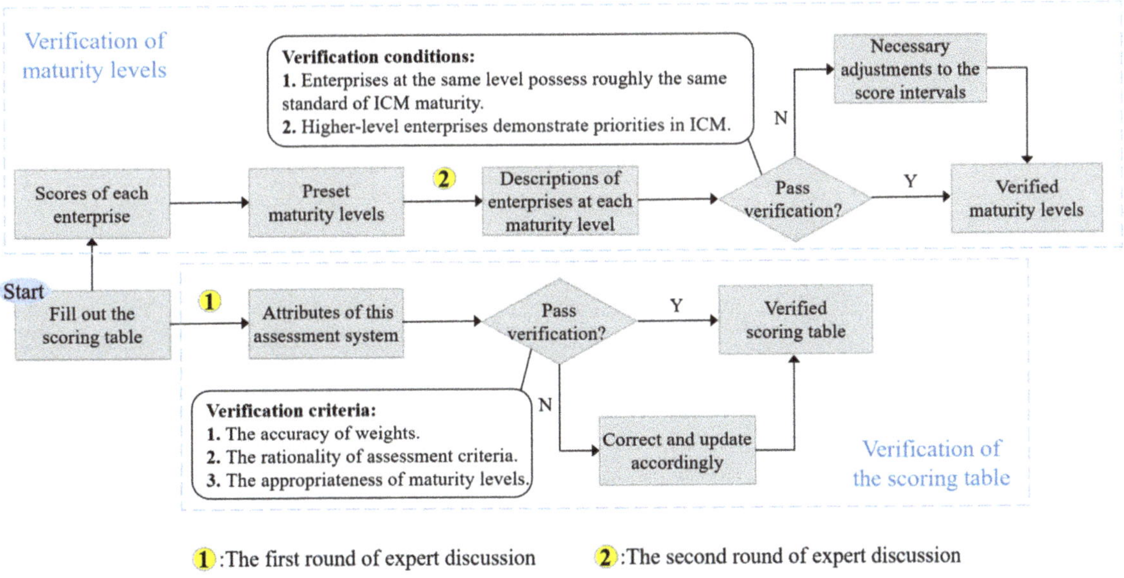

Figure 4. The overall verification process.

4.1. Verification of the Scoring Table

The scoring table was sent to leaders of many construction enterprises, and they filled out the scoring table according to the ICM maturities of their own enterprises. Except for collecting the scores of each enterprise, we discussed with the leaders the crucial attributes of this assessment system. Through the feedback of these leaders, related contents of the scoring table were verified and updated.

4.2. Verification of Maturity Levels

After the collection of the scores of each enterprise, the distribution of all scores was obtained. Rough score intervals were divided by clustering all the scores, and thus the maturity levels were preset. According to the score intervals of the preset maturity levels, participating enterprises were differentiated into different maturity levels. Through detailed discussions with leaders from several representative enterprises at each maturity level, summary descriptions of the overall ICM maturity of enterprises at each maturity level were formed. The setting of the maturity levels must satisfy two criteria: (1) enterprises at the same level possessed roughly the same standard of ICM maturity; (2) enterprises at higher levels demonstrated relatively obvious priorities in ICM maturity compared with those at lower levels. Necessary adjustments to the set score intervals were conducted to satisfy the two principles, and the appropriate corresponding relationship between maturity levels and the scoring results was finally obtained. Thus far, all contents in the whole assessment system have been verified to ensure their accuracy, rationality and appropriateness.

5. Assessment System

5.1. Assessment Indicators and Dimensions

To obtain more extensive responses, we set two criteria for selecting the potential respondents for the questionnaire: (1) respondents with ample work experience in construction management or with ample knowledge of intelligent technologies; (2) respondents from as many professional fields as possible. Leaders from many different construction enterprises were interested in our study, so they helped us select employees from their enterprises according to these two criteria to answer the questionnaires. They told us that older employees tended to have more work experience while younger employees tended to have more knowledge about intelligent technologies. We offered them a QR code, which could be scanned to access our questionnaire and these leaders assigned qualified employees to complete the questionnaire. The questionnaires were collected two weeks after sending the QR code to leaders, and incomplete ones were deleted. Of the remaining questionnaires, 706 were considered valid. The basic information of the respondents is shown in Table 6. Respondents thought that the existing indicators were proper, so there was no need for eliminations or supplements. Furthermore, respondents provided us with suggestions on setting the assessment criteria, such as taking the operability of IT into account.

Table 6. Basic information of respondents.

Professional Field	Amount	Professional Title	Amount	Working Post	Amount	Working Year	Amount
Roads and bridges	483	Primary	131	General supervisor	166	Within 5	73
Tunnels	60	Middle	337	Specific supervisor	346	5–10	140
Traffic engineering	84	Vice-senior	170	Supervisor	148	10–20	301
Electromechanics	10	Senior	68	Enterprise administrator	32	More than 20	192
Others [1]	69			Others [2]	14		

[1] Safety, structure, electric power, water transport, contract, experiment and engineering economy. [2] Vice-general supervisor, consultant and experimentalist.

There were twenty indicators in total, and their accurate weights were calculated using PCM. However, in order to make the data neat and easy to use, it was necessary to make adjustments within an appropriate range and keep the outcome as an integer. The scores of all assessment indicators and dimensions in the maturity scoring table were calculated by converting their weights, as shown in Table 7. They illustrated not only the present situation but also the future development trend of ICM.

(1) The first dimension had the highest weight, of nearly 40%, in the whole scoring table. It described the personnel organization, management and workflow of ICM. Effective

personnel organization and management are a crucial basis for every kind of enterprise and company to maintain competence, and it is the same for construction enterprises. Furthermore, as ICM is still rapidly developing, more and more suitable workflows will always be a key for construction management to develop more intelligence and for ITs and managerial approaches to maximize their superiorities. According to the expert discussion, nearly every enterprise has developed its own cloud platform. They work online, and their organizational framework was adjusted to adapt to the intelligent working mode.

(2) Indicators I-5, I-6, III-3, I-1 and V-2 have the top five highest weights. According to the expert discussion, they were all at present developing focuses for ICM. Most enterprises have been collaborating with researchers from institutes and universities to develop ITs and managerial approaches that these indicators describe. These ITs and managerial approaches have been realized to varying degrees among different enterprises. Due to the high weights these indicators possess, they are now decisive factors for a construction enterprise's ICM maturity.

(3) Indicators IV-3, V-3 and V-1 have relatively low weights. According to the expert discussion, only a few employees were contacted with the ITs and managerial approaches these indicators described. The development levels of these ITs and managerial approaches among different enterprises were pretty low. Publications about these ITs and managerial approaches were mostly limited to theoretical or prospective; they have not been comprehensively and maturely applied to ICM. However, as the publications imply, these ITs and managerial approaches have great benefit and considerable application potential to ICM, including VR [34], AR [35], prediction model [68], time-space conflicts management [67], etc. A number of construction enterprises have included these ITs and managerial approaches in their future development plans, and their weights will definitely increase in the future.

Table 7. Weights of dimensions and indicators.

Assessment Dimension	Assessment Indicator
I. Organizational framework and working process (38)	1. Working post setting (8), 2. Collaboration mode (7), 3. Personnel training (4), 4. Personnel assessment (1), 5. Workflow (9), 6. Transaction tracking (9)
II. Information collection and monitoring (12)	1. Collection range (6), 2. Collection frequency (3), 3. Equipment integration (3)
III. Information transmission and aggregation (22)	1. Transmission speed (6), 2. Information integration (7), 3. Information storage (9)
IV. Decision-making supported by visualization (12)	1. Data visualization (2), 2. Knowledge base management (4), 3. Expanding reality (1), 4. Comprehensive decision (5)
V. Intelligent analysis and deduction (16)	1. Auxiliary calculation (2), 2. Anomaly Identification (8), 3. Deduction and prediction (1), 4. Early warning and optimization (5)

5.2. Maturity Scoring Table

The scoring table is attached in Appendix A. Assessment criteria for each indicator were set according to the application status of the relevant IT or managerial approach.

5.3. Maturity Levels

Maturity levels and corresponding score intervals were set according to the two criteria, as shown in Table 8. Five maturity levels were set, among which there was a particular level called "Minimum Intelligent Maturity". During the expert discussions, we found that ITs and managerial approaches described by many indicators have been well developed. Therefore, more than half of the scores in the scoring table were easily acquired by every construction enterprise. Enterprises with scores less than 60 must demonstrate shortcomings in many aspects under this circumstance. Therefore, the lowest maturity level, "Minimum Intelligent Maturity", was set to conclude that these enterprises were not "intelligent" enough. As mentioned, many leaders of different construction enterprises filled out the scoring table and their scores were collected. The number of enterprises at level 2 was the largest. Few enterprises were located at the Minimum Intelligent Maturity level, and a few enterprises just entered level 3. There was not even one enterprise that entered level 4.

Table 8. Maturity levels and corresponding score intervals.

Maturity Level	Score Interval
Minimum Intelligent Maturity	<60
1	[60,70)
2	[70,80)
3	[80,90)
4	[90,100]

5.4. Radar Chart

The radar chart was designed to compare the relative development of each dimension, but the weights of each dimension were different so that their full scores in the scoring table were also different. When using the radar chart, scores of each dimension in the scoring table should be converted into a centesimal system:

$$R_i = \frac{T_i}{W_i} \times 100$$

In this function, R_i is the score of dimension i in the radar chart, T_i is the score of dimension i in the scoring table and W_i is the weight of dimension i (shown in Table 7). Then the pentagon representing the ICM maturity of the enterprise in these dimensions was drawn.

6. Case Study

The proposed assessment system was examined in a case study of two construction enterprises, A and B. These two enterprises are both located in Hangzhou, a city in southern China. Enterprise A mainly focuses on the construction of highways and canals; enterprise B mainly focuses on the construction of highways and railways. One leader from each of the two enterprises filled out the maturity scoring table according to the actual situations of their enterprises.

6.1. Assessment Results

Their assessment results, using the maturity scoring table, were 78 and 81, respectively. Therefore, the ICM maturities of enterprises A and B were very close. They both almost entirely satisfied the demands of level 2, and enterprise B had just entered level 3. Furthermore, to show their strengths and weaknesses in each dimension, a radar chart of their ICM maturities was drawn, as shown in Figure 5.

Then two discussions with the two leaders were conducted for detailed situations of their ICM maturities, as shown in Table 9. Enterprise A and B paid great attention to the development of ICM, and they made specific development plans for the introduction and application of several ITs and managerial approaches. However, their unsystematic development plans caused different kinds of deficiencies in each dimension and led to relatively elementary maturity levels thereafter. Enterprise A and B are located in the same city, and they have known each other for years. They both commented that their development situations in ICM are very close, and they admitted that their maturity levels are relatively elementary. Not enough new roles have been set in enterprise A, and existing roles have not been thoroughly adapted to ICM, leading to a burden on the newly developed workflow. It is the main reason, according to our assessment system, for enterprise A to have a slight gap in scores compared with enterprise B.

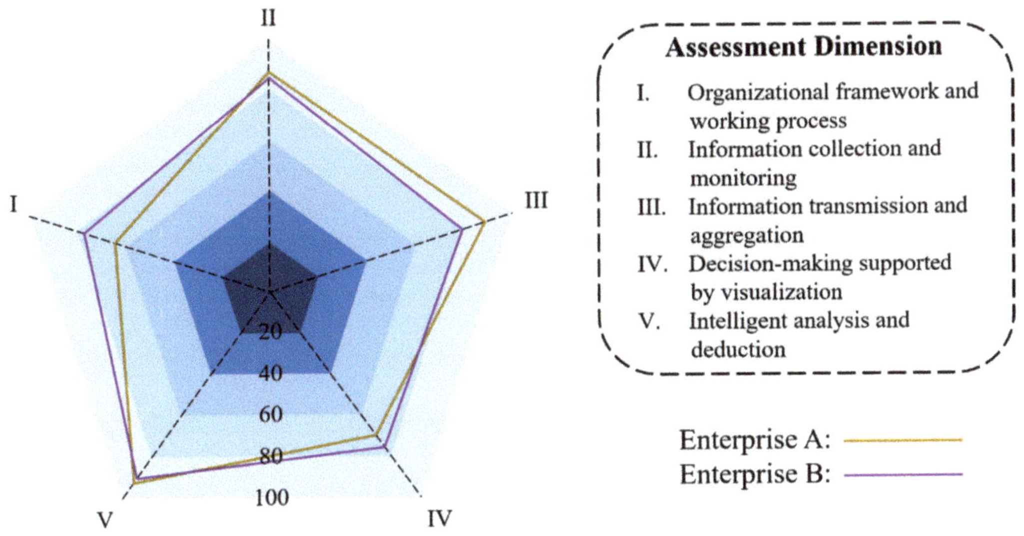

Figure 5. The radar chart of enterprises A and B.

Table 9. Detailed situations of the ICM maturities of enterprises A and B.

Assessment Dimension	Strength	Weakness
I. Organizational framework and working process	A and B: Develop a construction management platform, and transactions are strictly tracked. A: Monthly personnel assessment. B: Many new roles are set.	A and B: Insufficient personnel training. Some employees are still not skilled with the platform, and workflows are still not clear. A: Only a few new roles are set.
II. Information collection and monitoring	A and B: Many data collection devices are arranged on construction site. A: Enough mobile phones for management. B: A few attempts at equipment integration.	A: Almost no equipment integration, efficiency and accuracy of data collection are low.
III. Information transmission and aggregation	A and B: All data and information collected are stored. A: Efficient information aggregation (a new role was set for this).	A and B: Inadequate coverage of network signal on-site; real-time upload and receipt cannot be guaranteed.
IV. Decision-making supported by visualization	A and B: Many kinds of important data and information are displayed in real time by adequate display devices.	A and B: Poor interaction between different kinds of models, no application of VR, AR or MR.
V. Intelligent analysis and deduction	A and B: Many intelligent functions and algorithms are developed. A: Employed a software system team.	A and B: The frequency of use is unstable.

6.2. Validation of the Assessment System

Despite many existing weaknesses and deficiencies for these two construction enterprises, the biggest problem for them is the low capacities of organizational management, which are caused by the incompletely adjusted working posts. Even if there are ITs and managerial approaches developed, they cannot be efficiently applied. Therefore, we strongly recommend these two enterprises set more new roles first, and then other measures can be applied, such as training employees more often, improving the network signal on site, introducing VR, AR or MR, etc. After the discussions, these two leaders claimed that they would adjust their development plans according to our recommendations.

In total, these two leaders spoke highly of the usability of the assessment system. They commented that the assessment criteria helped them to reach deeper into their actual development situation with ICM, and the assessment dimensions are essential for them to find their deficiencies and weaknesses in detail. Therefore, the case study proved the validation of the proposed assessment system, and it can provide not only overall and specific representations of the ICM maturity of a certain enterprise, but also targeted development plans thereafter.

6.3. Discussion

There should also be a potential, anticipated or typical development path to the desired target state after evaluation [81]. The ultimate purpose of the maturity assessment of the ICM is to efficiently and accurately improve its maturity. After assessment using the maturity scoring table, construction enterprises can further understand details of their weaknesses through the indicators with lower scores, and these are the aspects that need to improve most. To provide specific improvement plans for construction enterprises, this study discussed improvement strategies from two perspectives.

The first perspective was the detailed ITs and managerial approaches. This study extracted and summarized ITs and managerial approaches that could improve the maturity of ICM during the extraction of the assessment indicators. They could be used to help construction enterprises that have been assessed to discover and fill the present gaps. These ITs and managerial approaches are listed in Table 1, where they are very detailed and specific. They were simply classified according to their effects on ICM, and their sources are contained in the table, which can be searched and consulted by enterprises to understand the development and application methods of any IT or managerial approach in detail.

The second perspective was the framework of enterprises at level 4. Organizations engaged in information, digitalization, or intelligence have similar frameworks, which include many layers. However, each layer in the framework of different kinds of organizations consisted of different components. The frameworks of highly developed organizations were powerful and comprehensive, where all layers had undergone extensive development. When a certain construction enterprise had the highest maturity of ICM, it must have the fourth maturity level and developed almost every IT and managerial approach in Table 1. Under this circumstance, each layer in its framework possessed adequate components. Representative components are summarized according to the sources of ITs and managerial approaches in Table 1, as shown in Figure 6. This framework can be used for reference by construction enterprises to supplement their own framework and finally continuously improve their ICM maturities.

In general, construction enterprises are able to obtain suitable improvement plans synthetically from these two perspectives after assessment, as shown in Figure 7.

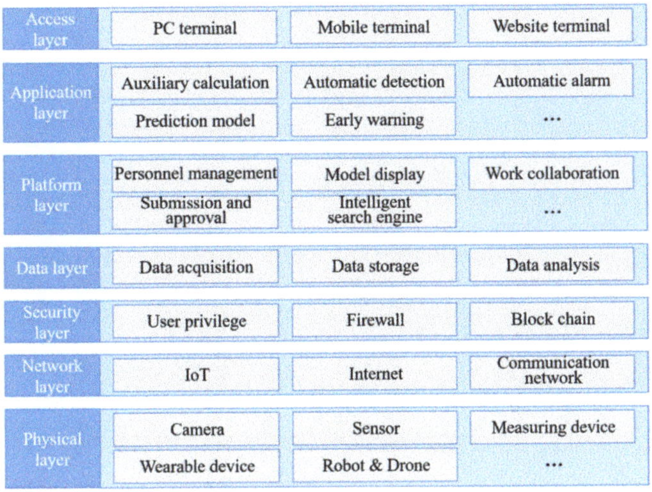

Figure 6. Framework of enterprises at level 4.

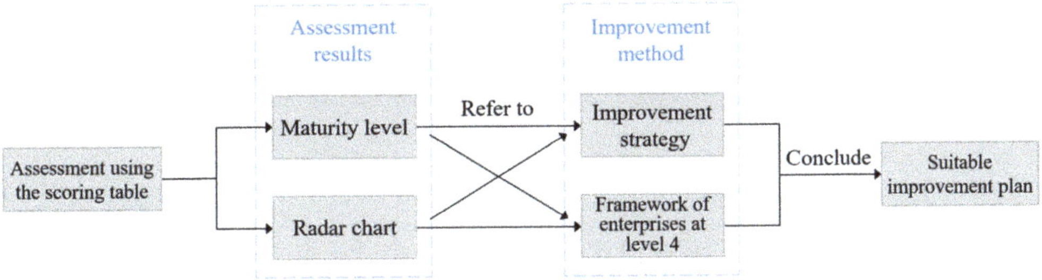

Figure 7. Method to obtain suitable improvement plans.

7. Conclusions

ICM is rapidly developing at present, but there is still a lack of systems, methods or even indicators to systematically assess the maturity of ICM. In this study, we developed a maturity assessment system through literature reviews, questionnaires, expert discussions and a case study. The maturity assessment system consisted of a maturity scoring table, maturity levels and a radar chart of dimensions, which can be used by construction enterprises to assess their ICM maturities and formulate suitable future improvement plans. First, fill out the maturity scoring table based on the ICM development situation of a certain construction enterprise. Then the score can be converted into a maturity level and a dimension radar chart. The position of the enterprise in the whole construction industry and its own strengths and weaknesses can be accurately understood. Finally, a suitable improvement plan for this construction enterprise can be created with reference to the improvement strategy and the framework of enterprises at level 4.

The maturity scoring table consists of five assessment dimensions and twenty assessment indicators. When using it, assessors need to score each indicator based on their subjective judgment of their own construction enterprises. Since the subjective difference is inevitable, it is strongly recommended that the ICM maturities of each enterprise are assessed by more than one leader, and the average of their scores is taken as the final result. According to the scoring table, developing ITs and managerial approaches, which support organizational framework and working process, are of great significance for construction enterprises to reach high ICM maturity. Many enterprises do not pay enough attention to advanced managerial approaches because they have not realized the unimagined progress that these approaches can bring to them. During our discussion with leaders in construction enterprises, we found that many enterprises had already developed adequate IT and managerial approaches. Unfortunately, a large number of ITs and managerial approaches remained unused or insufficiently used for a lack of suitable organizational frameworks and working processes, leading to a low ICM maturity with a huge amount of resource waste.

The study has two limitations. First, the study used the maturity scoring table as the assessment method because of the high complexity of the assessment process of ICM maturity. Although the scoring table is accurate and reliable, it is not efficient enough. Future research can use the assessment indicators in this study to establish a more efficient maturity assessment methods for ICM, such as the assessment methods consisting of yes or no questions, flowcharts or single-choice questions. Second, the assessment indicators and their weights in this study represented the developing situation of ICM in the present and short-term future. There is still a long way for construction management to thoroughly reach intelligence since most construction enterprises are located at level 2, and there is nearly none that have entered level 4. However, there will be a time when most enterprises have entered level 4 because of the efforts that the whole construction industry is making. Meanwhile, more ITs and managerial approaches will come into use and serve as indicators. The weights of all indicators must change with time. Future studies are recommended to add new indicators and correspondingly adjust the weights of all indicators. For example, the ITs and managerial approaches toward automation are rapidly developing and will occupy more and more weight [82,83].

Author Contributions: Conceptualization, J.-R.L.; methodology, C.L. and J.-R.L.; validation, Z.-Z.H. and Y.-C.D.; formal analysis, C.L., C.Y. and W.Z.; investigation, C.L. and J.-R.L.; resources, C.Y. and W.Z.; data curation, C.L.; writing—original draft preparation, C.L. and J.-R.L.; writing—review and editing, C.L., J.-R.L. and Y.-C.D.; visualization, C.L.; supervision, Z.-Z.H., J.-R.L. and Y.-C.D.; project administration, Z.-Z.H., J.-R.L. and Y.-C.D.; funding acquisition, J.-R.L. and Z.-Z.H. All authors have read and agreed to the published version of the manuscript.

Funding: This research was funded by the National Key R&D Program of China (grant number 2018YFD1100900) and the Science and Technology Plan Project of the Zhejiang Provincial Department of Transportation (grant number 2020061).

Data Availability Statement: Data sharing is not applicable.

Conflicts of Interest: The authors declare no conflict of interest.

Appendix A

Table A1. The maturity scoring table for ICM.

Dimension	Indicator	Criterion	Score
I. Organizational framework and working process	1. Working post setting: Adjust former posts and set specific responsibilities and corresponding organizational relationships for all posts.	Reasonable, perfect personnel allocation and gross wage.	8
	2. Collaboration mode: Conduct online personnel management and workflow interaction based on the management platform.	Real-time uploads, reminders and feedback.	7
	3. Personnel training: Train personnel in a variety of ways to adapt to the working mode of ICM.	Check regularly and trace the training data.	4
	4. Personnel assessment: Use a variety of data sources to assess the attendance and performance of personnel.	Quantitative, qualitative and objective assessments.	1
	5. Workflow: Assign designated, responsible personnel to complete the workflow of each task with a clear work sequence.	Smooth workflow with high efficiency.	9
	6. Transaction tracking: Record and track the processing flow and relevant responsible personnel for all transactions.	Clearly record the process and responsible person.	9

Table A1. *Cont.*

Dimension	Indicator	Criterion	Score
II. Information collection and monitoring	1. Collection range: Collect as many types of data and information as possible on-site and make the collection range as wide as possible.	Collect comprehensively and all key areas covered.	6
	2. Collection frequency: Collect data and information as frequently as possible to improve their continuity.	Avoid interruptions in data and information acquisition.	3
	3. Equipment integration: Develop equipment that collects multiple data and information simultaneously and efficiently.	Improve the accuracy of data and information collection.	3
III. Information transmission and aggregation	1. Transmission speed: Improve the transmission speed to ensure the timeliness of data and information transmission.	Ensure real-time uploading and receiving on-site.	6
	2. Information integration: Integrate, fuse, summarize and transform the collected data and information.	Automatic preprocessing of data and information.	7
	3. Information storage: Archive and save the collected data and information to support efficient utilization and security management.	Store all data for the whole life cycle of the project.	9
IV. Decision-making supported by visualization	1. Data visualization: Model, visualize and simulate using all kinds of construction data and information.	Concrete, intuitive and accurate visualizations.	2
	2. Knowledge base management: Upload construction data and information to the platform and set search functions for viewing.	Comprehensive categories and accurate search results.	4
	3. Expanding reality: Assist scheduling and management with the help of VR, AR, MR and other extended reality technologies.	Widely used in the complete workflow.	1
	4. Comprehensive decision: Real-time display of the data and information being monitored and collected.	Display on a variety of devices widely.	5
V. Intelligent analysis and deduction	1. Auxiliary calculation: Intelligently calculate the schedule, cost, etc., with collected data and information.	Introduce intelligent computing for all calculation processes.	2
	2. Anomaly identification: Identify occurring abnormal conditions, including automatic detection of construction results and automatic alarm of unsafe behaviors, etc.	Comprehensive categories, fast detection and identification speed with high accuracy.	8
	3. Deduction and prediction: Establish a prediction model based on collected data and information to predict the work focus and potential risks in the next stage.	The model considers all types of data and information, real-time update.	1
	4. Early warning and optimization: Adjust the management plan according to the prediction results to avoid possible risks and improve the project management ability.	Real-time optimize the management plan and implement it on time.	5

References

1. Ofori, G. Nature of the construction industry, its needs and its development: A review of four decades of research. *J. Constr. Dev. Ctries.* **2015**, *20*, 115–135.
2. National Bureau of Statistics of China. Available online: http://data.stats.gov.cn/easyquery.htm?cn=C01 (accessed on 27 September 2022).
3. Lu, Y.; Gong, P.; Tang, Y.; Sun, S.; Li, Q. BIM-integrated construction safety risk assessment at the design stage of building projects. *Autom. Constr.* **2021**, *124*, 103553. [CrossRef]
4. Ameh, O.J.; Soyingbe, A.A.; Odusami, K.T. Significant factors causing cost overruns in telecommunication projects in Nigeria. *J. Constr. Dev. Ctries.* **2010**, *15*, 49–67.
5. Construction Management Association of America. Available online: https://www.cmaanet.org/about-us/what-construction-management (accessed on 30 September 2022).
6. Lasi, H.; Fettke, P.; Kemper, H.-G.; Feld, T.; Hoffmann, M. Industry 4.0. *Bus. Inf. Syst. Eng.* **2014**, *6*, 239–242. [CrossRef]
7. Sawhney, A.; Riley, M.; Irizarry, J. *Construction 4.0—An Innovation Platform for the Built Environment*; Routledge: New York, NY, USA, 2020; ISBN 978-0-429-39810-0.
8. Wu, C.; Li, X.; Guo, Y.; Wang, J.; Ren, Z.; Wang, M.; Yang, Z. Natural language processing for smart construction: Current status and future directions. *Autom. Constr.* **2022**, *134*, 104059. [CrossRef]
9. Forcael, E.; Ferrari, I.; Opazo-Vega, A.; Pulido-Arcas, J.A. Construction 4.0: A literature review. *Sustainability* **2020**, *12*, 9755. [CrossRef]

10. García de Soto, B.; Agustí-Juan, I.; Joss, S.; Hunhevicz, J. Implications of Construction 4.0 to the workforce and organizational structures. *Int. J. Constr. Manag.* **2022**, *22*, 205–217. [CrossRef]
11. Cheng, G.-J.; Liu, L.-T.; Qiang, X.-J.; Liu, Y. Industry 4.0 development and application of intelligent manufacturing. In Proceedings of the 2016 International Conference on Information System and Artificial Intelligence (ISAI), Hong Kong, China, 24–26 June 2016; pp. 407–410.
12. De Bruin, T.; Rosemann, M.; Freeze, R.; Kaulkarni, U. Understanding the main phases of developing a maturity assessment model. In Proceedings of the 16th Australasian Conference on Information Systems (ACIS), Sydney, Australia, 29 November–2 December 2005; pp. 8–19.
13. Pfeffer, J.; Sutton, R.I. Knowing "what" to do is not enough: Turning knowledge into action. *Calif. Manag. Rev.* **1999**, *42*, 83–108.
14. Mettler, T. Maturity assessment models: A design science research approach. *Int. J. Soc. Syst. Sci.* **2011**, *3*, 81–98. [CrossRef]
15. National BIM Standard-United States. Available online: https://www.nationalbimstandard.org (accessed on 30 September 2022).
16. Westerman, G.; Calméjane, C.; Bonnet, D.; Ferraris, P.; McAfee, A. Digital Transformation: A Road-Map for Billion-Dollar Organizations. MIT Center for Digital Business and Capgemini Consulting. 2011. Available online: https://www.capgemini.com/wp-content/uploads/2017/07/Digital_Transformation__A_Road-Map_for_Billion-Dollar_Organizations.pdf (accessed on 30 September 2022).
17. Zheng, Z.; Lu, X.-Z.; Chen, K.-Y.; Zhou, Y.-C.; Lin, J.-R. Pretrained domain-specific language model for natural language processing tasks in the AEC domain. *Comput. Ind.* **2022**, *142*, 103733. [CrossRef]
18. Deng, H.; Xu, Y.; Deng, Y.; Lin, J. Transforming knowledge management in the construction industry through information and communications technology: A 15-year review. *Autom. Constr.* **2022**, *142*, 104530. [CrossRef]
19. Woo, S.; Chang, C.-K.; Lee, S.; Cho, C.-S. Comparison of efficiency and satisfaction level on different construction management methods for public construction projects in Korea. *KSCE J. Civ. Eng.* **2019**, *23*, 2417–2425. [CrossRef]
20. Sun, H.; Liu, Z. Research on Intelligent Dispatching System Management Platform for Construction Projects Based on Digital Twin and BIM Technology. *Adv. Civ. Eng.* **2022**, *2022*, 8273451. [CrossRef]
21. Xu, X.; Jin, F.; Fu, L.; Zhou, H. Construction information management system for rock-filled concrete dam (CIM4R). In Proceedings of the 2021 7th International Conference on Hydraulic and Civil Engineering & Smart Water Conservancy and Intelligent Disaster Reduction Forum (ICHCE & SWIDR), Nanjing, China, 6–8 November 2021; pp. 457–460.
22. Ho, S.-P.; Tserng, H.-P.; Jan, S.-H. Enhancing knowledge sharing management using BIM technology in construction. *Sci. World J.* **2013**, *2013*, 170498. [CrossRef]
23. Elghaish, F.; Chauhan, J.K.; Matarneh, S.; Rahimian, F.P.; Hosseini, M.R. Artificial intelligence-based voice assistant for BIM data management. *Autom. Constr.* **2022**, *140*, 104320. [CrossRef]
24. Mendoza, J.; de-la-Bandera, I.; Álvarez-Merino, C.S.; Khatib, E.J.; Alonso, J.; Casalderrey-Díaz, S.; Barco, R. 5G for Construction: Use Cases and Solutions. *Electronics* **2021**, *10*, 1713. [CrossRef]
25. Hu, Y. Design and Implementation of Using Intelligent Attendance System to Assess Human Resource Management. In Proceedings of the 2021 International Conference on Aviation Safety and Information Technology, Changsha, China, 18–20 December 2021; pp. 128–132.
26. Granda, J.C.; Nuño, P.; Suárez, F.J.; Pérez, M.A. E-pSyLon: A synchronous e-learning platform for staff training in large corporations. *Multimed. Tools Appl.* **2013**, *66*, 431–463. [CrossRef]
27. Detsimas, N.; Coffey, V.; Sadiqi, Z.; Li, M. Workplace training and generic and technical skill development in the Australian construction industry. *J. Manag. Dev.* **2016**, *35*, 486–504. [CrossRef]
28. Zhou, B.; Xu, M. Research on the construction of a comprehensive information platform for publishing enterprises with efficient resource reuse. In Proceedings of the 2021 2nd International Conference on Computers, Information Processing and Advanced Education, Ottawa, ON, Canada, 25–27 May 2021; pp. 1397–1402.
29. Zhang, M.; Shi, R.; Yang, Z. A critical review of vision-based occupational health and safety monitoring of construction site workers. *Saf. Sci.* **2020**, *126*, 104658. [CrossRef]
30. Yang, Z.; Yuan, Y.; Zhang, M.; Zhao, X.; Tian, B. Assessment of Construction Workers' Labor Intensity Based on Wearable Smartphone System. *J. Constr. Eng. Manag.* **2019**, *145*, 4019039. [CrossRef]
31. Simpeh, F.; Amoah, C. COVID-19 guidelines incorporated in the health and safety management policies of construction firms. *J. Eng. Des. Technol.* **2021**, *20*, 6–23. [CrossRef]
32. Tak, A.N.; Taghaddos, H.; Mousaei, A.; Bolourani, A.; Hermann, U. BIM-based 4D mobile crane simulation and onsite operation management. *Autom. Constr.* **2021**, *128*, 103766. [CrossRef]
33. Li, L.; Yuan, J.; Tang, M.; Xu, Z.; Xu, W.; Cheng, Y. Developing a BIM-enabled building lifecycle management system for owners: Architecture and case scenario. *Autom. Constr.* **2021**, *129*, 103814. [CrossRef]
34. dela Cruz, O.G.; Dajac, J.S. Virtual Reality (VR): A Review on Its Application in Construction Safety. *Turk. J. Comput. Math. Educ.* **2021**, *12*, 3379–3393.
35. Garbett, J.; Hartley, T.; Heesom, D. A multi-user collaborative BIM-AR system to support design and construction. *Autom. Constr.* **2021**, *122*, 103487. [CrossRef]
36. Kim, C.; Park, T.; Lim, H.; Kim, H. On-site construction management using mobile computing technology. *Autom. Constr.* **2013**, *35*, 415–423. [CrossRef]

37. Dave, B.; Kubler, S.; Främling, K.; Koskela, L. Opportunities for enhanced lean construction management using Internet of Things standards. *Autom. Constr.* **2016**, *61*, 86–97. [CrossRef]
38. Sheng, D.; Ding, L.; Zhong, B.; Love, P.E.D.; Luo, H.; Chen, J. Construction quality information management with blockchains. *Autom. Constr.* **2020**, *120*, 103373. [CrossRef]
39. Dominicis, C.M.D.; Depari, A.; Flammini, A.; Rinaldi, S.; Sisinni, E. Smartphone based localization solution for construction site management. In Proceedings of the 2013 IEEE Sensors Applications Symposium Proceedings, Galveston, TX, USA, 19–21 February 2013; pp. 43–48.
40. Hasan, S.M.; Lee, K.; Moon, D.; Kwon, S.; Jinwoo, S.; Lee, S. Augmented reality and digital twin system for interaction with construction machinery. *J. Asian Arch. Build. Eng.* **2021**, *21*, 564–574. [CrossRef]
41. Ahn, C.R.; Lee, S.; Peña-Mora, F. Application of low-cost accelerometers for measuring the operational efficiency of a construction equipment fleet. *J. Comput. Civ. Eng.* **2015**, *29*, 04014042. [CrossRef]
42. Kasim, N.; Latiffi, A.A.; Fathi, M.S. RFID Technology for materials management in construction projects–A Review. *Int. J. Constr. Eng. Manag.* **2013**, *2*, 7–12.
43. Gulghane, A.A.; Khandve, P.V. Management for construction materials and control of construction waste in construction industry: A review. *Int. J. Eng. Res. Appl.* **2015**, *5*, 59–64.
44. Kang, H.; Sung, S.; Hong, J.; Jung, S.; Hong, T.; Park, H.S.; Lee, D.-E. Development of a real-time automated monitoring system for managing the hazardous environmental pollutants at the construction site. *J. Hazard. Mater.* **2021**, *402*, 123483. [CrossRef]
45. KIM, N.; Denissen, L.J.A.; LOOI, E.; Yong, C.N.K.; Hong, F.C.; Ping, F.G.G.; He, Z.; Kamaruddin, N.S.B. Environment Monitoring Mesh System (EM2S). In Proceedings of the 2019 IEEE 5th International Conference on Mechatronics System and Robots (ICMSR), Singapore, 3–5 May 2019; pp. 97–101.
46. El-Omari, S.; Moselhi, O. Integrating automated data acquisition technologies for progress reporting of construction projects. *Autom. Constr.* **2011**, *20*, 699–705. [CrossRef]
47. Inoue, F.; Ohmoto, E. Development of high accuracy position marking system applying mark robot in construction site. In Proceedings of the 50th SICE Annual Conference, Tokyo, Japan, 13–18 September 2011; pp. 2413–2417.
48. Chen, L.; Luo, H. A BIM-based construction quality management model and its applications. *Autom. Constr.* **2014**, *46*, 64–73. [CrossRef]
49. Chun, P.J.; Yamane, T.; Maemura, Y. A deep learning based image captioning method to automatically generate comprehensive explanations of bridge damage. *Comput.-Aided Civ. Eng. Infrastruct. Eng.* **2021**, *37*, 1387–1401. [CrossRef]
50. Duan, R.; Deng, H.; Tian, M.; Lin, J. SODA: A large-scale open site object detection dataset for deep learning in construction. *Autom. Constr.* **2022**, *142*, 104499. [CrossRef]
51. Kim, J. A study to investigate using mobile devices in the construction management classroom as rationalized by the needs of industry. In Proceedings of the Creative Construction Conference, Ljubljana, Slovenia, 30 June–3 July 2018; pp. 121–128.
52. Omar, T.; Nehdi, M.L. Data acquisition technologies for construction progress tracking. *Autom. Constr.* **2016**, *70*, 143–155. [CrossRef]
53. Igwe, U.S.; Mohamed, S.F.; Azwarie, M.B.M.D.; Paulson Eberechukwu, N. Recent Developments in Construction Post Contract Cost Control Systems. *J. Comput. Theor. Nanosci.* **2020**, *17*, 1236–1241. [CrossRef]
54. Zhang, J.P.; Hu, Z.Z. BIM-and 4D-based integrated solution of analysis and management for conflicts and structural safety problems during construction: 1. Principles and methodologies. *Autom. Constr.* **2011**, *20*, 155–166. [CrossRef]
55. Abdelkhalek, H.A.; Refaie, H.S.; Aziz, R.F. Optimization of time and cost through learning curve analysis. *Ain Shams Eng. J.* **2020**, *11*, 1069–1082. [CrossRef]
56. Kim, K.; Lee, G.; Kim, S. A study on the application of blockchain technology in the construction industry. *KSCE J. Civ. Eng.* **2020**, *24*, 2561–2571. [CrossRef]
57. Zhang, Y.; Wang, T.; Yuen, K.-V. Construction site information decentralized management using blockchain and smart contracts. *Comput.-Aided Civ. Eng. Infrastruct. Eng.* **2021**, *37*, 1450–1467. [CrossRef]
58. Chen, G.; Luo, Y. A BIM and ontology based intelligent application framework. In Proceedings of the 2016 IEEE Advanced Information Management, Communicates, Electronic and Automation Control Conference (IMCEC, 2016), Xi'an, China, 3–5 October 2016; pp. 494–497.
59. Leng, S.; Lin, J.-R.; Li, S.-W.; Hu, Z.-Z. A Data Integration and Simplification Framework for Improving Site Planning and Building Design. *IEEE Access* **2021**, *9*, 148845–148861. [CrossRef]
60. Wu, L.-T.; Lin, J.-R.; Leng, S.; Li, J.-L.; Hu, Z.-Z. Rule-based information extraction for mechanical-electrical-plumbing-specific semantic web. *Autom. Constr.* **2022**, *135*, 104108. [CrossRef]
61. Guo, S.; Li, J.; Liang, K.; Tang, B. Improved safety checklist analysis approach using intelligent video surveillance in the construction industry—A case study. *Int. J. Occup. Saf. Ergon.* **2019**, *27*, 1064–1075. [CrossRef]
62. Luo, H.; Liu, J.; Fang, W.; Love, P.E.; Yu, Q.; Lu, Z. Real-time smart video surveillance to manage safety: A case study of a transport mega-project. *Adv. Eng. Inform.* **2020**, *45*, 101100. [CrossRef]
63. Thomas, S.; Teizer, J.; Reynolds, M. SmartHat: A Battery-free Worker Safety Device Employing Passive UHF RFID Technology. In Proceedings of the International Conference on RFID, Orlando, FL, USA, 12–14 April 2011; pp. 85–90.
64. Hwang, S. Ultra-wide band technology experiments for real-time prevention of tower crane collisions. *Autom. Constr.* **2012**, *22*, 545–553. [CrossRef]

65. Su, Y.; Mao, C.; Jiang, R.; Liu, G. Data-driven fire safety management at building construction sites: Leveraging CNN. *J. Manag. Eng.* **2021**, *37*, 04020108. [CrossRef]
66. Boje, C.; Guerriero, A.; Kubicki, S.; Rezgui, Y. Towards a semantic Construction Digital Twin: Directions for future research. *Autom. Constr.* **2020**, *114*, 103179. [CrossRef]
67. Dashti, M.S.; RezaZadeh, M.; Khanzadi, M. Taghaddos, H. Integrated BIM-based simulation for automated time-space conflict management in construction projects. *Autom. Constr.* **2021**, *132*, 103957. [CrossRef]
68. Hoła, B.; Topolski, M.; Szer, I.; Szer, J.; Blazik-Borowa, E. Prediction model of seasonality in the construction industry based on the accidentality phenomenon. *Arch. Civ. Mech. Eng.* **2022**, *22*, 30. [CrossRef]
69. Berghaus, S.; Back, A. Stages in Digital Business Transformation: Results of an Empirical Maturity Study. In Proceedings of the Tenth Mediterranean Conference on Information Systems (MCIS), Paphos, Cyprus, 4–6 September 2016.
70. Wernicke, B.; Stehn, L.; Sezer, A.A.; Thunberg, M. Introduction of a digital maturity assessment framework for construction site operations. *Int. J. Constr. Manag.* **2021**, *21*, 1–11. [CrossRef]
71. Succar, B.; Kassem, M. Macro-BIM adoption: Conceptual structures. *Autom. Constr.* **2015**, *57*, 64–79. [CrossRef]
72. Kassem, M.; Succar, B. Macro BIM adoption: Comparative market analysis. *Autom. Constr.* **2017**, *81*, 286–299. [CrossRef]
73. De Carolis, A.; Macchi, M.; Negri, E.; Terzi, S. A Maturity Model for Assessing the Digital Readiness of Manufacturing Companies. In Proceedings of the APMS Conference 2017–Advances in Production Management Systems, Hamburg, Germany, 3–7 September 2017; Springer: Berlin/Heidelberg, Germany, 2017; pp. 13–20.
74. Rolstadås, A.; Schiefloe, P.M. Modelling project complexity. *Int. J. Manag. Proj. Bus.* **2017**, *10*, 295–314. [CrossRef]
75. Sezer, A.A.; Thunberg, M.; Wernicke, B. Digitalization index: Developing a model for assessing the degree of digitalization of construction projects. *J. Constr. Eng. Manag.* **2021**, *147*, 04021119. [CrossRef]
76. You, Z.; Fu, H.; Shi, J. Design-by-analogy: A characteristic tree method for geotechnical engineering. *Autom. Constr.* **2018**, *87*, 13–21. [CrossRef]
77. Yu, X.; Ma, S.; Cheng, K.; Kyriakopoulos, G.L. An evaluation system for sustainable urban space development based in green urbanism principles—A case study based on the Qin-Ba mountain area in China. *Sustainability* **2020**, *12*, 5703. [CrossRef]
78. Li, J.; Chen, C.; Li, Y.; Wu, H.; Li, X. Difficulty assessment of shoveling stacked materials based on the fusion of neural network and radar chart information. *Autom. Constr.* **2021**, *132*, 103966. [CrossRef]
79. Blanco, J.; Mullin, A.; Pandya, K.; Sridhar, M. The new age of engineering and construction technology. *McKinsey Q.* **2017**, 1–16.
80. Allen, I.E.; Seaman, C.A. Likert scales and data analyses. *Qual. Prog.* **2007**, *40*, 64–65.
81. Pöppelbuß, J.; Röglinger, M. What makes a useful maturity model? A framework of general design principles for maturity models and its demonstration in business process management. In Proceedings of the 19th European Conference on Information Systems, ECIS 2011 Proceedings, Helsinki, Finland, 9–11 June 2011.
82. Zhou, Y.-C.; Zheng, Z.; Lin, J.-R.; Lu, X.-Z. Integrating NLP and context-free grammar for complex rule interpretation towards automated compliance checking. *Comput. Ind.* **2022**, *142*, 103746. [CrossRef]
83. Zheng, Z.; Zhou, Y.-C.; Lu, X.-Z.; Lin, J.-R. Knowledge-informed semantic alignment and rule interpretation for automated compliance checking. *Autom. Constr.* **2022**, *142*, 104524. [CrossRef]

buildings

MDPI

Article

Factors Affecting BIM Adoption in the Yemeni Construction Industry: A Structural Equation Modelling Approach

Ali Hamoud Mssoud Al-sarafi, Aidi Hizami Alias *, Helmi Zulhaidi Mohd. Shafri and Fauzan Mohd. Jakarni

Department of Civil Engineering, Faculty of Engineering, Universiti Putra Malaysia, Serdang 43400 UPM, Malaysia

* Correspondence: aidihizami@upm.edu.my

Abstract: The construction sector is one of Yemen's most important economic pillars. Building information modelling (BIM) is a new information technology implementation that can create an intelligent digital design of buildings to support a variety of tasks and provides a wide range of benefits throughout the project life cycle. However, BIM is not widely embraced in Yemeni construction firms. Compared with other countries, Yemen presents a unique case for BIM adoption due to the ongoing war in the country, which will assist in rapid rebuilding processes. Thus, a complete and systematic investigation of the factors affecting BIM adoption in the Yemeni construction industry is required. This study utilises five categories of impacting factors: Technology, Process, Policy, People, and the Environment to model the strategic implementation for BIM in the Yemeni construction industry. A random sample was used to achieve homogeneity and increase the consistency and quality of data. Purposive sampling was used to choose participants for the framework validation. The data were analysed using partial least squares structural equation modelling (PLS-SEM), and the key factors influencing BIM adoption were determined and modelled. The results show multivariate results indicate a high correlation within the measurement model for all factors affecting BIM adoption in Yemen. In addition, the developed model was deemed to fit because the analysis result of the model's coefficient of determination test (R^2) is BIM adoption having 0.437, Environment at 0.589, and People having 0.310, demonstrating high acceptance. Moreover, the results reveal a high correlation between policy and people (>0.50), while the environment significantly affected BIM adoption (0.304). Overall, the model illustrated how various factors influence BIM adoption. The created framework highlights the importance of understanding BIM adoption concepts and challenges in the Yemeni construction industry. It is believed that this study highlights the BIM implementation in developing countries such as Yemen and the possibility of implementing the proposed method in other countries to develop their own BIM implementation strategy.

Keywords: building information modelling; BIM adoption; construction industry; technology factors; Yemen; partial least square; PLS; SEM

Citation: Al-sarafi, A.H.M.; Alias, A.H.; Shafri, H.Z.M.; Jakarni, F.M. Factors Affecting BIM Adoption in the Yemeni Construction Industry: A Structural Equation Modelling Approach. *Buildings* **2022**, *12*, 2066. https://doi.org/10.3390/buildings12122066

Academic Editors: Hongling Guo, Jia-Rui Lin and Yantao Yu

Received: 25 August 2022
Accepted: 9 November 2022
Published: 25 November 2022

Publisher's Note: MDPI stays neutral with regard to jurisdictional claims in published maps and institutional affiliations.

Copyright: © 2022 by the authors. Licensee MDPI, Basel, Switzerland. This article is an open access article distributed under the terms and conditions of the Creative Commons Attribution (CC BY) license (https://creativecommons.org/licenses/by/4.0/).

1. Introduction

Construction industries worldwide use building information modelling (BIM) to plan, build, and monitor their projects. The BIM adoption rate is increasing in several countries in the public and private construction sectors [1]. However, BIM has not been studied adequately for building project management, and there is a lack of extensive evaluations that objectively analyse the advancements in BIM applications in the construction sector [2]. Succar [3] discusses the BIM framework's fields, stages, and lenses. BIM competencies include implementation maturity, activity domain, level/scope, and requirements assessment. Adopting BIM requires professionals and organisations, not software or technology; three knowledge models are described (i.e., in Technology, Process, and Policy developments).

The significance, drivers, obstacles, and factors for government policies on BIM adoption methods need to be determined and implemented [4]. The project lifecycle benefits greatly from BIM. These advantages are, nevertheless, compromised by challenges and the construction industry's failure to integrate BIM technology properly [5]. The primary elements influencing BIM adoption in the global construction sector are processes, people, and technology [6]. Babatunde [7] highlighted the barriers to BIM adoption and implementation in the Nigerian construction sector and discovered that enterprises are still underperforming in BIM adoption and implementation. Therefore, further work is needed to deepen the BIM adoption and acceptance strategies. The unpleasant precedent predicted by building professionals since BIM's inception in Malaysia has prompted more profound research into BIM adoption [8].

Despite several attempts to analyse and model the BIM framework in the construction industry, current research has flaws in how it views BIM as a separate technical and administrative process rather than a working process that is supposed to be instantaneous or simultaneous and emergent. The BIM knowledge framework, practical implementation at the industry level, BIM application, and BIM acceptance are some of the frameworks proposed to disperse BIM application in the construction sector [8]. Numerous studies looked at technical integration, the usage of BIM tools, and model-sharing concerns, which dwell on a technical aspect of BIM technology. However, this research was limited to the use of BIM in the building sector in general. There is a shortage of in-depth examination of the obstacles and success factors for effective BIM deployment among local organisations or at the organisational level in building projects [9].

The Yemeni construction industry experienced noticeable growth in early 2011. After the war started in 2015, there was a considerable decline in the construction industry, and many projects were suspended. In early 2016, people got used to the instability within the war environment and started to adopt the new norms of life, resulting in the unsustainable building of their houses despite the shortage of materials and machinery. Additionally, ongoing projects were funded by different non-governmental agencies and the Social Fund for Development [9].

Complexity, instability, and time constraints are among the significant issues affecting project delivery in the construction industry. Yemen's construction industry is no exception [10–13]. Using 2D CAD for many years has not enhanced collaboration or project performance. Yemen's construction sector is growing. Yemeni construction industry stakeholders are working to improve construction efficiency. Still, the sector has technical obstacles, such as a lack of acceptable building materials, labour construction technology and a lack of BIM awareness and knowledge [12]. Therefore, techniques for BIM application in the Yemeni construction sector should be studied to improve project cooperation and performance [12,13].

There are limited studies on BIM in the Yemen construction industry, and construction stakeholders are resistant to embracing changes, which encourages traditional building practices linked with incorrect planning and monitoring resulting in cost overruns, schedule overruns, low quality, and project failure [12]. Ineffective regulation and law, limited utilisation of local construction technology, inadequate financial structure, and incorrect use of local building resources were other problems [13]. An extensive review of the relevant literature shows that most construction industries in Yemen still use 2D CAD. Holistic research on BIM adoption, especially in Yemen, is absent. This research integrated and examined BIM implementation aspects across the building process to produce an effective and complete implementation plan in Yemen. For effective BIM adoption, local organisations and the entire building industry need in-depth analyses of the challenges and success factors.

This study fills this gap and narrows the scope by focusing on the Yemeni industry scenario. Moreover, this study investigated the extent of BIM adoption in construction projects, particularly among local organisations in Yemen, and has contributed to the body of knowledge due to the limited literature on BIM adoption in the Yemeni construction industry. It also examines the factors affecting BIM adoption, knowledge, and awareness in the Yemeni construction sector. As a result of the thorough literature review, in-depth discussion during the interviews, and factor analysis evaluation, new factor groupings are also identified. In addition, the study also created an SEM model that detailed the correlations between the factors that influence the adoption of BIM. The model helps to better understand how independent factors impact the adoption of BIM. The created framework is an excellent example of the significance of understanding the BIM adoption ideas and challenges and assessing the elements and BIM technology motivators to accomplish effective adoption in building projects. Other researchers might use the suggested approach to evaluate its value in promoting BIM adoption in construction projects.

2. Literature Review

The need for BIM in the construction industry became apparent after reviewing papers and studies. A substantial amount of information on BIM, including definitions and relevance to the construction industry, has been published in the academic and non-academic literature. Some frameworks suggested for implementing BIM in the construction industry include the BIM knowledge framework, industry-level implementation, and BIM acceptance. Other studies investigated the concerns of model sharing, BIM tool use, and technology integration. Most construction industries in underdeveloped nations are BIM infant industries that struggle with adoption and implementation [14]. Therefore, BIM infants are confronted with obstacles varying from innovative features to internal and exterior settings. As noted in several developed nations, the lack of official backing for BIM in most countries is a substantial barrier.

Recent research by [15] in "macro-BIM adoption: comparative market analysis" contributes to comparative market research. That paper offers suggestions for newly adopting countries seeking to deploy macro-BIM. In this expanding industry, precedent is crucial for education and acceptability. This study evaluated BIM adoption trends in the United States, the United Kingdom, and Australia to serve as a model for early adopting countries. The study demonstrates: government engagement increases BIM adoption; government mandates enable widespread BIM adoption and integrate a country's industry into the global marketplace; the ruling also supports BIM research and training, which leads to revenue development through training and workforce export; diffusion dynamics vary throughout time based on a country's propensity to absorb innovation; and the dynamics also alter as the culture and regulations of a sector evolve.

Moreover, the study by [16] evaluates and defines the usefulness and inefficiency of BIM technology in construction infrastructure projects and presents a comparative and exhaustive examination of academic literature and industry reports. Its implementation provides a framework solution to profit and utilise BIM to overcome inefficiencies and obstacles. The study intended to develop a method for identifying difficulties. The excessive nature of BIM (acronyms and competing acronyms) also results in a gap. People need a framework for applying an objective emphasis on BIM methodology, requirements, goal achievement, and agreed-upon measurements, as well as an objective focus on what to deploy and when (standards and technology) concerning the project's aims and advantages.

Information collecting relies on the expert analysis provided by conventional storage. The Internet of Things (IoT) and smart devices generate vast quantities of live data from several sources; hence, IoT–BIM integration is essential. Replace semantic information with internal conditions to construct Service-Oriented-Architecture (SOA). Connect static to real-time models using SOA. It is crucial to develop two-way communication to imitate human thought. Cloud computing is required for IoT device connectivity. Integrating BIM with Internet of Things (IoT) real-time data enhances construction and operational efficiency and produces high-fidelity BIM models. The study addresses IoT concerns connected to BIM. Cloud computing eliminates interoperability problems. The document investigates and identifies new BIM–IoT application areas, followed by enhanced procedures [17].

Due to country-specific socio-cultural, economic, and legal conditions, marketing and implementing BIM for building projects varies. Cambodia's building sector's BIM adoption is unknown. This study investigates BIM industry obstacles. Detailed survey responses and professional architects and contractors. In the final datasets, 13 key drivers were identified. The use of technology enhances project visibility, and technology alters project timelines. The future of an industry is influenced by the information stakeholders share. Technology adoption's is the most significant obstacle that pushes toward industrial resistance to change, especially, reluctance towards inadequate BIM conversion from 2D to 3D, which is expensive. Study implementation can adapt and apply technology to improve Cambodia's construction progress and project success through socio-cultural, economic, and regulatory parallels [17].

Enegbuma [18] conducted a study in Malaysia to investigate BIM adoption, focusing on BIM interpretation (factors influencing BIM) and sources for successful BIM adoption. This collaborative approach mediated the interaction between strategic IT planning and BIM adoption. It identified the factors that have the most significant impact on BIM perception in Malaysia. Another development was made in Singapore by Attarzadeh and Tiong [19], likely to interest many researchers and industries looking to implement the BIM technique. This study was to see what factors impact BIM adoption and application in the AEC sector in Singapore. The study results aid AEC firms in ensuring BIM acceptance during the project life cycle. The study also recommended that government agencies develop standard, comprehensive functional guidelines, models, and BIM public libraries for various areas to promote new technologies.

Similarly, Rosli et al. [8] investigated the link between numerous constructs that influence BIM adoption. The Structural Equation Modelling (SEM) model fit indices and the association strength within the components were used to investigate this link. It is advised to employ ongoing BIM-friendly policy formulation, individuals, procedures, and technology to primarily address the issues impacting BIM adoption in the worldwide construction sector. Hosseini et al. [20] introduced some results of a study effort in Australia where they employed a questionnaire survey to target SMEs in the construction sector. The research provides the most up-to-date information on BIM in Australia's small- and medium-sized enterprises. It offers and expands upon a framework based on the innovation diffusion concept (IDT).

Yemen has suffered and is continuously experiencing mass structural destruction from a war that has been happening for many years. Many vital structures have been destroyed, and reconstruction is inevitable [21]. Usually, the reconstruction of destroyed buildings, such as hospitals, schools, universities, factories, highways, etc., requires a substantial amount of time, money, and effort. An effective and efficient construction approach such as building information modelling (BIM) is essential for rebuilding efficiency and cost-effectiveness.

Gamil et al. [9] noticed that Yemen's construction industry had substantially declined and failed. The sector's growth has been halted, and most projects have collapsed. Several factors have played a significant role in the industry's downfall. It is problematic because the study views BIM as a discrete technical and administrative procedure rather than an interactive, continuous, and emergent working process. Alaghbari [22] indicates that construction project costs and time overruns are caused by various factors, including poor labour productivity.

Moreover, according to Kassem [13], the economics of Yemen prepends on heavily the gas sector. Any active building project has a unique set of risk concerns. As a result, external risk factors have the most significant impact on Yemen's oil industry. The greatest risk indicators for cost and schedule overruns were those related to project management. According to Dahmas [21], Yemen's construction industry is pressured to reduce production time and project costs. Yemen needs to use concurrent engineering (CE) to speed up the reconstruction of its facilities. CE focuses on the design stage and gets it done right the first time. However, delays in implementing construction projects, especially public projects, have become common in Yemen.

3. Methodology

3.1. Identifying and Evaluating the Factors That Affect the Adoption of BIM

This study aims to identify the factors influencing the adoption of BIM in Yemen's construction industry. The initial technique for conducting research is to go through several sources, such as scholarly journal articles, conference proceedings, and books, to determine all aspects of the concepts [23]. This study conducted intensive research of the previous literature to study the adoption of BIM in the construction industry. Two essential techniques are employed to extract and filter the components from the literature: similarity analysis and frequency analysis. Analysis of similarity is a technique used to avoid duplication of variables with similar meanings and distinct phrases; it also aids in establishing a collection of factors that differ in terms of purpose and intent [24]. Frequency analysis is the number of repetitions from the various literature sources of BIM adoption in the construction industry [19]. The list of factors connected to the construction industry is given to five Yemeni professionals; each has over 20 years of experience working in Yemen's construction industry. Using expert opinion in Yemen's construction industry proved extremely valuable in identifying the most critical challenges of implementing BIM in Yemen [3]. Then, the experts were asked to classify these components conceptually into categories, and ambiguous elements were improved. Figure 1. Summarises all stages of the methodology.

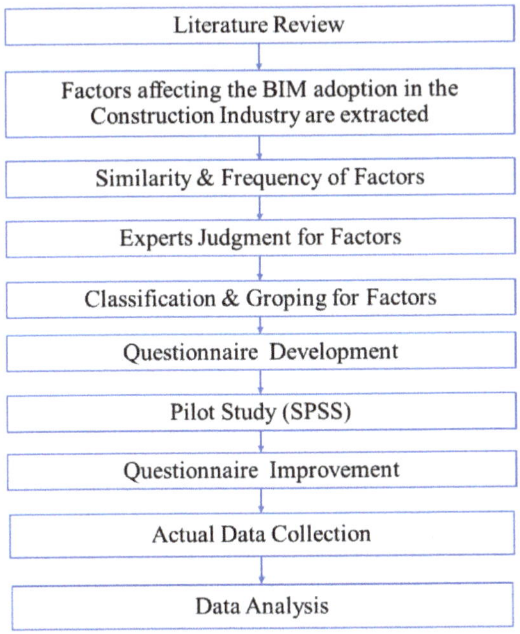

Figure 1. Research method.

A random sample was used to achieve homogeneity and increase the consistency and quality of data. Purposive sampling was used to choose participants for the expert's evaluation depending on criteria such as years of experience within the Yemeni construction industry, BIM experience, organisation size, and job description. Table 1. lists the final assessment items for these constructs.

Table 1. Assessment of the factors that influence BIM adoption in the Yemeni construction industry.

Code	Category 1: Technology (TEC)	References
TEC01	Lack of full automation in the construction industry	[8,18,25,26]
TEC02	Lack of BIM knowledge within the project	[27–33]
TEC03	Visualisation of construction sequences	[6,24,34–39]
TEC04	Trialability (Possibility of risk reduction with the try-out before adopting BIM in practice, and trying out various BIM features in my work to verify its effects)	[6,24,29,34–36]
TEC05	The usefulness of digital transfer of data	[24,26,40–42]
Code	**Category 2: Process (PR)**	**References**
PR01	Information availability and sharing	[35,41,42]
PR02	Providing guidance on the use of BIM	[40,42]
PR03	The leadership of senior management	[18,42–45]
PR04	Contractual sharing norm	[35,41,42]
PR05	Shared norms and collective expectations diffused through information exchange activities	[35,42]
PR06	Shared liability between project participants	[41–46]
PR07	Production of drawings and schedules	[27,47,48]
PR08	Desire to speed up the design process	[24,42]
PR09	Collaboration (project) management tools	[42]
PR10	Standard and rules	[42]
PR11	Companies' collaboration experience with project partners	[27,42,47,49,50]
PR12	Developing data exchange standards	[24,41,42,45,46,51,52]
PR13	Greater collaboration with consultants and other project team members	[46]
Code	**Category 3: Policy (PL)**	**References**
PL01	Financial resources of the organisation	[6,35,42,47,53,54]
PL02	Regulation and policy	[35,42,47,55]
PL03	Organisational readiness	[6,8,26,29,34,35,46,55–59]
PL04	Weak legal institutions	[60,61]
PL05	Guidance on the use of BIM	[40,42]
PL06	The increased demand for design and building	[42,47,51]
PL07	Lack of government incentives	[29,33,41,45,51,62]
PL08	Lack of construction codes	[9,22,24,53,57,63,64]
Code	**Category 4: People (PPL)**	**References**
PPL 01	Lack of skills and knowledge of one of the partners	[65–70]
PPL 02	Lack of cooperative concept	[4,18,21,24,26,41,71–75]
PPL 03	Lack of BIM expertise	[29,32,41]
PPL 04	Lack of top management support	[28,74–81]
PPL 05	Errors by a design team in construction projects	[13,33,56,82–85]
PPL 06	Weak supervision and control	[50,86–90]
PPL 07	Lack of demand by clients	[20,32,33,45,47,53,62,84,91–94]

Table 1. *Cont.*

Code	Category 5: Environment (ENV)	References
ENV 01	Security of information on project data	[22,24,42,46,51,52,54,62,94,95]
ENV 02	Poor Internet connectivity	[50,64,96]
ENV 03	Allows coordination and collaboration between disciplines	[46,47,51,53,57,97]
ENV 04	BIM readiness by project consultants.	[50,64,96]
ENV 05	Poor economic condition	[5,13,55]
ENV 06	Method of communication between the team	[18,20,24,26,32,35,36,41,52,92]
ENV 07	Market demand, size, and competition increase	[98–101]
ENV 08	Risk management	[2,34,72,102–107]
ENV 09	Facility management and building operation	[17,108,109]
Code	**Intention to Adopt the BIM**	**References**
ADBIM1	Encourage the staff to use BIM in regular workflow, even without BIM being the official workflow process at the organisation	[94]
ADBIM2	Implement BIM in future projects, regard less of its implementation level	[94]
ADBIM3	Invite other partner organisations to use BIM for project communication purposes	[94]

3.2. Questionnaire Design

A questionnaire is a comprehensive set of instruments presenting respondents with questions to answer by choosing responses that match their ideas [110]. This study uses the literature review results and expert interviews to improve the questionnaire design. The factors affecting the adoption of BIM in the construction industry were extracted and then categorized into groups. Using a Likert-style scale, these factors determine the elements' degree of importance and seriousness.

A pilot study observes the perspectives and feedback of construction industry experts. It also aids in identifying issues, evaluating questions for clarity, confirming quality, and validating measurement scales. The second objective of the pilot research is to assess and improve the questionnaire's content [111]. This study surveyed 30 Yemeni construction professionals for a pilot study to examine the internal accuracy of the questionnaire regarding data evaluation and assess the variables' importance.

3.3. Data Collection

Surveys are often used to collect research field sample data. Despite a poor response rate and bias, they can examine essential topics. This survey is based on earlier research that led to government guidelines, suggestions, and principles for determining research data requirements [112]. A two-part quantitative questionnaire was developed and utilised for data collection. The first part comprised respondents' demographic information, including their age, education, position, BIM experience, and work experience. Table 2 shows that most participants have more than ten years of experience in the construction industry, and their career is strongly tied to civil/structural engineering. There were more designers or consultants in the research than in the public sector, with fewer participants. The rest are in the private sector and (Mix) public and private.

Table 2. Factors affecting demographics.

		Frequency	Percent %
Qualification	High School	1	0.4
	Diploma	5	2.1
	Bachelor	137	58.3
	Masters	58	24.7
	PhD	34	14.5
Specialisation	Designer or Consultant	160	68.1
	Contractor/Construction	64	27.2
	Client	11	4.7
Organisation	Public	35	14.9
	Private	94	40
	Public and Private (Mix)	106	45.1
Profession	Architecture	33	14
	Civil/Structural Engineering	147	62.6
	Electrical Engineering	13	5.5
	Mechanical Engineering	2	0.9
	Project Management	14	6
	Construction Management	11	4.7
	Quantity Surveying	3	1.3
	Technical in panning team	5	2.1
	Others	7	3

The second part of the questionnaire had 45 items (see Table 1 for details). Using a Likert scale of 1 to 5, the respondents' attitudes and comprehension of BIM adoption factors in the Yemeni construction industry were evaluated (1: strongly disagree; 2: disagree; 3: neutral; 4: agree; and 5: strongly agree). The online surveys were open to a broad public. The most efficient method of communication during the COVID-19 pandemic was online; hence, the Ministry of Public Works and Highways and Yemeni Engineers Syndicates (YES) were contacted repeatedly to distribute the survey to all registered engineers. Access to the study was permitted for four months. Despite receiving 235 survey responses, the intended sample size for the study was 475 people. The questionnaire was answered by 49% of the respondents that participated in the study. A quantitative technique was used to investigate the factors affecting BIM adoption in the Yemeni construction industry.

3.4. Structural Equation Modelling (SEM)

A measurement model (confirmatory factor analysis) and a structural model are combined in the SEM test. In formulas, all evaluation component connections are specified. Since SEM captures the structure of latent variable relationships, the measuring method must be validated. Scale reliability is the dependability of an internal element. It is computed using Cronbach's alpha coefficient, with a minimum value of 0.70 and a higher value suggesting more accurate measurement scales for latent variables. The analyses include concept validity, reliability, convergent and discriminant validity, and structural model evaluation.

Estimating and quantifying relationships for interactions among components/latent variables distinguishes structural equation modelling from other data analysis methodologies [113]. Over the last decade, SEM has captivated the interest of a rising number of scholars in psychology, social science, and strategic management [114]. SEM is used to explain a wide range of empirical data to evaluate the validity of statistical models' underlying ideas. On the other hand, the researcher employs the SEM technique to estimate a specific model. Hypotheses can be tested using SEM, including both latent and observable variables. SEM's aggregate topographies of factor analysis and multiple regression are used to examine the structural properties of both theoretical and measurement models.

Many academics have resorted to SEM as an alternative to first-generation data analysis approaches, such as regression analysis and defining multi-layer correlations between dependent and independent variables [115]. SEM concurrently examines structural models and data. It concurrently models several dependent and independent variables. SEM must be understood before usage. PLS-SEM and CB-SEM are examples of methods. Smart-PLS software created a conceptual measurement model for examining observable characteristics. PLS replicates the model by calculating and measuring item loading, reliability, and validity. To estimate PLS model parameters, first, solve the measurement model's blocks, then compute the structural model's path coefficients [116]. Even though individual item reliability was satisfactory, construct reliability was nevertheless advised for observing group item reliability within the same construct. The internal relationship between items belonging to the exact constructions is more remarkable, as seen by construct-level dependability [117]. The commonly used "Average Variance Extracted" strategy was used in this study to examine convergent validity [118]. This method is considered comparable to that of Fornell and Lacker. The HTMT number must be less than 0.90 [119].

Because PLS does not need distribution assumptions, bootstrapping was utilised to generate t statistics and confidence ranges. Route estimates based on the inner path model or hypothetical relations demonstrated the correct connection. It was utilised to assess each framework path. PLS bootstrap was used to determine structural model hypotheses. According to research, the path coefficient must be at least 0.1 for a model to have an effect. The mediating analysis uses a rigorous bootstrapping method. Some scholars believe that mediation analysis diminishes the significance of the direct impact. Inadequate sample size or predictive ability may limit the detection of a relevant direct correlation. As a result, the mediation analysis is the most important part of observing the indirect impact [120].

4. Results and Findings

This research method investigates BIM acceptance and usage, as well as how the perspectives of BIM drivers, advocates, and early adopters may be utilised to develop a contextualised BIM adoption framework. The conceptual framework supports fundamental research methodologies. This model integrates Policy, Process, Technology, People, and the Environment for BIM adoption in Yemen's construction industry.

In this study, eleven hypotheses were formulated based on the theoretical model illustrated in Figure 2; the potential for BIM's further adoption in the construction sector:

H1. *Environment (ENV) has a significant effect on BIM adoption (ADBIM).*

H2. *People (PPL) have a significant effect on BIM adoption (ADBIM).*

H3. *Policy (PL) has a significant effect on BIM adoption (ADBIM).*

H4. *Policy (PL) has a significant effect on Environment (ENV).*

H5. *Policy (PL) has a significant effect on People (PPL).*

H6. *Process (PR) has a significant effect on BIM adoption (ADBIM).*

H7. *Process (PR) has a significant effect on Environment (ENV).*

H8. *Process (PR) has a significant effect on People (PPL).*

H9. *Technology (TEC) has a significant effect on BIM adoption (ADBIM).*

H10. *Technology (TEC) has a significant effect on Environment (ENV).*

H11. *Technology (TEC) has a significant effect on People (PPL).*

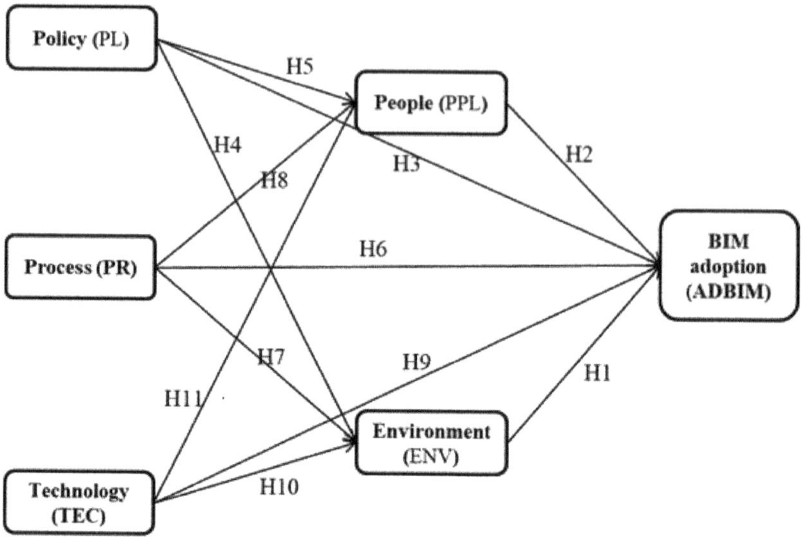

Figure 2. The hypotheses of the research.

A partial least square estimate technique was used to investigate the theoretical model. The measurement and the structural model parameters were estimated using Smart PLS 3.0.

4.1. Experts' Assessment of the Factors

Establishing a model for final variables focuses this investigation on defining features that benefit the Yemeni construction industry. Expert opinion was precious in identifying the most crucial issues in implementing BIM. Table 3 illustrates the respondent's demography, indicating that 40% of the experts are above 55 years, while those between 36–45 years represent 60%. Moreover, respondents that spent more than 20 years in the industry represent 60%, whereas those between 11 to 15 years represent 40%.

Table 3. Demographic characteristic analysis for the experts.

Demographic Characteristics	Frequency	%
Age group:		
Above 55 years	2	40%
36–45 years	3	60%
Experience in the construction industry:		
Above 20 years	3	60%
11 to 15 years	2	40%
Qualification:		
PhD	5	100%
Organisation:		
Private	2	40%
Private (Mix)	3	60%
Job description:		
Commercial Buildings; Industrial Buildings	2	40%
Governmental Buildings; Roads and Transportation; Water and Sanitation Projects	1	20%
Residential Buildings	2	40%

One hundred percent of the respondent have a Ph.D. The majority projects undertaken by the respondents are residential and commercial buildings, with 40% each. Respondents from the private sector represent 40%.

4.2. Pilot Survey

The pilot survey is aimed to test the questionnaire's accuracy, completeness, and ease of understanding by the respondent. It helps uncover flaws, assess whether questions are straightforward, and check to measure scales' reliability and validity. The pilot research helped improve the questionnaire's content and find unclear or complicated questions. After explaining and clarifying questions to the respondents, the researcher collected 30 complete responses from the respondents who were emailed the pilot study questionnaire.

Reliability Test: This section calculates the first Cronbach alpha values based on five BIM adoption factors affecting the construction industry. SPSS is used to calculate Cronbach's alpha; the result of the original Cronbach's alpha value is less than the minimum [121]. Using Cronbach's alpha, which ranges from 0 to 1, a reliability analysis determines if the data obtained are consistent. If Cronbach's alpha value is less than 0.3, the reliability is poor, and the data cannot be trusted. A higher Cronbach's alpha implies better internal consistency in the data [122]. The data have a high and respectable level of consistency if Cronbach's alpha value exceeds 0.7. The pilot study's Cronbach alpha is shown in Table 4.

Table 4. Cronbach Alpha (Pilot Study) Constant factors affecting BIM adoption.

Construct	No of Items	Cronbach Alpha Value
Technology (TEC)	5	0.838
Proses (PR)	13	0.825
Policy (PL)	8	0.826
People (PPL)	7	0.925
Environment (ENV)	9	0.800
The extent of BIM adoption in the Yemeni construction industry (All Categories)	42	0.930

All the items are trustworthy, and the real test is internally consistent according to the overall model's Cronbach alpha value being substantially higher than 0.7.

4.3. Assessment of Measurement Model

Figure 3 shows the model development. The first stage in examining the model is to evaluate the measurement model, which involves assessing Cronbach's alpha and composite reliability for construct reliability, convergent and discriminant reliability, and discriminant validity for composite and discriminant validity. The outer model, also known as the measurement model, is used in factor analysis to determine how loaded observed variables are on their underlying construct. To confirm the underlying relationship between the observable variables and the hidden components, an outer model/CFA is advised. Figure 3 shows each item's factor loadings/outer loadings, and the Cronbach alpha (CA) for each constant derived using the PLS-Algorithm. Moreover, Table A1 indicates some descriptive analyses resulting from Smart PLS.

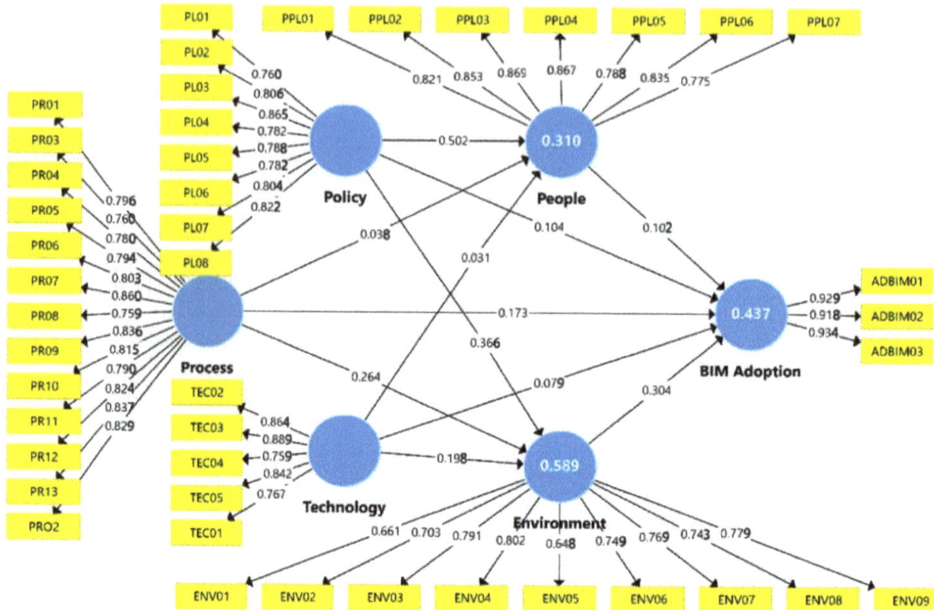

Figure 3. Factor loadings/outer loadings of each item calculated through the PLS algorithm.

4.3.1. Validity and Reliability of Constructs

The construct validity and reliability findings demonstrate that the absolute correlation between the construct and its measuring items is between 0.661 to 0.934, which is higher than the minimum threshold criteria.

4.3.2. Convergent Validity

The Average Variation Extracted (AVE) for each latent variable was more significant than the required threshold of 0.5 (50 percent), indicating that each construct could explain more than half of the variance to its measuring items on average, as shown in Table 5.

Table 5. Internal Consistency and Convergence Validity Results.

Constructs/Items	Code	F. L	CA	CR	AVE
BIM Adoption	**AD-BIM**		**0.918**	**0.948**	**0.859**
Encourage employees to utilise BIM in their daily work, even if it is not the organisation's formal workflow process	ADBIM 01	0.929			
Implement BIM in future projects, no matter how advanced it is	ADBIM 02	0.918			
Invite additional collaborative partners to utilise BIM for project communication	ADBIM 03	0.934			
Environment Factors	**ENV**		**0.896**	**0.916**	**0.548**
Security of information on project data	ENV 01	0.661			
Poor Internet connectivity	ENV 02	0.703			
Allows coordination and collaboration between disciplines.	ENV 03	0.791			

Table 5. *Cont.*

Constructs/Items	Code	F. L	CA	CR	AVE
BIM readiness by project consultants.	ENV 04	0.802			
Poor economic condition	ENV 05	0.648			
Method of communication between the team	ENV 06	0.749			
Market demand, size, and competition increase	ENV 07	0.769			
Risk management	ENV 08	0.743			
Facility management and buildings operation	ENV 09	0.779			
People Factors	**PPL**		**0.925**	**0.940**	**0.690**
Lack of skills and knowledge of one of the partners	PPL 01	0.820			
Lack of cooperative concept	PPL 02	0.854			
Lack of BIM expertise	PPL 03	0.871			
Lack of top management support	PPL 04	0.868			
Errors by the design team in construction projects	PPL 05	0.789			
Weak supervision and control	PPL 06	0.834			
Lack of demand by clients	PPL 07	0.773			
Policy Factors	**PL**		**0.920**	**0.935**	**0.643**
Financial resources of the organisation	PL01	0.760			
Regulation and policy	PL02	0.806			
Organisational readiness	PL03	0.866			
Strong legal institutions	PL04	0.782			
Guidance on the use of BIM	PL05	0.788			
The increased demand for design and building	PL06	0.782			
Government incentives	PL07	0.804			
Construction codes	PL08	0.822			
Process Factors	**PR**		**0.955**	**0.960**	**0.651**
Information availability and sharing	PR01	0.796			
Guiding the use of BIM	PR02	0.829			
The leadership of senior management	PR03	0.760			
Contractual sharing norm	PR04	0.780			
Information-sharing activities disseminate shared norms and community expectations	PR05	0.794			
Shared liability between project participants	PR06	0.803			
Production of drawings and schedules	PR07	0.860			
Desire to have the design process go faster	PR08	0.759			
Collaboration (project) management tools	PR09	0.836			
Standard and rules	PR10	0.815			
Collaboration experience of companies with project partners	PR11	0.790			
Creating data interchange standards	PR12	0.824			
Greater collaboration with consultants and other project team members.	PR13	0.837			
Technology Factors	**TEC**		**0.882**	**0.914**	**0.682**
Full automation in the construction industry	TEC01	0.767			

Table 5. *Cont.*

Constructs/Items	Code	F. L	CA	CR	AVE
BIM knowledge within the projects	TEC02	0.864			
Visualisation of construction sequences	TEC03	0.889			
Trialability (possibility of risk reduction by experimenting with BIM before implementing it in practice and experimenting with various BIM features in my work to validate their impact)	TEC04	0.759			
The usefulness of digital transfer of data	TEC05	0.842			

Hints: (AVE) Average Variance Extracted; (CA) Cronbach's alpha; (CR) Composite reliability.

4.3.3. Measurement of Discriminant Validity

Table 6 shows that the square roots of the AVE are more significant than their comparable inter-correlations. As a consequence, the validity and reliability of the measurement model is established.

Table 6. Discriminant Validity—Fornell and Lacker Criterion.

Constructs	BIM Adoption	Environment	People	Policy	Process	Technology
BIM Adoption	0.927					
Environment	0.614	0.740				
People	0.447	0.560	0.831			
Policy	0.585	0.730	0.556	0.802		
Process	0.588	0.721	0.481	0.837	0.807	
Technology	0.532	0.665	0.424	0.726	0.763	0.826

The diagonal represents the square root of AVE, while the off-diagonal values are correlations between latent variables.

As shown in Table 7, the discriminant findings demonstrate that most of the Heterotrait–Monotrait (HTMT) values are less than 0.9, which is extremely good and meets the discriminant validity criteria since the value is less than 0.90.

Table 7. Heterotrait–Monotrait Ratio Results (HTMT).

Constructs	BIM Adoption	Environment	People	Policy	Process	Technology
BIM Adoption						
Environment	0.668					
People	0.483	0.622				
Policy	0.631	0.797	0.596			
Process	0.623	0.770	0.509	0.893		
Technology	0.587	0.738	0.463	0.803	0.828	

4.4. *The Structural Model's Assessment*

The structural model is a theoretical model that analyses the inner path model using structural equations. Statistical measures such as path coefficient, predictive relevance (Q^2), effect size (f^2), and coefficient of determination (R^2) were used to verify the structural model. Once the measurement model was fit, the structural model's validity was evaluated. The next step was to create a causal route between independent (exogenous) and dependent (endogenous) variables to develop a linear covariance connection. The path coefficient, coefficient of determination (R^2) for the endogenous prediction relevance (Q^2), variable, effect size (f^2), and multicollinearity were used to evaluate the structural model in this study [123].

4.4.1. Coefficient of Determination (R^2)

The coefficient of determination (R^2) is the most significant criterion for assessing structural models and determining R^2 values. If the R^2 value is 0.26 or higher, effective results are expected. It is moderate if the R^2 value is between 0.13 and 0.25, and it is weak if the R^2 value is between 0.02 and 0.12 [124]. The R^2 results are presented in Table 8, with BIM adoption having 0.437, Environment at 0.589, and People having 0.310, demonstrating high acceptance.

Table 8. Result of R-square.

Endogenous Variables	R Square	R Square Adjusted
BIM Adoption	0.437	0.424
Environment	0.589	0.584
People	0.310	0.301

4.4.2. Effect Size (f^2)

The f^2 measures the influence of a predictive construct on an endogenous construct. According to [125], R^2 looks at how much one external construct helps explain a particular endogenous component. Significant, medium, and minor impact sizes are defined by f^2 values of 0.35, 0.15, and 0.02. Table 9 shows that policy on people and the environment has significant effects considering a value of 0.103 and 0.092, respectively. Other values indicate medium and small size effects.

Table 9. F-square Result.

Exogenous Variables	BIM Adoption	Environment	People
BIM Adoption			
Environment	0.062		
People	0.012		
Policy	0.005	0.092	0.103
Process	0.013	0.042	0.001
Technology	0.004	0.037	0.001

4.4.3. Result of Multicollinearity (Inner VIF)

The presence of two or more independent but highly connected entities is referred to as multicollinearity. It is a multicollinearity problem if there are common indicators across multiple constructs. Before moving further with model testing, we strongly suggest the researcher looks at multicollinearity [126]. The variables are assumed to have a collinearity problem when the correlation coefficient values are more than 0.9. The Variance Infant Tolerance (VIF) can detect collinearity concerns instead of the correlation coefficient. The VIF value in Smart-PLS must not be greater than five, indicating that the variables in the model are not collinear. This investigation did not consider multicollinearity because the inner VIF values were less than 5. Table 10 shows that the maximum VIF is 4.196, and the lowest is 1.561, indicating no multicollinearity at the site as the VIF is less than 10.

Table 10. Multicollinearity—Inner VIF Values.

Exogenous Variables	BIM Adoption	Environment	People
BIM Adoption			
Environment	2.630		
People	1.567		
Policy	4.102	3.559	3.559
Process	4.196	4.022	4.022
Technology	2.644	2.547	2.547

4.4.4. Predictive Relevance (Q^2 Value)

The Q^2 value was calculated using a blindfolding test to measure the model's predictive effectiveness. The blindfolding Q^2 test assesses endogenous variables' predictive capabilities and the structural model's predictive abilities. It is also a sample process strategy for assessing cross-validation in a model. The model is accurate in its predictions. The model's predictive significance is insufficient if the Q^2 value is more than zero [47]. As shown in Table 11, because the Q^2 values are more than zero, the model establishes a good fit and vital predictive significance. All matters are greater than zero ranging from 0.210 to 0.672, which indicates that the model is significant.

Table 11. Predictive Relevance Results.

Endogenous Variables	CCC Q^2 (=1-SSE/SSO)	CCR Q^2 (=1-SSE/SSO)
BIM Adoption	0.672	0.350
Environment	0.433	0.314
People	0.583	0.210
Policy	0.540	
Process	0.586	
Technology	0.522	

(CCC), Construct cross-validated communality; (CCR), construct cross-validated redundancy.

4.5. *Analysis of Direct Effect Path Coefficients*

The path coefficient results, as shown in Table 12 and Figure 4, indicate that the most significant path (t = 5.276) was found between Policy (PL) and People (PPL), which Policy (PL) and Environment (ENV) follow, and then Environment (ENV) and BIM adoption (ADBIM) with *t* values of 4.050 and 2.889, respectively, with all having *p* significance values of 0.000. The minor significance paths are those between Process (PR) and BIM adoption, Process and People, Technology (TEC) and BIM adoption (ADBIM), and Technology and People, all having a P-value above 0.05, and hence their hypotheses are not supported.

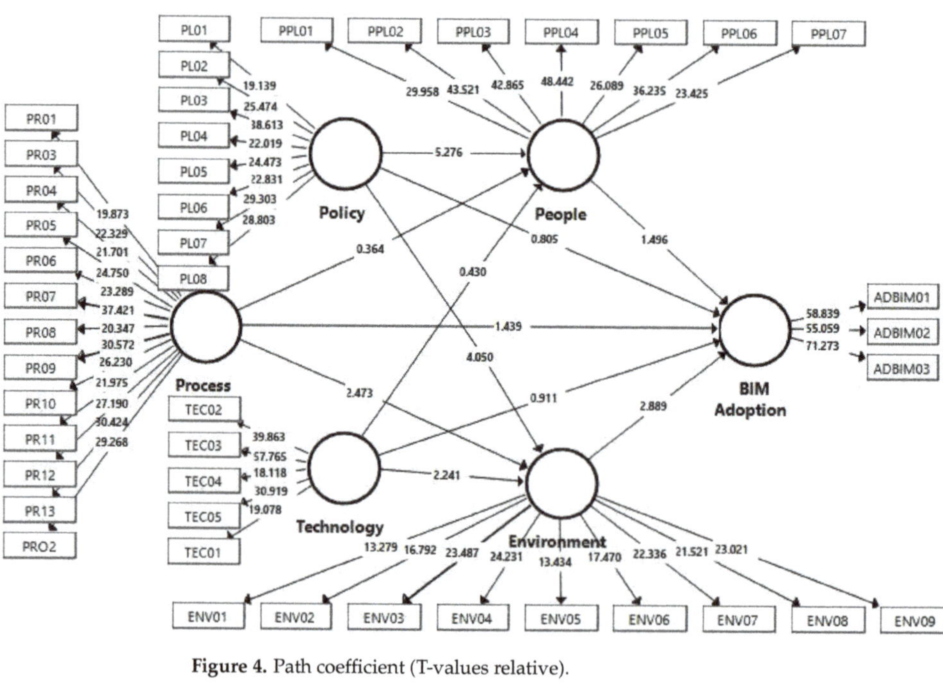

Figure 4. Path coefficient (T-values relative).

Table 12. Path Coefficient Result.

	Original Sample (O)	Sample Mean (M)	Standard Deviation (STDEV)	T Statistics (|O/STDEV|)	*p* Values	Decision
Environment → BIM Adoption	0.304	0.304	0.105	2.889	0.004	Significant
People → BIM Adoption	0.102	0.097	0.068	1.496	0.135	Not Significant
Policy → BIM Adoption	0.104	0.107	0.129	0.805	0.421	Not Significant
Policy → Environment	0.366	0.365	0.090	4.050	0.000	Significant
Policy → People	0.502	0.503	0.095	5.276	0.000	Significant
Process → BIM Adoption	0.173	0.169	0.120	1.439	0.151	Not Significant
Process → Environment	0.264	0.257	0.107	2.473	0.014	Significant
Process → People	0.038	0.036	0.104	0.364	0.716	Not Significant
Technology → BIM Adoption	0.079	0.079	0.087	0.911	0.363	Not Significant
Technology → Environment	0.198	0.205	0.088	2.241	0.025	Significant
Technology → People	0.031	0.034	0.071	0.430	0.668	Not Significant

Significant: $p < 0.05$.

Table 12 shows the study's path coefficient results, which show that five hypotheses were supported and six were not, with a *p*-value of less than 0.05 for the supported hy potheses.

4.6. Indirect (Mediation) Effect Analysis

The bootstrapping results for the indirect effect are shown in Table 13, where the bootstrapping analysis was used to indicate the indirect effect of PL, PR, and technology (TEC) on BIM adoption (ADBIM). The mediation impact of independent variables on dependent variables was statistically significant using PL, PR, and TEC. The findings of the mediation analysis show that two of the three mediating hypotheses were supported, while the third was not. The mediating path Policy (PL) → Environment (EMV) is significant, having a $p = 0.000$ and $t = 4.050$. Moreover,, Policy (PL) → People (PL) is significant with $p = 0.000$ and $t = 5.276$, Environment → BIM adoption is significant with $p = 0.004$ and $t = 2.889$, and Process → Environment is significant with $p = 0.025$ and $t = 2.241$.

Table 13. Mediation Result.

Hypothesis	OS	SM	SD	T	*p* Values	Decision	Mediation
Policy (PL) → BIM adoption (ADBIM)	0.162	0.155	0.055	2.964	0.003 *	Sig.	Full Mediation
Process (PR) → BIM adoption (ADBIM)	0.083	0.076	0.046	1.804	0.045 *	Sig.	Full Mediation
Technology (TEC) → BIM adoption	0.063	0.064	0.039	1.604	0.109	Not Sig.	No Mediation

Significant; * $p < 0.05$.

4.7. Hypotheses Testing Result

The summary of the hypotheses testing is presented in Table 14, which shows that five hypotheses are accepted and six are rejected. This indicates Environment, People, and Policy are the most influencing factors on BIM adoption in the Yemeni construction industry. It also shows that other factors, such as Technology and Process, can be crucial in achieving the said objectives. The findings conform with the studies conducted by previous researchers.

Table 14. Hypotheses Results.

No.	Hypotheses	Results
H1	ENV has a significant effect on ADBIM	Accepted
H2	PPL have a significant effect on ADBIM	Rejected
H3	PL has a significant effect on ADBIM	Rejected
H4	PL has a significant effect on ENV	Accepted
H5	PL has a significant effect on PPL	Accepted
H6	PR has a significant effect on ADBIM	Rejected
H7	PR has a significant effect on ENV	Accepted
H8	PR has a significant effect on PPL	Rejected
H9	TEC has a significant effect on ADBIM	Rejected
H10	TEC has a significant effect on ENV	Accepted
H11	TEC has a significant effect on PPL	Rejected

The path between policy (PL) and Environment is the next meaningful relationship (ENV). It was discovered via structural equation modelling evaluating this link that there is a sizeable direct relationship between Policy and People. The Environment mediates the relationship between increased BIM adoption in the Yemeni construction industry. Previous studies have repeatedly emphasised this desired transformation.

5. Discussion

The structural equation model path analysis shows that the five variables, Policy, Process, Technology, People, and Environment, affect BIM adoption. Specifically, the relationship between Policy and People was found to be the most significant (t-values = 5.276, p-values = 0.000; significant), the relationship between Policy and the Environment was shown to be the second most important (t-values = 4.050, p-values = 0.000; significant). Following that, in terms of the significance of the association, the Process (which is an independent variable) and the Environment (which acts as a mediator) come in with t-values = 2.473 and p-values = 0.014; significant). Only the direct affect environment has a significant active impact on the rate of BIM adoption (t-values = 2.89, p-values = 0.004; significant). Consequently, the other direct-effect constants do not contribute considerably to the relationships. The relationship between Technology and the Environment was the last one to reach the significant level (t-values = 2.241, p-values = 0.025; significant). The other correlations lack statistical significance because the p-value is more than 0.05, and the t has a considerably lower value. As shown in Figure 5, the BIM adoption model includes two mediation paths: PL and PR→PPL→ADBIM and PR and TEC→ENV→ADBIM. In the first path, PPL acts as a link between PL→ADBIM. Such findings indicate that the construction industry's comprehensive understanding of policy BIM implementation factors (particularly construction codes) encourages the People (PPL) with a positive attitude to implement BIM in an existing workflow; this will eventually influence the organisation's decision to adopt BIM. Rogers' (2003) innovation process has five stages: agenda-setting, decision-making, implementation, and evaluation. Diffusion theory argues that organisations start implementing innovations by identifying issues and suggesting solutions. After analysing the innovation's viability, a decision can be taken about its implementation. This study focuses on initiation and decision stages, not a five-stage process. The analysis results indicate that the Yemeni construction industry shows fewer considerations of negative factors, including People (PPL). This reveals that the Yemeni construction industry idealises the BIM adoption process. Such results are not limited to Yemen. For instance, a study conducted in Qatar [127] revealed that more than half of the interviewees understood BIM to be the collaboration, cooperation, and digital data management that modifies the traditional manner of work. Despite this, most respondents (71%) stated that the industry lacks a sufficient understanding of BIM. Another mediation path identified in this study is from TEC and PR to ENV and eventually to ADBIM. Organisations whose teams have better capabilities in using BIM tools and processes tend to advance more in understanding the work environment, which will finally contribute to BIM adoption. Based on the above logic, conventional organisations with a lack of focus on improving personnel's technical knowledge and simplifying the way of work are less likely to adopt BIM. The establishment of a suitable environment is profoundly affected by process (PR) or technology (TEC). The stronger the BIM process or experiences of staff on technology, the more influential the environment the Yemeni construction industry will establish [26]. Moreover, a test of the hypothesised components suggests that the model can explain 24.6% of BIM adoption based on the sample size. The most significant influence of Technology was on Processes, confirming the belief that technology facilitates strategic innovation and alters traditional business processes. Non-challenge attitudes of the authorities toward adopting BIM are eminent in developing countries. The construction industry's stakeholders in most developing countries are still in the early stage of BIM adoption and implementation. Hence, they face numerous issues ranging from qualities of innovation inside and outside environments. Small- and medium-scale construction organisations contributing significantly to Yemen's development are the most affected due to their peculiar nature. The industry faces insufficient human resources, limited resources, and a lack of technological innovation, which has always been a significant setback to BIM adoption and implementation. This study appraises the factors affecting BIM adoption in the Yemeni construction industry using a structural equation modelling approach. Factors affecting BIM adoption were identified, reviewed, and synthesised into groups. Professionals within the Yemeni construction industry were consulted to determine the relationship among the factors affecting BIM adoption using the structural equation modelling (SEM) technique. As a result, Yemeni construction experts can investigate, examine, pinpoint, and assess the challenges associated with implementing BIM in construction projects. This study fills a gap and narrows the scope by focusing on the Yemeni construction industry scenario. A schematic relationship model of effective BIM adoption was also developed in the research. The government of Yemen is making several efforts to promote BIM among local groups. As a result, there is an opportunity to investigate, examine, identify, and assess the constraints of poor BIM adoption in construction projects among Yemeni construction professionals. In Yemen, almost all projects struggle to accomplish their goals. The government should use this study's results to enhance the construction sector's state. This is necessary to investigate the previous projects to identify the leading causes of issues and draw lessons for new initiatives. The Yemeni government should use the results of this study to enhance the state of the construction sector currently [63]. This systematic research on BIM adoption in the Yemeni construction industry has increased the literature and describes the research's originality. The research filled the knowledge gap regarding identifying and evaluating barriers to and impacts on BIM adoption in Yemeni construction projects. The overall results of this study are anticipated to boost and succeed in BIM adoption.

The research's findings are essential for the construction sector for the reasons listed below. Firstly, review the variables influencing the adoption of BIM in the construction industry. Secondly, research the elements influencing BIM adoption in the Yemeni construction industry. Investigate the level of BIM adoption, awareness, and knowledge in the Yemeni construction industry. Finally, the study created a framework for enhancing BIM adoption in the Yemeni construction industry. This framework can be used as a visual aid to comprehend the requirements for BIM adoption and potential obstacles. The research has paved the way for further study in various fields, including an international application. The findings of this research can be expanded and updated to support and improve construction practices in other countries.

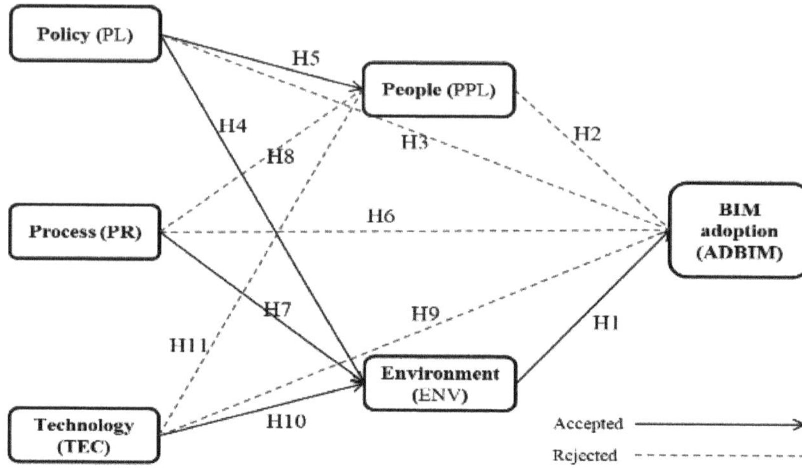

Figure 5. The final hypotheses of the research.

6. Conclusions

This study looks at the factors that drive BIM adoption in the construction sector. Its purpose is to provide an integrated framework for BIM adoption in the Yemeni construction industry, where additional research is needed. Making information more accessible to project members is an awareness factor that is more significant among the responses received. Structural equation modelling was employed, and essential factors loading was observed, which led to the development of the BIM adoption framework, which was successfully validated. This study's framework is represented diagrammatically with essential information embedded within. The findings show that the most critical factor for BIM adoption in the Yemeni construction industry is Policy, which would include regulation and policy, organisational readiness, government incentives, and construction codes. Visualisation of a sequence is the most significant technological factor toward BIM adoption. Greater collaboration between consultants and contractors is the most significant process factor. In contrast, BIM adoption is a policy-driven factor that lacks top management support as a people factor in addition to examining BIM adoption determinants and awareness in the Yemeni construction industry in order to establish a strategy that enables the development of a practical framework to proceed smoothly.

This research contributes originally to knowledge and the Yemeni construction industry. According to the literature review, there has never been academic research in Yemen on BIM adoption for the construction industry that has raised or increased the literature on sustainable construction. The study framework will provide consultants and contractors with a systematic and realistic technique for encouraging collaboration and consultation in the BIM adoption decision-making process. The findings of this study contribute to a better understanding of the factors affecting BIM adoption in the Yemeni construction industry. It is believed that these factors will help the construction industry improve the effectiveness of BIM implementation, achieve full benefits, and maximise the advantages for each project stakeholder with the existing tools and technologies available. A research framework is developed as the main contribution of this research, in which are attributes for BIM adoption in the construction industry. Particular attention is given to the challenging requirements of the Yemen construction industry, together with the need for government support for BIM adoption and implementation across all disciplines throughout the project lifecycle.

This study is extensive, and the findings are valuable to construction stakeholders. Nonetheless, there are certain drawbacks to this study. The literature supporting BIM adoption in the Yemeni construction sector was limited. As a result, this study could provide a solution to bridging this gap. Furthermore, it investigated BIM adoption factors in the Yemeni construction sector and built a strategy that allows for the smooth development of a practical framework. The framework's design and development were limited to Yemen, and possibly other countries needed to be studied. The usefulness of this study remains, however, because it does not detract from the limitations but allows for future research.

The following recommendations for improving BIM adoption were derived from the findings of this study:

- This study aimed to create a BIM adoption model in Yemen that could be expanded to include the operational and destruction steps and investigations into nations other than Yemen. More research may be conducted to examine the parameters of their impact on different types of infrastructure.
- The built environment curriculum in Yemeni tertiary institutions should be studied to include BIM education to produce a stream of BIM-oriented professionals.
- Similar to other developed countries, the Yemeni government should adopt construction policies to promote the use of BIM on every construction project. These policies would stimulate the implementation of BIM in Yemen.
- Due to the high cost of BIM infrastructure, the government might implement a loan scheme to aid construction companies in acquiring it.
- It would be interesting to investigate the level of BIM adoption in developed and developing nations. As a result, benchmark data and best practices for addressing problems with worldwide BIM adoption should be established.

Author Contributions: Conceptualisation, A.H.M.A.-s.; data curation, A.H.M.A.-s.; formal analysis, A.H.M.A.-s.; funding acquisition, A.H.A.; methodology, A.H.M.A.-s. and H.Z.M.S.; supervision, A.H.A., H.Z.M.S. and F.M.J.; validation, A.H.M.A.-s. and A.H.A.; visualisation, F.M.J.; writing—original draft, A.H.M.A.-s. and A.H.A.; writing—review and editing, H.Z.M.S. and F.M.J. All authors have read and agreed to the published version of the manuscript.

Funding: This research received no external funding.

Informed Consent Statement: Informed consent was obtained from all subjects involved in the study.

Data Availability Statement: The data used in the study can be obtained from the authors upon request.

Conflicts of Interest: The authors declare no conflict of interest.

Appendix A

Table A1. MV Descriptive (from Smart PLS).

No.		Mean	Median	Min	Max	Standard Deviation	Excess Kurtosis	Skewness	Number of Observations Used
1	ADBIM01	4.000	4.000	1.000	5.000	1.023	1.603	−1.320	235.000
2	ADBIM02	3.885	4.000	1.000	5.000	1.031	0.911	−1.103	235.000
3	ADBIM03	4.021	4.000	1.000	5.000	1.000	1.751	−1.329	235.000
4	ENV01	3.485	4.000	1.000	5.000	1.077	−0.290	−0.568	235.000
5	ENV02	3.523	4.000	1.000	5.000	1.211	−0.341	−0.771	235.000
6	ENV03	3.749	4.000	1.000	5.000	1.092	0.614	−1.027	235.000
7	ENV04	3.902	4.000	1.000	5.000	1.041	1.265	−1.191	235.000
8	ENV05	3.672	4.000	1.000	5.000	1.248	−0.317	−0.814	235.000
9	ENV06	3.813	4.000	1.000	5.000	0.931	1.269	−1.018	235.000
10	ENV07	3.796	4.000	1.000	5.000	1.044	0.605	−0.982	235.000
11	ENV08	3.706	4.000	1.000	5.000	1.049	0.150	−0.790	235.000
12	ENV09	3.762	4.000	1.000	5.000	1.008	0.980	−1.039	235.000
13	PL01	3.715	4.000	1.000	5.000	1.076	0.640	−1.001	235.000

Table A1. *Cont.*

No.		Mean	Median	Min	Max	Standard Deviation	Excess Kurtosis	Skewness	Number of Observations Used
14	PL02	3.753	4.000	1.000	5.000	1.035	0.607	−0.905	235.000
15	PL03	3.851	4.000	1.000	5.000	1.031	1.077	−1.128	235.000
16	PL04	3.681	4.000	1.000	5.000	1.078	0.090	−0.793	235.000
17	PL05	3.974	4.000	1.000	5.000	0.993	1.145	−1.129	235.000
18	PL06	3.800	4.000	1.000	5.000	1.063	0.824	−1.068	235.000
19	PL07	3.800	4.000	1.000	5.000	1.166	0.080	−0.949	235.000
20	PL08	3.991	4.000	1.000	5.000	1.126	1.005	−1.261	235.000
21	PPL01	3.528	4.000	1.000	5.000	1.153	−0.561	−0.612	235.000
22	PPL02	3.498	4.000	1.000	5.000	1.165	−0.509	−0.597	235.000
23	PPL03	3.570	4.000	1.000	5.000	1.227	−0.734	−0.595	235.000
24	PPL04	3.609	4.000	1.000	5.000	1.265	−0.509	−0.729	235.000
25	PPL05	3.455	4.000	1.000	5.000	1.142	−0.584	−0.511	235.000
26	PPL06	3.532	4.000	1.000	5.000	1.186	−0.556	−0.585	235.000
27	PPL07	3.477	4.000	1.000	5.000	1.168	−0.514	−0.564	235.000
28	PR01	3.996	4.000	1.000	5.000	1.033	1.827	−1.388	235.000
29	PR03	3.791	4.000	1.000	5.000	1.021	0.640	−0.950	235.000
30	PR04	3.723	4.000	1.000	5.000	1.042	0.264	−0.835	235.000
31	PR05	3.817	4.000	1.000	5.000	0.974	0.909	−0.931	235.000
32	PR06	3.889	4.000	1.000	5.000	1.000	1.028	−1.085	235.000
33	PR07	4.085	4.000	1.000	5.000	1.011	2.233	−1.514	235.000
34	PR08	3.877	4.000	1.000	5.000	1.043	0.605	−1.020	235.000
35	PR09	3.898	4.000	1.000	5.000	0.953	1.794	−1.218	235.000
36	PR10	3.813	4.000	1.000	5.000	1.043	0.946	−1.089	235.000
37	PR11	3.826	4.000	1.000	5.000	0.980	1.052	−0.980	235.000
38	PR12	4.000	4.000	1.000	5.000	0.932	2.222	−1.303	235.000
39	PR13	4.132	4.000	1.000	5.000	1.012	2.016	−1.456	235.000
40	PRO2	3.889	4.000	1.000	5.000	1.058	1.215	−1.209	235.000
41	TEC01	3.672	4.000	1.000	5.000	1.103	0.389	−0.970	235.000
42	TEC02	3.864	4.000	1.000	5.000	1.051	1.254	−1.229	235.000
43	TEC03	3.936	4.000	1.000	5.000	1.048	1.369	−1.256	235.000
44	TEC04	3.783	4.000	1.000	5.000	1.035	1.016	−1.152	235.000
45	TEC05	3.911	4.000	1.000	5.000	1.058	0.830	−1.100	235.000

References

1. Manzoor, B.; Othman, I.; Gardezi, S.; Harirchian, E. Strategies for Adopting Building Information Modeling (BIM) in Sustainable Building Projects—A Case of Malaysia. *Buildings* **2021**, *11*, 249. [CrossRef]
2. Baarimah, A.O.; Alaloul, W.S.; Liew, M.S.; Kartika, W.; Al-Sharafi, M.A.; Musarat, M.A.; Alawag, A.M.; Qureshi, A.H. A bibliometric analysis and review of building information modelling for post-disaster reconstruction. *Sustainability* **2022**, *14*, 393. [CrossRef]
3. Succar, B. Building information modelling framework: A research and delivery foundation for industry stakeholders. *Autom. Constr.* **2009**, *18*, 357–375. [CrossRef]
4. Sacks, R.; Eastman, C.; Lee, G.; Teicholz, P. Facilitators of BIM Adoption and Implementation. In *BIM Handbook*; John Wiley & Sons, Inc.: Hoboken, NJ, USA, 2018; pp. 323–363.
5. Ahmed, S. Barriers to Implementation of Building Information Modeling (BIM) to the Construction Industry: A Review. *J. Civ. Eng. Constr.* **2018**, *7*, 107–113. [CrossRef]
6. Elghdban, M.G.; Azmy, N.B.; Bin Zulkiple, A.; Al-Sharafi, M.A. A Systematic Review of the Technological Factors Affecting the Adoption of Advanced IT with Specific Emphasis on Building Information Modeling. *Recent Adv. Intell. Syst. Smart Appl.* **2020**, *29*, 29–42.
7. Babatunde, S.O.; Udeaja, C.; Adekunle, A.O. Barriers to BIM implementation Barriers and ways forward to improve its adoption in the Nigerian AEC firms. *Int. J. Build. Pathol. Adapt.* **2020**, *39*, 48–71. [CrossRef]
8. Enegbuma, W.I.; Aliagha, G.U.; Ali, K.N. Effects of Perceptions on BIM Adoption in Malaysian Construction Industry. *J. Teknol.* **2015**, *77*, 1–6. [CrossRef]
9. Gamil, Y.; Rahman, I.A.; Nagapan, S.; Nasaruddin, N.A.N. Nasaruddin. Exploring the failure factors of Yemen construction industry using PLS-SEM approach. *Asian J. Civ. Eng.* **2020**, *21*, 967–975. [CrossRef]

10. Mishmish, M.; El-Sayegh, S.M. Causes of claims in road construction projects in the UAE. *Int. J. Constr. Manag.* **2016**, *18*, 26–33. [CrossRef]
11. El-Sayegh, S.M.; Mansour, M.H. Risk assessment and allocation in highway construction projects in the UAE. *J. Manag. Eng.* **2015**, *31*, 04015004. [CrossRef]
12. Issa, U.H.; Farag, M.A.; Abdelhafez, L.M.; Ahmed, S.A. A risk allocation model for construction projects in Yemen. *Civ. Environ. Res.* **2015**, *7*, 78–89.
13. Kassem, M.A.; Khoiry, M.A.; Hamzah, N. Risk factors in oil and gas construction projects in developing countries: A case study. *Int. J. Energy Sect. Manag.* **2019**, *13*, 846–861. [CrossRef]
14. Saka, A.B.; Chan, D.W.M.; Siu, F.M.F. Drivers of sustainable adoption of building information modelling (BIM) in the nigerian construction small and medium-sized enterprises (SMEs). *Sustainability* **2020**, *12*, 3710. [CrossRef]
15. Hamma-Adama, M.; Kouider, T. Comparative Analysis of BIM Adoption Efforts by Developed Countries as Precedent for New Adopter Countries. *Curr. J. Appl. Sci. Technol.* **2019**, *36*, 1–15. [CrossRef]
16. Pidgeon, A.; Dawood, N. BIM Adoption Issues in Infrastructure Construction Projects: Analysis and Solutions. *J. Inf. Technol. Constr.* **2021**, *26*, 263–285. [CrossRef]
17. Altohami, A.; Haron, N.; Ales@alias, A.; Law, T. Investigating approaches of integrating BIM, IoT, and facility management for renovating existing buildings: A review. *Sustainability* **2021**, *13*, 3930. [CrossRef]
18. Enegbuma, W.I.; Aliagha, U.G.; Ali, K.N. Preliminary building information modelling adoption model in Malaysia A strategic information technology perspective. *Constr. Innov.* **2014**, *14*, 408–432. [CrossRef]
19. Attarzadeh, M.; Nath, T.; Tiong, R.L.K. Identifying key factors for building information modelling adoption in Singapore. *Inst. Civ. Eng. Manag. Procure. Law* **2015**, *168*, 220–231.
20. Hosseini, M.; Banihashemi, S.; Chileshe, N.; Namzadi, M.O.; Udaeja, C.; Rameezdeen, R.; McCuen, T. BIM adoption within Australian Small and Medium-sized Enterprises (SMEs): An innovation diffusion model. *Constr. Econ. Build.* **2016**, *16*, 71–86. [CrossRef]
21. Dahmas, S.; Li, Z.; Liu, S. Solving the difficulties and challenges facing construction based on concurrent engineering in Yemen. *Sustainability* **2019**, *11*, 3146. [CrossRef]
22. Alaghbari, W.; Al-Sakkaf, A.A.; Sultan, B. Factors affecting construction labour productivity in Yemen. *Int. J. Constr. Manag.* **2017**, *19*, 79–91. [CrossRef]
23. Bahamid, R.A.; Doh, S.I.; Al-Sharafi, M.A.; Rahimi, A.R. Risk Factors Influencing the Construction Projects in Yemen from Expert's Perspective. *IOP Conf. Ser. Mater. Sci. Eng.* **2020**, *712*, 012007. [CrossRef]
24. Shirowzhan, S.; Sepasgozar, S.M.E.; Edwards, D.J.; Li, H.; Wang, C. BIM compatibility and its differentiation with interoperability challenges as an innovation factor. *Autom. Constr.* **2020**, *112*, 103086. [CrossRef]
25. Enegbuma, W.I.; Aliagha, G.U.; Ali, K.N.; Badiru, Y.Y. CConfirmatory strategic information technology implementation for building information modelling adoption model. *J. Constr. Dev. Ctries.* **2016**, *21*, 113–129.
26. Enegbuma, W.I.; Aliagha, U.G.; Ali, K.N. Measurement of Theoretical Relationships in Building Information Modelling Adoption in Malaysia. In Proceedings of the 31st International Symposium on Automation and Robotics in Construction and Mining (ISARC), Sydney, Australia, 9–11 July 2014.
27. Pickup, J. BIM Adoption & Precast Concrete: Design and Implementation of a Strategic Guide. Bachelor's Thesis, University of Sydney, Sydney, Australia, 2013.
28. Elhendawi, A.; Omar, H.; Elbeltagi, E.; Smith, A. Practical approach for paving the way to motivate BIM non-users to adopt BIM. *Int. J. BIM Eng. Sci.* **2019**, *2*, 1–22. [CrossRef]
29. Ahuja, R.; Sawhney, A.; Jain, M.; Arif, M.; Rakshit, S. Factors influencing BIM adoption in emerging markets–the case of India. *Int. J. Constr. Manag.* **2018**, *20*, 65–76. [CrossRef]
30. Alhumayn, S.A. Developing A Framework for BIM Implementation in the Saudi Arabian Construction Industry. Ph. D. Thesis, University of Wolverhampton, Wolverhampton, UK, 2018.
31. Sodangi, M.; Fouad, A.; Muhammad, S. Building Information Modeling: Awareness Across the Subcontracting Sector of Saudi Arabian Construction Industry. *Arab. J. Sci. Eng.* **2017**, *43*, 1807–1816. [CrossRef]
32. Enshassi, A.; Ayyash, A.; Choudhry, R.M. BIM for construction safety improvement in Gaza strip: Awareness, applications and barriers. *Int. J. Constr. Manag.* **2016**, *3599*, 249–265.
33. Gamil, Y.; Rahman, I.A.R. Awareness and challenges of building information modelling (BIM) implementation in the Yemen construction industry. *J. Eng. Des. Technol.* **2019**, *17*, 1077–1084. [CrossRef]
34. Hochscheid, E.; Halin, G. Generic and SME-specific factors that influence the BIM adoption process: An overview that highlights gaps in the literature. *Front. Eng. Manag.* **2019**, *7*, 119–130. [CrossRef]
35. Ahmed, A.L.; Kawalek, J.P.; Kassem, M. A comprehensive identification and categorisation of drivers, factors, and determinants for BIM adoption: A systematic literature review. In *Computing in Civil Engineering*; American Society of Civil Engineers (ASCE): Reston, VA, USA, 2017; pp. 220–227.
36. Ahmed, A.; Kawalek, P.; Kassem, M. A Conceptual Model For Investigating BIM Adoption by Organisations. In Proceedings of the Joint Conference on Computing in Construction (JC3), Heraklion, Greece, 4–12 July 2017; Volume 1, pp. 447–455.
37. Muhammad, H.; Shehzad, F.; Ibrahim, R.B.; Fadhil, A.; Anwar, K.; Shawkat, S. Recent developments of BIM adoption based on categorization, identification and factors: A systematic literature review. *Int. J. Constr. Manag.* **2020**, *17*, 1–13.

38. Shehzad, H.M.F.; Ibrahim, R.B.; Yusof, A.F.; Khaidzir, K.A.M. Building Information Modeling: Factors Affecting the Adoption in the AEC Industry. In Proceedings of the 2019 6th International Conference on Research and Innovation in Information Systems (ICRIIS), 2–3 December 2019; pp. 1–6.
39. Ahmed, A.L.; Kassem, M. A unified BIM adoption taxonomy: Conceptual development, empirical validation and application. *Autom. Constr.* **2018**, *96*, 103–127. [CrossRef]
40. Zhang, L.; Chu, Z.; Song, H. Understanding the Relation between BIM Application Behavior and Sustainable Construction: A Case Study in China. *Sustainability* **2019**, *12*, 306. [CrossRef]
41. Oesterreich, T.D.; Teuteberg, F. Behind the scenes: Understanding the socio-technical barriers to BIM adoption through the theoretical lens of information systems research. *Technol. Forecast. Soc. Chang.* **2019**, *146*, 413–431. [CrossRef]
42. Ahmed, S.H.; Suliman, S.M. A structure equation model of indicators driving BIM adoption in the Bahraini construction industry. *Constr. Innov.* **2019**, *20*, 61–78. [CrossRef]
43. Gurevich, U.; Sacks, R. Longitudinal Study of BIM Adoption by Public Construction Clients. *J. Manag. Eng.* **2020**, *36*, 05020008. [CrossRef]
44. Gu, N.; London, K. Understanding and facilitating BIM adoption in the AEC industry. *Autom. Constr.* **2010**, *19*, 988–999. [CrossRef]
45. Rogers, J.; Chong, H.Y.; Preece, C. Adoption of Building Information Modelling technology (BIM): Perspectives from Malaysian engineering consulting services firms. *Eng. Constr. Archit. Manag.* **2015**, *22*, 424–445. [CrossRef]
46. Alsaeedi, F.; Lafta, M.J.; Ahmed, A. State of Building Information Modelling (BIM) adoption in Iraq. *IOP Conf. Ser. Mater. Sci. Eng.* **2020**, *737*, 012007. [CrossRef]
47. Ayinla, K.; Adamu, Z. Bridging the digital divide gap in BIM technology adoption. *Eng. Constr. Archit. Manag.* **2018**, *25*, 1398–1416. [CrossRef]
48. Patil, S.D. Application of BIM for Scheduling and Costing of Building Project. *Int. J. Res. Appl. Sci. Eng. Technol.* **2018**, *6*, 1609–1615. [CrossRef]
49. Khan, S.U.; Niazi, M.; Ahmad, R. Factors influencing clients in the selection of offshore software outsourcing vendors: An exploratory study using a systematic literature review. *J. Syst. Softw.* **2011**, *84*, 686–699. [CrossRef]
50. Ogunlana, S.; Charoenngam, C.; Herabat, P.; Hadikusumo, B.H.W. International Symposium on Globalisation and Construction Proceedings CIB W107 (Construction in Developing Economies) and CIB TG23 (Culture in Construction) Joint Symposium Sponsored by Edited by [Internet]. academia.edu. 2004. Available online: https://www.academia.edu/download/3477348/cib5911.pdf#page=394 (accessed on 24 August 2022).
51. Herr, C.M.; Fischer, T. BIM adoption across the Chinese AEC industries: An extended BIM adoption model. *J. Comput. Des. Eng.* **2018**, *6*, 173–178. [CrossRef]
52. Oraee, M.; Hosseini, M.R.; Edwards, D.J.; Li, H.; Papadonikolaki, E.; Cao, D. Collaboration barriers in BIM-based construction networks: A conceptual model. *Int. J. Proj. Manag.* **2019**, *37*, 839–854. [CrossRef]
53. Almuntaser, T.; Sanni-Anibire, M.O.; Hassanain, M.A. Adoption and implementation of BIM–case study of a Saudi Arabian AEC firm. *Int. J. Manag. Proj. Bus.* **2018**, *11*, 608–624. [CrossRef]
54. Kassem, M.A.; Khoiry, M.A.; Hamzah, N. Evaluation of Risk Factors Affecting on Oil and Gas Construction Projects in Yemen. *Int. J. Eng. Technol.* **2019**, *8*, 6–14.
55. Succar, B.; Kassem, M. Macro-BIM adoption: Conceptual structures. *Autom. Constr.* **2015**, *57*, 64–79. [CrossRef]
56. Yuan, H.; Yang, Y. BIM Adoption under Government Subsidy: Technology Diffusion Perspective. *J. Constr. Eng. Manag.* **2020**, *146*, 04019089. [CrossRef]
57. Ma, G.; Jia, J.; Ding, J.; Shang, S.; Jiang, S. Interpretive structural model based factor analysis of BIM adoption in Chinese construction organizations. *Sustainability* **2019**, *11*, 1982. [CrossRef]
58. Banawi, A. *Barriers to Implement Building Information Modeling (BIM) in Public Projects in Saudi Arabia*; Springer: Berlin/Heidelberg, Germany, 2018; Volume 3, pp. 119–125.
59. Researcher, A. BIM: A Technology Acceptance Model In Peru. *J. Inf. Technol. Constr.* **2020**, *25*, 99–108.
60. Republic of Yemen. National Report Habitat III. In Proceedings of the Third United Nations Conference on Housing and Sustainable Urban Development-HABITAT III, Quito, Ecuador, 17–20 October 2016; pp. 1–67. Available online: http://habitat3.org/wp-content/uploads/Yemen-National-Report-September-2016.pdf (accessed on 24 August 2022).
61. Sultan, B.; Alaghbari, W. Political instability and the informal construction sector in Yemen. *Int. J. Civ. Eng. Technol.* **2018**, *9*, 1228–1235.
62. Delgado JM, D.; Oyedele, L.; Ajayi, A.; Akanbi, L.; Akinade, O.; Bilal, M.; Owolabi, H. Robotics and automated systems in construction: Understanding industry-specific challenges for adoption. *J. Build. Eng.* **2019**, *26*, 100868. [CrossRef]
63. Gamil, Y.; Rahman, I.A.; Nagapan, S.; Alemad, N. Qualitative Approach on Investigating Failure Factors of Yemeni Mega Construction Projects. *MATEC Web Conf.* **2017**, *103*, 03002. [CrossRef]
64. Gamil, Y.; Rahman, I.A. Assessment of critical factors contributing to construction failure in Yemen. *Int. J. Constr. Manag.* **2018**, *20*, 429–436. [CrossRef]
65. McCuen, T.L. BIM and Cost Estimating: A Change in the Process for Determining Project Costs. *Build. Inf. Model. Appl. Pract.* **2015**, 63–81.

66. Yahya Al-Ashmori, Y.; Bin Othman, I.; Bin Mohamad, H.; Rahmawati, Y.; Napiah, M. Establishing the Level of BIM implementation-A Case Study in Melaka, Malaysia. *IOP Conf. Ser. Mater. Sci. Eng.* **2019**, *601*, 012024. [CrossRef]
67. Awwad, K.A.; Shibani, A.; Ghostin, M. Exploring the critical success factors influencing BIM level 2 implementation in the UK construction industry: The case of SMEs. *Int. J. Const.r Manag.* **2022**, *22*, 1894–1901. [CrossRef]
68. CIDB. BIM Guide 1: Awareness. Construction Industry Development Board Malaysia. 2016. Available online: http://www.mybimcentre.com.my (accessed on 24 August 2022).
69. Joblot, L.; Paviot, T.; Deneux, D.; Lamouri, S. Automation in Construction Building Information Maturity Model specific to the renovation sector. *Autom. Constr.* **2019**, *101*, 140–159. [CrossRef]
70. Kassem, M.; Vukovic, V.; Dawood, N.; Hafeez, M.A.; Chahrour, R.; Naji, K. Approaches for Assessing BIM Adoption in Countries: A Comparative Study within Qatar. *World Build. Congr.* **2016**, *1*, 695–705.
71. Ahmed, S.; Dlask, P.; Shaban, M.; Selim, O. Possibility of applying bim in syrian building projects. *Eng. Rural. Dev.* **2018**, *17*, 524–530.
72. Babatunde, S.O.; Perera, S.; Ekundayo, D.; Adeleye, T.E. An investigation into BIM-based detailed cost estimating and drivers to the adoption of BIM in quantity surveying practices. *J. Financ. Manag. ofof Prop. Constr.* **2019**, *25*, 61–81. [CrossRef]
73. Li, P.; Zheng, S.; Si, H.; Xu, K. Critical Challenges for BIM Adoption in Small and Medium-Sized Enterprises: Evidence from China. *Adv. Civ. Eng.* **2019**, *2019*, 9482350. [CrossRef]
74. Alemayehu, S.; Nejat, A.; Ghebrab, T.; Ghosh, S. A multivariate regression approach toward prioritizing BIM adoption barriers in the Ethiopian construction industry. *Eng. Constr. Archit. Manag.* **2021**, *29*, 2635–2664. [CrossRef]
75. Qin, X.; Shi, Y.; Lyu, K.; Mo, Y. Using a tam-toe model to explore factors of building information modelling (BIM) adoption in the construction industry. *J. Civ. Eng. Manag.* **2020**, *26*, 259–277. [CrossRef]
76. Shi, Q.; Ding, X.; Zuo, J.; Zillante, G. Automation in Construction Mobile Internet based construction supply chain management: A critical review. *Autom. Constr.* **2016**, *72*, 143–154. [CrossRef]
77. Chien, K.-F.; Wu, Z.-H.; Huang, S.-C. Identifying and assessing critical risk factors for BIM projects: Empirical study. *Autom. Constr.* **2014**, *45*, 1–15. [CrossRef]
78. Lancashire, C. *Implementation of the Lean Approach in Sustainable Construction: A Conceptual Framework by Oyedolapo Ogunbiyi*; University of Central Lancashire: Preston, UK, 2014.
79. Ding, Z.; Zuo, J.; Wu, J.; Wang, J.Y. Key factors for the BIM adoption by architects: A China study. *Eng. Constr. Archit. Manag.* **2015**, *22*, 732–748. [CrossRef]
80. Bui, N.; Merschbrock, C.; Munkvold, B.E. A review of Building Information Modelling for construction in developing countries. *Procedia Eng.* **2016**, *164*, 487–494. [CrossRef]
81. Saka, A.B.; Chan, D.W.M. Profound barriers to building information modelling (BIM) adoption in construction small and medium-sized enterprises (SMEs): An interpretive structural modelling approach. *Constr. Innov.* **2020**, *20*, 261–284. [CrossRef]
82. Aigbavboa, C.; Thwala, W. *The Construction Industry in the Fourth Industrial Revolution*; Springer: Berlin/Heidelberg, Germany, 2020.
83. Liao, L.; Teo Ai Lin, E.; Low, S.P. Assessing building information modeling implementation readiness in building projects in Singapore: A fuzzy synthetic evaluation approach. *Eng. Constr. Archit. Manag.* **2019**, *27*, 700–724. [CrossRef]
84. Aibinu, A.; Venkatesh, S. Status of BIM Adoption and the BIM Experience of Cost Consultants in Australia. *J. Prof. Issues Eng. Educ. Pract.* **2014**, *140*, 04013021. [CrossRef]
85. Antwi-Afari, M.F.; Li, H.; Pärn, E.A.; Edwards, D.J. Critical success factors for implementing building information modelling (BIM): A longitudinal review. *Autom. Constr.* **2018**, *91*, 100–110. [CrossRef]
86. Al-Fadhali, N.; Zainal, R.; Kasim, N.; Dodo, M.; Kim-Soon, N.; Hasaballah, A.H.A. The desirability of Integrated Influential Factors (IIFs) Model of internal stakeholder as a panacea to project completion delay in Yemen. *Int. J. Constr. Manag.* **2017**, *19*, 128–136. [CrossRef]
87. Al-Fadhali, N.; Mansir, D.; Zainal, R. Validation of an integrated influential factors (IIFs) model as a panacea to curb projects completion delay in Yemen. *J. Sci. Technol. Policy Manag.* **2019**, *10*, 793–811. [CrossRef]
88. Alaghbari, W.; Sultan, B. Delay Factors Impacting Construction Projects in Sana' a-Yemen. *PM World J.* **2018**, *VII*, 1–28.
89. Purushothaman, K.; Ahmad, R. Integration of Six Sigma methodology of DMADV steps with QFD, DFMEA and TRIZ applications for image-based automated inspection system development: A case study. *Int. J. Lean Six Sigma* **2022**, *13*, 1239–1276. [CrossRef]
90. Li, C.Z.; Hong, J.; Xue, F.; Shen, G.Q.; Xu, X.; Mok, M.K. Schedule risks in prefabrication housing production in Hong Kong: A social network analysis. *J. Clean. Prod.* **2016**, *134*, 482–494. [CrossRef]
91. Cao, Y.; Zhang, L.H.; McCabe, B.; Shahi, A. The Benefits of and Barriers to BIM Adoption in Canada. *Int. Symp. Autom. Robot. Constr.* **2019**, *36*, 152–158.
92. Hong, Y.; Hammad, A.W.; Sepasgozar, S.; Nezhad, A.A. BIM adoption model for small and medium construction organisations in Australia. *Eng. Constr. Archit. Manag.* **2018**, *26*, 154–183. [CrossRef]
93. Ullah, K.; Lill, I.; Witt, E. An overview of BIM adoption in the construction industry: Benefits and barriers. In Proceedings of the 10th Nordic Conference on Construction Economics and Organization, Tallinn, Estonia, 7–8 May 2019; Emerald Publishing Limited: Bingley, UK, 2019; Volume 2, pp. 297–303. Available online: https://www.emerald.com/insight/content/doi/10.1108/S2516- (accessed on 24 August 2022).

94. Doan, D.T.; Ghaffarianhoseini, A.; Naismith, N.; Ghaffarianhoseini, A.; Zhang, T.; Tookey, J. Examining Green Star certification uptake and its relationship with Building Information Modelling (BIM) adoption in New Zealand. *J. Environ. Manag.* **2019**, *250*, 1–11. [CrossRef] [PubMed]
95. Fitriani, H.; Budiarto, A.; Ajayi, S.; Idris, Y. Implementing BIM in architecture, engineering and construction companies: Perceived benefits and barriers among local contractors in palembang, Indonesia. *Int. J. Constr. Supply Chain Manag.* **2019**, *9*, 20–34. [CrossRef]
96. Zafar, I.; Shen, G.; Ahmed, S.; Yousaf, T. Stakeholders Responsibilities for Time Overrun Risks of Highway Projects in Terrorism Affected Areas. In Proceedings of the 13th International Postgraduate Conference (IPGRC 2017), London, UK, 19–21 February 2017; pp. 504–515.
97. Hore, A.; Kuang, S.; McAuley, B.; West, R.P. Development of a Framework to Support the Effective Adoption of BIM in the Public Sector: Lessons for Ireland. In *Conference Papers; by the School of Multidisciplinary Technologies at ARROW@TU Dublin, Hong Kong, China, 17–21 June 2019*; Technological University Dublin: Dublin, Ireland, 2019; pp. 1–17. Available online: https://arrow.tudublin.ie/schmuldistcon/25%0Ahttps://arrow.tudublin.ie/schmuldistcon/25 (accessed on 24 August 2022).
98. Hong, Y.; Hammad, A.W.A.; Akbarnezhad, A. Impact of organization size and project type on BIM adoption in the Chinese construction market. *Constr. Manag. Econf.* **2019**, *37*, 675–691. [CrossRef]
99. Kassem, M.; Succar, B. Macro BIM adoption: Comparative market analysis. *Autom. Constr.* **2017**, *81*, 286–299. [CrossRef]
100. Bew, M.; Underwood, J. Delivering BIM to the UK Market. In *Handbook of Research on Building Information Modeling and Construction Informatics: Concepts and Technologies*; IGI Global: Hershey, PA, USA, 2010; pp. 30–64.
101. Construction, M.H. The Business Value of BIM for Construction in Major Global Markets. In *Smart Market Report*; McGraw Hill Construction: Bedford, MA, USA, 2014.
102. Alhumayn, S.A.; Saka, A.B.; Chan, D.W.M. A Scientometric Review and Metasynthesis of Building Information Modelling (BIM) Research in Africa. *Buildings* **2019**, *9*, 85.
103. Fadhil, A.; Khaidzir, K.; Husain, O. The Evolution of Technology Adoption Theories in Building Information Modelling Research Building Information Modelling Adoption: Systematic Literature Review. In Proceedings of the 5th International Conference of Reliable Information and Communication Technology 2020, London, UK, 26–28 July 2020; Yemeni Scientists Reseaarch Group & (ISSIRG) in Universiti Teknologi Malaysia (Malaysia): Johor Bahru, Malaysia, 2020; pp. 1–12.
104. Hochscheid, E.; Halin, G. Micro BIM adoption in design firms: Guidelines for doing a BIM implementation plan. *Proc. Creat. Constr. Conf.* **2019**, *119*, 864–871.
105. Hochscheid, E.; Halin, G. A model to approach BIM adoption process and possible BIM implementation failures. In Proceedings of the Creative Construction Conference 2018, CCC 2018, Ljubljana, Slovenia, 30 June–3 July 2018; Diamond Congress Ltd.: Budapest, Hungary, 2018; pp. 257–264.
106. Prabhakaran, A.; Mahamadu, A.-M.; Mahdjoubi, L.; Andric, J.; Manu, P.; Mzyece, D. An investigation into macro BIM maturity and its impacts: A comparison of Qatar and the United Kingdom. *Archit. Eng. Des. Manag.* **2021**, *17*, 496–515. [CrossRef]
107. Van Tam, N.; Diep, T.N.; Toan, N.Q.; Quy, N.L.D. Factors affecting adoption of building information modeling in construction projects: A case of Vietnam. *Cogent Bus. Manag.* **2021**, *8*, 1918848. [CrossRef]
108. Liu, Z. Feasibility Analysis of BIM Based Information System for Facility Management at WPI. Master's Thesis, Worcester Polytechnic Institute, Worcester, MA, USA, 2010.
109. Kang, T.W.; Hong, C.H. A study on software architecture for effective BIM/GIS-based facility management data integration. *Autom. Constr.* **2015**, *54*, 25–38. [CrossRef]
110. Vriens, R.G.M. *The Handbook of Marketing Research: Uses, Misuses, and Future Advances*; Sage: New York, NY, USA, 2006.
111. Bhattacherjee, A. *Social Science Research: Principles, Methods, and Practices*, 2nd ed.; Creative Commons Attribution-NonCommercial-ShareAlike 3.0 Unported License; University of South Florida Tampa: Tampa, FL, USA, 2012.
112. Fellows, R.; Liu, A.M.M. Use and misuse of the concept of culture. *Constr. Manag. Econ.* **2013**, *31*, 401–422. [CrossRef]
113. Weston, R.; Gore, P.A. A brief guide to structural equation modeling. *Couns. Psychol.* **2006**, *34*, 719–751. [CrossRef]
114. Chin, W.W. The partial least squares approach to structural equation modeling. *Mod. Methods Bus. Res.* **1998**, *295*, 295–336.
115. Haenlein, M.; Kaplan, A.M. A beginner's guide to partial least squares analysis. *Underst. Stat.* **2004**, *3*, 283–297. [CrossRef]
116. Latan, H.; Noonan, R.; Matthews, L. Partial least squares path modeling. In *Basic Concepts, Methodol*; Springer: Berlin/Heidelberg, Germany, 2017; Volume 3.
117. Bagozzi, R.P.; Baumgartner, H. The evaluation of structural equation models and hypothesis testing. *Princ. Mark. Res.* **1994**, *1*, 386–422.
118. Tabachnick, B.G.; Fidell, L.S. *Using Multivariate Statistics*, 5th ed.; Allyn Bacon: Boston, MA, USA, 2007.
119. Henseler, J.; Ringle, C.M.; Sarstedt, M. A new criterion for assessing discriminant validity in variance-based structural equation modeling. *J. Acad. Mark. Sci.* **2015**, *43*, 115–135. [CrossRef]
120. Hayes, A.F. *Introduction to Mediation, Moderation, and Conditional Process Analysis: A Regression-Based Approach*; Guilford publications: New York, NY, USA, 2017.
121. Kim, H.; Ionides, E.; Almirall, D. A sample size calculator for SMART pilot studies. *SIAM Undergrad. Res.* **2016**, *9*, 229. [CrossRef]
122. Sürücü, L.; Maslakçi, A. Validity and reliability in quantitative research. *Bus. Manag. Stud. An Int. J.* **2020**, *8*, 2694–2726. [CrossRef]
123. Götz, O.; Liehr-Gobbers, K.; Krafft, M. Evaluation of structural equation models using the partial least squares (PLS) approach. In *Handbook of Partial Least Squares*; Springer: Berlin/Heidelberg, Germany, 2010; pp. 691–711.

124. Cohen, A. Comparison of correlated correlations. *Stat. Med.* **1989**, *8*, 1485–1495. [CrossRef]
125. Shmueli, G.; Sarstedt, M.; Hair, J.F.; Cheah, J.-H.; Ting, H.; Vaithilingam, S.; Ringle, C.M. Predictive model assessment in PLS-SEM: Guidelines for using PLSpredict. *Eur. J. Mark.* **2019**, *53*, 2322–2347. [CrossRef]
126. Hair, J.F.; Anderson, R.E.; Babin, B.J.; Black, W.C. *Multivariate Data Analysis: A Global Perspective*; Pearson: Bergen, NJ, USA, 2010.
127. Vukovic, V.; Hafeez, M.A.; Chahrour, R.; Kassem, M.; Dawood, N. BIM Adoption in Qatar: Capturing High Level Requirements for Lifecycle Information Flow. In Proceedings of the 15th International Conference on Construction Applications of Virtual Reality (CONVR), Banff, AB, Canada, 5–7 October 2015; pp. 1–11.

MDPI
St. Alban-Anlage 66
4052 Basel
Switzerland
Tel. +41 61 683 77 34
Fax +41 61 302 89 18
www.mdpi.com

Buildings Editorial Office
E-mail: buildings@mdpi.com
www.mdpi.com/journal/buildings

www.ingramcontent.com/pod-product-compliance
Lightning Source LLC
Chambersburg PA
CBHW061807210926
51850CB00006B/637
9783036581507